Resisting the Rule of Law in Nineteenth-Century Ceylon

This book offers in-depth insights on the struggles implementing the rule of law in nineteenth-century Ceylon, introduced into the colonies by the British as their "greatest gift." The book argues that resistance can be understood as a form of negotiation to lessen oppressive colonial conditions, and that the cumulative impact caused continual adjustments to the criminal justice system, weighing it down and distorting it.

The tactical use of rule of law is explored within the three bureaucracies: the police, the courts and the prisons. Policing was often "governed at a distance" due to fiscal constraints and economic priorities and the enforcement of law was often delegated to underpaid Ceylonese. Spaces of resistance opened up as Ceylon was largely left to manage its own affairs. Villagers, minor officials, as well as senior British government officials, alternately used or subverted the rule of law to achieve their own goals. In the courts, the imported system lacked political legitimacy and consequently the Ceylonese undermined it by embracing it with false cases and information, in the interests of achieving justice as they saw it. In the prisons, administrators developed numerous biopolitical techniques and medical experiments in order to punish prisoners' bodies to their absolute lawful limit. This limit was one which prison officials, prisoners and doctors negotiated continuously over the decades.

The book argues that the struggles around rule of law can best be understood not in terms of a dualism of bureaucrats versus the public, but rather as a set of shifting alliances across permeable bureaucratic boundaries. It offers innovative perspectives, comparing the Ceylonese experiences to those of Britain and India, and where appropriate to other European colonies. This book will appeal to those interested in law, history, postcolonial studies, cultural studies, cultural and political geography.

James S. Duncan was Reader in Cultural Geography, University of Cambridge until his retirement. He is now Emeritus Fellow of Emmanuel College. His research interests are cultural and historical geography, South Asian history and history of law.

Routledge Research in Historical Geography

Series Editors: **Simon Naylor**, *University of Glasgow, UK* and
Laura Cameron, *Queen's University, Canada*

This series offers a forum for original and innovative research, exploring a wide range of topics encompassed by the sub-discipline of historical geography and cognate fields in the humanities and social sciences. Titles within the series adopt a global geographical scope and historical studies of geographical issues that are grounded in detailed inquiries of primary source materials. The series also supports historiographical and theoretical overviews, and edited collections of essays on historical-geographical themes. This series is aimed at upper-level undergraduates, research students and academics.

Historical Geographies of Anarchism
Early Critical Geographers and Present-Day Scientific Challenges
Edited by Federico Ferretti, Gerónimo Barrera de la Torre, Anthony Ince and Francisco Toro

Cultural Histories, Memories and Extreme Weather
A Historical Geography Perspective
Edited by Georgina H. Endfield and Lucy Veale

Commemorative Spaces of the First World War
Historical Geographies at the Centenary
Edited by James Wallis and David C. Harvey

Architectures of Hurry
Mobilities, Cities and Modernity
Edited by Phillip Gordon Mackintosh, Richard Dennis and Deryck W. Holdsworth

Anarchy and Geography
Reclus and Kropotkin in the UK
Federico Ferretti

Twentieth Century Land Settlement Schemes
Edited by Roy Jones and Alexandre M. A. Diniz

Resisting the Rule of Law in Nineteenth-Century Ceylon
Colonialism and the Negotiation of Bureaucratic Boundaries
James S. Duncan

For more information about this series, please visit: www.routledge.com/ Routledge-Research-in-Historical-Geography/book-series/RRHGS

Resisting the Rule of Law in Nineteenth-Century Ceylon

Colonialism and the Negotiation of Bureaucratic Boundaries

James S. Duncan

Routledge
Taylor & Francis Group

LONDON AND NEW YORK

First published 2021
by Routledge
2 Park Square, Milton Park, Abingdon, Oxon OX14 4RN

and by Routledge
52 Vanderbilt Avenue, New York, NY 10017

Routledge is an imprint of the Taylor & Francis Group, an informa business

British Library Cataloguing-in-Publication Data
A catalogue record for this book is available from the British Library

Library of Congress Cataloging-in-Publication Data
Names: Duncan, James Stuart, 1945- author.
Title: Resisting the rule of law in nineteenth century Ceylon : colonialism and the negotiation of bureaucratic boundaries / James S. Duncan.
Description: Abingdon, Oxon ; New York, NY : Routledge, 2020. | Series: Routledge research in historical geography | Includes bibliographical references and index.
Identifiers: LCCN 2020008577 (print) | LCCN 2020008578 (ebook)
Subjects: LCSH: Criminology–Sri Lanka–History–19th century. | Criminal justice, Administration of–Sri Lanka–History–19th century. | Prisons–Sri Lanka–History–19th century. | Bureaucracy–Sri Lanka–History–19th century.
Classification: LCC HV6022.S75 D86 2020 (print) | LCC HV6022.S75 (ebook) | DDC 364.9549309/034–dc23
LC record available at https://lccn.loc.gov/2020008577
LC ebook record available at https://lccn.loc.gov/2020008578

ISBN: 978-0-367-51551-5 (hbk)
ISBN: 978-1-003-05438-2 (ebk)

Typeset in Times New Roman
by Swales & Willis, Exeter, Devon, UK

For Nancy, As Ever

Contents

PART IV
The prison and the arts of dark biopower   155

Figures

Tables

Acknowledgements

I owe a debt of gratitude to my colleagues in the Geography Department at the University of Cambridge and would particularly like to single out David Nally and Phil Howell for their intellectual stimulation and friendship. I also gained immeasurably over the years from discussing ideas with my postgraduates and later reading their published research, some of which I have drawn on in my analysis.

I am greatly indebted to Nuala Johnson, David Nally, Maruja Jackman, Stephen Legg and two anonymous reviewers, not only for their meticulous editorial suggestions, but especially for their analytical insights and for steering me towards related works that I had missed. My greatest debt, as ever, is to Nancy Duncan who has discussed the ideas in this book with me at length and did a heroic job of editing my sometimes laboured prose. So involved has she been over the years with this project that at times it becomes difficult to know which ideas are mine and which are hers.

I would like to thank the staffs of the British Library, the National Archives, Kew, the Royal Commonwealth Society Library, the Munby Rare Book Room and the Manuscript Room of the University of Cambridge for access to their holdings.

Glossary of terms

Appa   A thin pancake of rice flour and coconut milk.

Arrack renter   A licenced and regulated distiller of arrack, an alcoholic drink made from the fermented sap of the coconut flower.

Assistant Government Agent (AGA)   Official in charge of a district.

Batta   Maintenance or travel expenses for a government employee or witness.

Burgher   A descendant of the Dutch rulers of Ceylon.

Ceylonese   A native of Ceylon.

Chena   Common land used for shifting cultivation, especially important during times of hunger.

Chetty   A South Indian merchant caste.

Estate   The name given to a coffee or tea plantation.

Gansabhava   A village council or tribunal.

Government Agent (GA)   Official in charge of a province.

Kangany   A Tamil supervisor of a labour gang on a plantation.

Kachcheri   A district administration office.

Korale   The administrative subdivision of a district. Also the administrator of a district.

Malay   A person of Malayan descent.

Moor   A non-Malay Muslim living in Ceylon.

Mudaliyar   A chief headman.

Pattu   Subdivision of a district.

Peon   A labourer, servant, or very low-level official.

Police Vidane   A village policeman under the supervision of the Government Agent.

Ratemahatmaya   Chief headman of each subdivision of a Pattu.

Regular Police   Member of the police force under the supervision of the Inspector General of Police.

Sinhalese   The numerically dominant ethnic group in Ceylon and their language.

Tamil   A person from the south of India or a Ceylonese of South Indian descent.

Veddah   The inhabitants of the island before the arrival of the Sinhalese and Tamils.

Vidane   A headman or village official.

Vidane Arachchi   Chief headman of each subdivision of a Pattu. Supervises other headmen.

Part I

Introduction

1 Introduction

This book is about struggles over the implementation of rule of law during the second half of the nineteenth century in Ceylon, a British Crown colony willing to spend only very limited resources on the criminal justice system and a subject people who, while not openly rebelling against that system, systematically undermined it by appropriating it for their own purposes. The rule of law was introduced into the colonies by the British as their "greatest gift" and the reformed English criminal justice system was an important justification for their rule. Charles Hay Cameron wrote with reference to Ceylon,

> among all the duties incumbent on the British rulers in the east, it is impossible to name one more imperative than that of providing for the effectual decision by public authority of the disputes arising among the poorer classes ... and it is a benefit which none but an European government can confer.[1]

The criminal justice system was part of a larger bureaucratic administrative structure whose role was not only to secure the spread of British capitalism, but to establish firmer control over local populations and ultimately to inculcate British bourgeois beliefs about individualism, personal discipline, time management and a "proper" work ethic.[2] As Scott says, "the utopian, immanent, and continually frustrated goal of the modern state is to reduce the chaotic, disorderly, constantly changing social reality beneath it to something more closely resembling the administrative grid of its observations."[3]

---

1 Report upon the Judicial Establishments and Procedures in Ceylon, 31 January, 1832. *Parliamentary Papers*, 1831/32, XXXVII, 274, p. 65.
2 Yang, A.A. 1987. "Disciplining 'natives': prisons and prisoners in early nineteenth century India." *South Asia*, Vol. 10, No. 2, p. 30; Freitag, S.B. 1985. "Collective crime and authority in north India," in Yang, A.A. (ed.), *Crime and Criminality in British India*. Tucson: University of Arizona, p. 141; Comaroff, J. and J. Comaroff. 1991. *Of Revelation and Revolution: Christianity, Colonialism and Consciousness*. Chicago: University of Chicago Press; Engel, D.M. 2015. "Rights as wrongs: legality and sacrality in Thailand." *Asian Studies Review*, Vol. 39, No. 1, pp. 38–52.
3 Scott, J.C. 1998. *Seeing Like a State: How Certain Schemes to Improve the Human Condition Have Failed.* New Haven: Yale University Press, p. 82.

These bureaucratic goals, as Dwyer and Nettlebeck point out, were intended to supplement force with "an adaptable system of colonial practices that maintained the fundamental imbalance of power structuring colonial relations" and promote the development of capitalism.[4] In Ceylon, as in the colonies, where there was thought to be little respect for British institutions, an authoritarian administration of English law was considered necessary for security.[5] Security according to Foucault included a wide range of mechanisms, such as reports, commissions, regulations, surveillance strategies, penal techniques, statistical surveys of populations and partitioning of territory, all of which were deployed in Ceylon.[6] However, due to fiscal constraints and economic priorities, much of rural Ceylon was left to manage its own affairs. Given the small number of British officials on the island, the enforcement of law was delegated to underpaid Ceylonese who had no real stake in the system they were to administer. This lack of a strong British presence in Ceylon opened up physical as well as metaphorical spaces for resistance. As a consequence there were many "zones of darkness" and an uneven topography of law and defiance. Villagers, minor officials, as well as senior British officials, alternatively used or subverted rule of law to achieve their own goals.

The English criminal justice system has been described as the "cutting edge of colonialism."[7] It was an attempted inculcation of utilitarian moral principles and liberal economic values,[8] for as Garland put it, this constellation of ideas not only formed the taken-for-granted basis of Victorian British common sense, it was "commonly taken for universal logic and reason itself."[9] The idea of rule of law, along

4 Dwyer, P. and A. Nettelbeck. 2018. "'Savage wars of peace': violence, colonialism and empire in the modern world," in Dwyer, P. and A. Nettelbeck (eds.), *Violence, Colonialism and Empire in the Modern World*. London: Palgrave Macmillan, p. 11. Hindess, B. 2001. "The liberal government of unfreedom." *Alternatives*, Vol. 26, p. 99, says that the market was "seen as promoting the capacity for autonomous, self-directing activity—first, by encouraging individuals to calculate the costs and benefits of their own decisions and thereby ... fostering the cultivation of prudential virtues, and secondly, by undermining the relations of dependence and subservience."

5 Fitzpatrick, P. 2008. *Law as Resistance: Modernism, Imperialism, Legalism*. Aldershot: Ashgate, p. 31. Also see Stokes, E. 1959. *The English Utilitarians and India*. Delhi: Oxford University Press.

6 Here and elsewhere in this book I use the term population in the Foucauldian sense of an entity with quantifiable characteristics such as rates of crime, sickness, death, cultural and racial difference. On laissez faire as a strategy of governance see Foucault, M. 2007. *Security, Territory, Population: Lectures at the College de France, 1977–78*. Edited by M. Senellart. New York: Palgrave Macmillan.

7 Chanock, M. 1985. *Law, Custom and the Social Order: The Colonial Experience in Malawi and Zambia*. Portsmouth: Heinemann, p. 4.

8 See Ambirajan, S. 2008. *Classical Political Economy and British Policy in India*. Cambridge: Cambridge University Press; Mehta, U. 1999. *Liberalism, and Empire: A Study in Nineteenth-Century British Liberal Thought*. Chicago: University of Chicago Press.

9 Garland, D. 1985. *Punishment and Welfare: A History of Penal Strategies*. Aldershot: Gower, pp. 41–42.

with the market itself, were seen by the British as the foundation of the colonial civil-ising mission.[10]

This loosely amalgamated body of ideas was no mere disembodied discourse of course; it was materialised and given power in the day-to-day workings of colonial bureaucracies. The three bureaucracies in Ceylon that concern us here are the police, the courts and the prisons. My study is an example of the "every day state" as it was encountered at the local level. I show that in Ceylon there was no clear dichotomy between the state and those oppressed by it.[11] The social and spatial boundaries were always permeable and subject to negotiation by various parties despite their highly unequal positions within the colonial power structure. I have not written a history of these bureaucracies, for histories of the first two, at least, already exist.[12] Rather I have explored the tactics the Ceylonese, both inside and outside these bureaucracies, employed to resist bureaucratic goals and the punitive biopolitical and space-time surveillance strategies senior colonial government offi-cials employed to counter subaltern resistance.[13] In this regard, my study is indebted to Rogers' fine monograph on the impact of social change on crime and the government's response to it in nineteenth-century Ceylon.[14] The present study extends Rogers' analysis to include the colony's prison system and expands upon his work by focusing more squarely on how resistance to rule of law helped shape

10 David Nally makes the point that "the law and the market were preferred by liberals because they deploy force without being coercive." Personal communication. Hindess, "The liberal government of unfreedom," argues that liberal political reason endorses paternalistic, authoritarian government when subject populations are seen as not yet ready for self-direction—the ultimate goal being limited government and individual liberty. A self-regulating market and rule of law with the police function limited to peace-keeping and the protection of property were both seen as instrumental to achieving non-coercive rule. As I have found in Ceylon, non-coercive colonial rule was not only an ideological ideal, but was financially expedient, as British colonies were expected to be self-funded. Consequently, maintaining a system of direct state intervention throughout the island would have been considered too expensive.

11 On the everyday state see Fuller, C.J. and V. Benei. 2001. *The Everyday State and Society in Modern India*. London: Hurst & Co Ltd.

12 Pippet, G.K. 1938. *A History of the Ceylon Police, Volume 1, 1795–1870*. Colombo: The Times of Ceylon; Dep, A.C. 1969. *A History of the Ceylon Police, Volume 2, 1866–1913*. Colombo: Police Amenities Fund; Nanaraja, T. 1972. *The Legal System of Ceylon in its Historical Setting*. Leiden: E.J. Brill.

13 This volume is intended to be a companion to my earlier monograph Duncan, J.S. 2016. *In the Shadows of the Tropics: Climate Race and Biopower in Nineteenth Century Ceylon*. London: Rou-tledge, where I explore the themes of domination and resistance on the coffee plantations of nine-teenth-century Ceylon. The concepts of biopower and biopolitics are drawn from Foucault and others. Foucault introduces these ideas in his *History of Sexuality* and later in his lectures at the College de France (see Foucault, *Security, Territory, Population*). Foucault defines biopolitics as the "calculated management of life," the expert administration of collective bodies at the level of populations, a nineteenth-century "explosion of numerous and diverse techniques for achieving the subjugation of bodies and the control of populations." Foucault, M. 1990. *The History of Sexuality. Volume 1: An Introduction*. Translated by R. Hurley. New York: Vintage, p. 140.

14 Rogers, J.D. 1987. *Crime, Justice and Society in Colonial Sri Lanka*. London: Curzon Press.

the criminal justice system in Ceylon.[15] It should be noted, however, that during the nineteenth century this shaping brought only incremental change, for the Ceylonese had little power to overturn the conditions of disparity and exploitation which the rule of law presupposed and supported.[16] As I hope to show, resistance in the case of nineteenth-century Ceylon can at best be understood as a form of "negotiation" to lessen oppressive conditions by resetting bureaucratic norms, in the sense of lowering British expectations.[17] Here I follow Uday Chandra who reminds his readers of the "Latin root of resistance *re* + *sistere*, literally enduring or withstanding, to re-orient the older emphasis on opposition or negation towards a logic of negotiation." He employs the idea of resistance in a "narrower but arguably more robust sense of the term" which entails apprehending "the conditions of one's subordination, to endure or withstand those conditions in everyday life, and to act with sufficient intention and purpose to negotiate power relations from below in order to rework them in a more favourable or emancipatory direction."[18]

The time period I cover in the book is from the beginning of British rule into the early twentieth century. However, my particular focus is upon the period beginning in the mid-1860s when the government in Britain sent out to all the colonies an order to reform their prisons. As my study concentrates on how administrators in Ceylon managed the security of Ceylon through the prevention of "ordinary crime," rather than more overtly political crime, I end the study at the time when the nationalist movements of the twentieth century were beginning in earnest.

My methodology is to engage in a close and critical reading of the annual administration reports of the island's police, courts and prisons from 1867–1915. This is supplemented by the reports of the government agents and assistant government agents, the debates and papers laid before the Legislative Council and the three principal island newspapers from the period, the *Ceylon Examiner*, the *Ceylon Observer*, and the *Times of Ceylon*. These newspapers, through their editorials and letters to the editor, provide a range of opinion about the criminal justice system at the time.[19] Throughout the volume I point

---

15 There was of course resistance by the Ceylonese to other elements of the colonial administration such as the use of western medicine, especially the smallpox vaccination programme, and public works. See Sivasundaram, S. 2013. *Islanded: Britain, Sri Lanka and the Bounds of an Indian Ocean Colony.* Chicago: University of Chicago Press, pp. 247–82; 209–42.

16 Katz uses the term resilience to refer to individuals enduring oppressive conditions without fundamentally altering the overarching systems of power. Katz, C. 2004. *Growing Up Global: Economic Restructuring and Children's Everyday Lives.* Minneapolis: University of Minnesota Press. In the case of the Ceylonese and British rule of law, I show that short of bringing about emancipatory change, many Ceylonese were able to tactically manipulate and negotiate the conditions of the British colonial justice system.

17 Chandra, U. 2015. "Rethinking subaltern resistance." *Journal of Contemporary Asia*, Vol. 45, No. 4, p. 565.

18 *Ibid.*

19 As Sivasundaram points out, the press in Ceylon was both an organ of resistance and conservatism. He provides an important review of the press in the first half of the nineteenth century. Sivasundaram, *Islanded*, pp. 305–17. My focus is upon the second half of the nineteenth century.

to work on the rule of law in other colonial contexts and the circulation of ideas across empires. Much of this comparative material can be found in footnotes so as not to unduly interrupt the flow of my argument.

The reformed criminal justice system in Britain

At the heart of the nineteenth-century movement to reform the criminal justice system in England lay the idea of uniformity. Every person, regardless of their social position, was, in theory at least, to receive equal justice under the law.[20] Alongside this universalising theory of criminal justice was the utilitarian model of criminal behaviour based on rational decision-making.[21] The key to ensuring universal security was thus seen to be "rule of law." Liberal government by rule of law was deemed necessary to the development and spread of capitalism as it offered protection of property.[22] Rule of law was further justified as a check on arbitrary power at all levels including the police, providing legal liberty so that individuals could predict under what circumstances they would be subject to coercion by the state.[23] As individuals were conceptualised as free, rational, legally equal and responsible, those who committed crimes were deemed to have violated their implicit "contract" with society and consequently could be punished by "contractual" right. Furthermore, punishment was to be certain and proportionate to the crime.

The focus of penal practice both in England as well as in the colonies after the 1860s remained deterrence through the biopolitical targeting of prisoners

---

20 As Hay points out, "justice, in the sense of rational, bureaucratic decisions made in the common interest, is a particularly modern conception and would not have been widely shared in eighteenth century England." Hay, D. 1975. "Property, authority and the criminal law," in Hay, D., P. Linebaugh, J.G. Rule, E.P. Thompson and C. Winslow (eds.), *Albion's Fatal Tree: Crime and Society in Eighteenth-Century England*. New York: Pantheon, p. 39.

21 Utilitarianism had underpinned the reforms of the courts and the police in Britain in the 1820s and 1830s and the prison reforms of the 1850s. While Jeremy Bentham systematised the utilitarian doctrine, it was left to his followers John Stuart Mill, James Kay-Shuttleworth, Edwin Chadwick and Frederick Hill to propagate these ideas and implement them in state bureaucracies. See Roberts, D.F. 2002. *The Social Conscience of the Early Victorians*. Stanford: Stanford University Press, pp. 433–55.

22 Ellen Wood claims that capitalism is not just trade or a market economy; it is not only a quantitative increase in commercial activity, but a qualitatively new form of exploitation which entails legal enclosure, dispossession and the concentration of social power in the hands of private capital and a judicial system that supported such capitalist property relations. Wood, E. 1999. *The Origin of Capitalism*. New York: Monthly Review Press.

23 Legal liberty may be distinguished from political liberty understood as freedom from domination. The former requires that laws be declared publicly, applied equally and with certainty. Tamanaha, B.Z. 2004. *On the Rule of Law: History, Politics, Theory*. Cambridge: Cambridge University Press, p. 34. The utilitarians were very keen on legal liberty, which they associated closely with order and security, but not necessarily on political liberty and herein lay the problem of race in the colonies, which led some utilitarians to question the appropriateness, or at least the expediency, of equality before the law. It was capriciousness and a lack of rationality that was of greater concern.

with the hope of reducing recidivism and instilling fear in criminally inclined sub-populations.[24] Such deterrence entailed creating a harsh prison environment where any rational person would conclude that it was not worth choosing crime because the precisely adjusted pain of the punishment would outweigh the gain derived from the unlawful act.[25]

In line with utilitarian thinking, nineteenth-century England saw the development of the bureaucratic form of administration which, in its idealised form, was a highly rational, neutral and impersonal form of rule-governed organisation that subjugated the personal interests of its officials to the goals of the bureaucracy.[26] Uniformity was the guiding principle underpinning all the state bureaucracies and laws. A well-structured, rationally organised police force was expected to ensure that people followed the laws and regulations. Likewise, the job of the courts was to impose a proportionate, legally defined punishment based on deterrence rather than retribution. And so prisons were organised to manage penal servitude by delivering scientifically measured punishments as ordered by the courts.

The utilitarian view of freedom was considered a right to be earned and enjoyed by individuals and populations only as long as they behaved in a manner that was socially rational, but as the century wore on the British found such abstract ideas of limited use in their day-to-day practices of administration.[27] Utilitarianism, in the case of the poor in England and certainly in the colonial context, tended towards an illiberal authoritarian philosophy in which personal freedom was a secondary consideration.[28] Bureaucratic practice

24 While utilitarian theories of rational criminal behaviour waned somewhat in England after the mid-century, they survived in the colonies at least until the end of the century. See Wiener, M.J. 1990. *Reconstructing the Criminal: Culture, Law and Policy in England, 1830–1914.* Cambridge: Cambridge University Press, p. 12.

25 Stokes, *The English Utilitarians and India*, pp. 66–67. As we will see, however, toward the end of the nineteenth century penal theorists, but not Ceylon jailers, tended to give up on the idea of personal autonomy and sought alternative explanations of criminal behaviour.

26 Tijsterman, S.P. and P. Overeem. 2008. "Escaping the iron cage: Weber and Hegel on bureaucracy and freedom." *Administrative Theory & Praxis*, Vol. 30, No. 1, p. 74.

27 See Mantena, K. 2010. *Alibis of Empire: Henry Maine and the Ends of Imperialism.* Princeton: Princeton University Press. Also see Ambirajan, *Classical Political Economy*, on the distance between classical political economy and the implementation of policy and ad hoc solutions devised by administrators "on the spot" in India, many of whom had divergent views.

28 See Mantena, K. 2009. "The crisis of liberal imperialism," in Bell, D. (ed.), *Victorian Visions of the Global Order: Empire, and International Relations in Nineteenth Century Political Thought.* Cambridge: Cambridge University Press, pp. 113–35. Mantena argues that liberalism was qualified as "not quite or not yet universal." Also see Crimmins, J. 1996. "Contending interpretations of Bentham's utilitarianism." *Canadian Journal of Political Science*, Vol. 29, No. 4, pp. 751–77 on whether Bentham's utilitarianism was more authoritarian or liberal individualistic. Crimmins warns against the myth of coherence and acknowledges tensions within Bentham's work. In the context of utilitarianism imported into Ceylon by lightly educated administrators, of course there is far less reason to expect consistency. Liberalism and utilitarianism did not always mix well and when issues of racial difference were added into the mix there would inevitably have been many contradictions. Also see Losurdo, D. 2011. *Liberalism: A Counter-History.* London: Verso Books.

began to draw less on liberalism as a justification of imperial rule and more on the growing nineteenth-century enthusiasm for rule by experts and the quantification of all things in the management of life. Human behaviour was to be measured with the same rigour as applied to the physical and natural sciences.[29] To put it broadly, the task of government was to administer what Foucault has termed biopower, that is to say, to scientifically manage the life and welfare of populations. In the case of the colonies, the targets of this biopolitical approach were principally the populations that the British were most directly and legally responsible for, including inmates of state-run institutions such as prisons, asylums and hospitals. In Ceylon, as I have described in an earlier volume, it included a large number of migrant Indian labourers working on the plantations.[30] The responsibility for the welfare of the general population in the colonies was largely left to the local communities with the surveillance and punishment of criminal activity being the main focus of British policing.

Bringing the reformed criminal justice system to Ceylon

Liberal rule of law was a set of principles guaranteeing certainty, generality and equality. It was supposed to apply to everyone with no arbitrary distinctions among people. Laws were to be laid out so that everyone knew what the law was and understood what the consequences of disobeying it were. Rule of law was thought by the British to be an advance over what they considered the arbitrary rule of "oriental despots." The 1833 Ceylon Constitution, based on suggested reforms of the Colebrooke Commission, took significant steps towards securing rule of law in Ceylon. The Constitution ended the autocratic powers of the Governor, which had included imprisoning a person without a trial. It installed a common system of law courts and a Supreme Court in which all would be considered equal before the law. It provided for the separation of powers by introducing executive and legislative councils. It also did away with caste-based *rajakaria* (a feudal system of service originally for the king and later the government) in order to protect private ownership of property and to open the country to *laissez faire* capitalism. With the abolishment of *rajakariya*, peasants gained legal ownership of land, and with the ability to sell their land, they could be more mobile. Thus the increased mobility of populations through the ending of obligatory bonds to the land was seen as a benefit to the growth of capitalism, as was the ending of occupations based on caste that had also been an important component of the *rajakariya* system. The Constitution also abolished all special racial and caste privileges, such as excluding Ceylonese from the Civil Service.

29 Hacking, I. 1975. *The Emergence of Probability: A Philosophical Study of Early Ideas about Probability, Induction and Statistical Inference*. Cambridge: Cambridge University Press; Hacking, I. 1990. *The Taming of Chance*. Cambridge: Cambridge University Press.
30 Duncan, *In the Shadows of the Tropics*.

However, in seeing the actual working out of rule of law, "on the ground," so to speak, it was very clear that the British and Ceylonese officials as well as the general Ceylonese population resisted it in many ways.[31] It was also clear that the rule of law was no guarantee that the content of the law was just.[32] Given the nature of Ceylonese caste and class hierarchies and the privileges Europeans had secured for themselves, the impartial application of the law could never have brought about effective equality or rid the system of inherent injustices and inequities. As we will see, there were cases where Ceylonese officials felt constrained by the inflexibility of the rule of law, which took away their discretionary powers to achieve justice locally, that had been allowed under the traditional systems of justice and conflict resolution. For example, sometimes partiality and positive discrimination on the part of headmen was required in order to redress inequities within their communities. Most damaging to the idea of rule of law in Ceylon, however, were the many times when the idea of racial difference led the British to largely abandon the liberal values they saw as the foundation of rule of law.

Civil law in Ceylon was a hybrid mix of English law and various forms of customary law, themselves a product of ever-evolving societal relations.[33] Criminal law on the other hand was largely imported from England, albeit in a piecemeal fashion, and in 1883 Ceylon enacted a new penal code based upon

---

31 There is a large and growing literature on legal geographies. See Jeffrey, A. 2019. *The Edge of Law: Legal Geographies of a War Crimes Court*. Cambridge: Cambridge University Press; Blomley, N. 1994. *Law, Space, and the Geographies of Power*. New York: Guilford Press; Blomley, N., D. Delaney and R. Ford, eds. 2001. *The Legal Geographies Reader*. Oxford: Blackwell; Blomley, N. 2007. "Making private property: enclosure, common right and the work of hedges." *Rural History*, Vol. 18, pp. 1–21; Delaney, D. 2010. *The Spatial, the Legal, and the Pragmatics of World-Making: Nomospheric Investigations*. London: Glass House Books; Blomley, N. 2013. "Performing property, making the world." *Canadian Journal of Law and Jurisprudence*, Vol. 26, pp. 23–48; Braverman, I., N. Blomley and D. Delaney, eds. 2014. *The Expanding Spaces of Law: A Timely Legal Geography*. Palo Alto: Stanford University Press; Von Benda-Beckmann, F., K. Von Benda-Beckmann and A. Griffiths. 2009. "Space and legal pluralism: an introduction," in Von Benda-Beckmann, F., K. Von Benda-Beckmann and A. Griffiths (eds.), *Spatializing Law: An Anthropological Geography of Law in Society*. Farnham: Ashgate, pp. 1–30.

32 See Baxi, U. 2004. "Rule of law in India: theory and practice," in Peererboom, R. (ed.), *Discourses of Rule of Law: Theories and Implementation of Rule of Law in Twelve Asian Countries, France and India*. New York: Routledge, pp. 324–45.

33 See Hussain, N. 2003. *The Jurisprudence of Emergency: Colonialism and the Rule of Law*. Ann Arbor: University of Michigan, and Chanock, *Law Custom*, p. 5 who says of customary law in Africa that "far from being a received form *of* indigenous law, it *was* in fact the constructed product of colonial knowledge and of specific historical transactions between colonizers, local elites, and subject groups." In the case of Ceylon, this complex ongoing constructedness was especially true of the areas of the country that had been colonised by the Portuguese and Dutch long before the arrival of the British and was also complicated by the existence of several ethnic groups with different religious legal traditions. Merry points out that it is important to explore the inequalities as well as the mutually constitutive nature of such hybrid systems. Merry, S.E. 1992. "Anthropology of law and transnational processes." *Annual Review of Anthropology*, Vol. 21, pp. 357–79.

the Indian Penal Code originally drawn up by Macaulay.[34] Although in theory criminal laws applied equally to all, unquestionably they were framed in such a way as to support capitalist structures, facilitate market relations, and protect such property as could be legally documented and thus to disproportionately benefit the European and Ceylonese elites. While, in practice, rule of law was routinely violated, both by officials and non-officials alike, its theoretical value lay in defining such violations as illegitimate and punishable. Having said that, Shinar argues that while rule of law exists to stop government officials from making up their own rules, officials often violate it because they see themselves as fulfilling a public good that they consider to be in conflict with the spirit of the law. Out of hubris, they place either bureaucratic expediency or their beliefs concerning justice and racial difference over law.[35] In another context, Merry argues that rule of law is a "mythic vision" that constantly pressures the extra-legal to hide under cover.[36] While my focus in this book is upon the struggles over rule of law in the criminal justice bureaucracies, one must not lose sight of the fact that legal imperialism played a major role in facilitating the exploitation of Ceylon as it did in other British colonies.[37]

In the mid-1860s the British Parliament declared that the colonies be brought in line with the reformed criminal justice system in Britain. One of the key ideological justifications for colonialism in Asia was the provision of English justice through the moral legitimacy of the rule of laws, which, it was claimed, liberated the colonies from the capriciousness of "oriental despotism." As Hussain points out, in the absence of consent through democratic process, which for the British would never be an option in the colonies, "legality became the preeminent signifier of state legitimacy and of 'civilization,' the term that united politics and morality."[38] As I will show, in practice there was a tension

---

34 The Indian penal code, drawn up by Macaulay as a modified version of English law, was enacted in India in 1862.

35 Shinar, A. 2013. "Dissenting from within: why and how public officials resist the law." *Florida State University Law Review*, Vol. 40, No. 3, pp. 615–17, 656.

36 Merry, S.E. 1990. *Getting Justice and Getting Even: Legal Consciousness among Working-Class Americans*. Chicago: University of Chicago Press, p. 12. We will address some of the debates over rule of law in Chapter 4.

37 For example, English property law provided the legal rationale for much of the dispossession of common lands and lands that were held privately but with no documentation. On the role of law in legitimating plunder, see Mattei, U. and L. Nader. 2008. *Plunder: When the Rule of Law is Illegal*. Oxford: Blackwell. See also Bhandar, B. 2018. *Colonial Lives of Property*. London: Duke University Press. On legal imperialism see Kirkby, D. and C. Coleborne, eds. 2001. *Law, History and Colonialism: The Reach of Empire*. Manchester: Manchester University Press. Kearns describes colonialism as "always a violent rearrangement of property and person." Kearns, G. 2006. "Bare life, political violence, and the territorial structure of Britain and Ireland," in Gregory, D. and A. Pred (eds.), *Violent Geographies: Fear, Terror, and Political Violence*. New York: Routledge, p. 13.

38 Hussain, N. *The Jurisprudence of Emergency*, p. 4.

between the liberal ideal of uniformity and the competing idea of colonial difference which called for local specificity.

In reality, the actual implementation of the reformed system would have required increased governmental expenditures to enlarge and further professionalise colonial bureaucracies. Such a commitment to public spending was severely undercut by Whitehall's policy that the colonies should pay for themselves. This meant that public policy focused on the expansion of British capitalism in the colonies to increase their tax base rather than to invest directly in the safety and welfare of colonised peoples.[39] This had the effect of structurally undermining and largely invalidating any good intentions of the more humanitarian-minded administrators.[40] I should also point out that by invoking the rule of law, the colonial government emphasised the security functions of the police and criminal justice system and with a few exceptions neglected broader police powers.[41]

The Ceylon Government, although trumpeting the great gift of English justice, only grudgingly spent money on the criminal justice system. For example, although it was believed that well-educated, professionally trained Europeans could more effectively run the bureaucracies, few were hired because they commanded higher wages than the Ceylonese. Consequently, the bureaucracies were overwhelmingly staffed by miserably paid Ceylonese with inadequate training, little commitment to the institutions and virtually no oversight by colonial administrators. The British expected loyalty to bureaucracy and more broadly to English law, and they hoped that civil servants and lower level employees would follow the British bourgeois code where one's personal interests were secondary to that of the bureaucracy in which one serves. However, in the absence of genuine political legitimacy coupled with inadequate salaries and failure to impart the ideals of bureaucracy, resistance was widespread.[42] It is important to note that, as with many other aspects of the criminal justice system in Ceylon, the ideology of the bureaucrat whose highest loyalty is to the state was a relatively recent development in Britain itself.[43]

39 Wiener, M.J. 2009. *An Empire on Trial: Race, Murder and Justice under British Rule, 1870–1935.* Cambridge: Cambridge University Press, p. 13.

40 On humanitarianism and empire see Lester, A. and F. Dussart. 2014. *Colonization and the Origins of Humanitarian Governance: Protecting Aborigines across the Nineteenth-Century British Empire.* Cambridge: Cambridge University Press.

41 Exceptions included irrigation and later sanitation which were important for economic development and security benefiting the British as well as the Ceylonese.

42 For a comprehensive study of everyday corruption by subaltern officials in early twentieth-century Burma see Saha, J. 2013. *Law, Disorder and the Colonial State: Corruption in Burma c. 1900.* London: Palgrave Macmillan. On the importance in nineteenth century Britain of the inculcation of the cultural model of bourgeois virtue and character, see Brown, M. 2014. *Penal Power and Colonial Rule.* Abingdon: Routledge, pp. 41–43; Bellamy, R. 1992. *Liberalism and Modern Society: An Historical Argument.* Cambridge: Polity Press.

43 Osborne, T. 1994. "Bureaucracy as a vocation: governmentality and administration in nineteenth century Britain." *Journal of Historical Sociology*, Vol. 7, No. 3, pp. 289–313.

The struggles within the criminal justice system, which I describe in the following chapters, call into question the extent to which the British were truly committed to the rule of law in the colonies. I will argue that the very notion of rule of law itself was undermined by the idea of racial difference and that it was often in the interstices between the rule of law and belief in racial difference that resistance was able to flourish. It had become abundantly clear in the British attempts to administer their colonies that there was an irresolvable tension between the liberal desire for uniformity in the treatment of all people and prevailing racial theories which posited that different races were mentally and physically unequal. For example, British administrators in Ceylon were concerned that Europeans and "tropical races" such as the Ceylonese reacted differently to certain punishments such as solitary confinement, hard labour, flogging and even incarceration in the tropics, and therefore to give them the same sentences would be to impose a greater punishment on one than the other. Moreover, liberalism as an ideal was also at odds with the optics of racial domination, for Europeans feared that punishing their own in front of local populations would result in all Europeans losing face and consequently their authority being diminished.[44]

The tension between a belief in universal equality and a belief in racial difference underlay the British attempts to design new biopolitical technologies appropriate to the tropics. The British frequently argued that tropical peoples were naturally lazy, deceitful, irrational and cruel. Whether or not such traits were innate or acquired through the impress of the tropical environment was a matter of debate. However, that debate was largely conducted elsewhere at a more nuanced and abstract level of discourse than that in which the "men on the spot" in Ceylon engaged. The latter tended to hold uncritically racist interpretations of their own experiences in the colonies. Given their view that the Ceylonese either could not or would not display sufficient self-control, they assumed that deterrence would best be achieved through the strict application of scientifically calibrated technologies. Experiments were devised so as to tailor appropriate punishment to the Ceylonese body and psyche. The prison system became the site of much experimentation as the management of the daily life of racialised bodies was endlessly quantified.[45] New penal practices, as we shall see, were also imported from Britain where criminology was becoming

---

44 Stoler, A.L. 1995. *Race and the Education of Desire: Foucault's History of Sexuality and the Colonial Order of Things*. Durham: Duke University Press; Stoler, A.L. 2002. *Carnal Knowledge and Imperial Power: Race and the Intimate in Imperial Rule*. Berkeley: University of California Press.

45 Colonies have been conceptualised as laboratories of modernity. Robert Peel expressed this in 1812. He compared Trinidad with "a subject in an anatomy school or rather a poor patient in a country hospital and on whom all sorts of surgical experiments are tried, to be given up if they fail, and to be practiced on others if they succeed." Quoted in Phillips, R. 2017. *Sex, Politics and Empire: A Postcolonial Geography*. Oxford: Oxford University Press, p. 139. Also see Cooper, F. and A. Stoler, eds. 1997. *Tensions of Empire: Colonial Cultures in a Bourgeois World*. Berkeley: University of California Press, p. 5.

increasingly scientific in orientation drawing on the multiple emerging fields of penology, nutrition, bodily mechanics, psychology, criminology, medicine, climatology, and anthropology.

Resistance to the rule of law

Of course, the Ceylonese were not passive ciphers in this project of imperial control. This book details the manner in which they co-opted and surreptitiously resisted the bureaucratic structures at every level and in the process helped to re-shape them. Weber famously observed that an impartial, impersonal, rationally organised bureaucracy is an "iron cage" that controls individuals and limits their freedom. Yet in this book I will explore how the Ceylonese resisted the criminal justice bureaucracies from within as well as without and in doing so not only "rattled" the iron cage, but "rattled" their imperial rulers. The anxieties and frustrations of the British administrators are well documented in the administration reports sent back to London.

Although local people in Ceylon during the nineteenth century resisted the criminal justice system in various ways, at times they embraced it. Some had inculcated a sense of duty, but they were a minority and in many cases members of an educated elite with a greater stake in the system. This is not to claim that resistance was formally organised or centrally coordinated or that it was part of a larger strategy to drive the British from the island, for it was none of those things. A majority of Ceylonese peasants and labourers lived too close to the margins of survival to engage in any formal political activity at least during the nineteenth century. But their resistance wasn't entirely individual or unorganised either. Individuals could rarely act completely on their own without some degree of community support or at the very least tacit acceptance. As I will show, their actions were collective in the sense of including standardised sets of practices, which necessitated the complicity of fellow villagers who maintained barriers of silence and dissimulation.

In the few instances when overt resistance had taken place, it was easier for the British to counter. Leaders were rounded up and imprisoned or killed and concerted resistance snuffed out by the army and police, as happened in the early years of British rule and again during a small rebellion in the highlands in 1848.[46] On the contrary, most resistance was headless, and organic, welling up in spatially dispersed locations throughout the country without direction in a manner similar to what Foucault calls a "swarm of points of resistance" that rarely announced itself as such.[47] Scott refers to this as a kind of shadow

---

46 For an interesting account of unsuccessful open resistance to the British through the re-appropriation of ceremonies associated with Kandyan kingship during the first half of the nineteenth century see Wilson, J. 2017. "Re-appropriation, resistance, and British autocracy in Sri Lanka, 1820–1850." *The Historical Journal*, Vol. 60, No. 1, pp. 47–69.

47 Foucault, *The History of Sexuality*, p. 96.

society that existed under the colonial order of things; a structure of normative opposition to ruling elites. Such a culture of opposition, he writes, can produce "a form of tacit coordination that mimics or substitutes for formal organization." To the extent that the coordination of individual acts existed, Scott argues, it was the product of "dense informal networks" and "historically deep subcultures of resistance to outsider claims" rather than the coordination of formal institutional settings.[48] However, I argue that even the least organised, individual resistance can be considered political.[49] While recognising subversive acts as failures to submit to their rules, the British effectively depoliticised them by treating them not so much as political opposition, but as expressions of the inherent mendaciousness, incompetence and backwardness of the Ceylonese. This racist attitude of low expectations pervaded the British dealings with the local populations and, as I will show, opened up spaces of resistance allowing resetting by the Ceylonese of the norms of what was expected of them.

During the second half of the nineteenth century, it is unlikely that the peasants who resisted the rule of law believed that the British could be driven out of the island. For, after all, Europeans had controlled the coasts of Ceylon for centuries and the political revolts in the interior in 1818 and 1848 were brutally crushed.[50] Visible, organised and violent political challenges to the British presence on the island were a loser's game as the Ceylonese had discovered. Rather, their resistance was at the level of ignoring when possible laws that curtailed their traditional activities such as grazing on common lands, deceiving or threatening representatives of the state such as non-local policemen, and appropriating the court system by using it in ways that benefitted themselves or their community. The Ceylonese learned from experience that the British punished what they perceived to be racial shortcomings much less harshly than they did open opposition. Consequently, the Ceylonese adopted what Scott has termed "the weapons of the weak"; the tactics of those who cannot openly challenge overwhelming power, and therefore employ covert strategies which ingeniously

48 Scott, J.C. 2013. *Decoding Subaltern Politics: Ideology, Disguise, and Resistance in Agrarian Politics*. London: Routledge, pp. 10, 70, 94.

49 Scott refers to this type of politics which occurs outside of the recognised channels of politics as "infrapolitics." (Scott, J.C. 1990. *Domination and the Arts of Resistance*. New Haven: Yale University Press). Guha and the subaltern studies school also considered that poor peasants engaging in atomistic and uncoordinated resistance was a political act. They rejected the interpretations of Eurocentric historians who have tended to see these as merely "pre-political." Guha, R. 1999. *Elementary Aspects of Peasant Insurgency in Colonial India*. Durham: Duke University Press, p. 5. See also Legg, S. 2016. "Empirical and analytical subaltern space? Ashrams, brothels and trafficking in colonial Delhi." *Cultural Studies*, Vol. 30, No. 5, pp. 793–815; Legg, S. and T. Jazeel. 2019. *Subaltern Geographies*. Athens: University of Georgia Press.

50 Likewise in 1915 Muslim-Sinhalese riots were misinterpreted by the British as a political revolt and brutally suppressed. Martial law was declared and lasted for three months and there were numerous arrests of middle class Sinhalese on political grounds. The reaction of the British served as an impetus to the nascent nationalist movement. See Fernando, P.T.M. 1970. "The post-riots campaign for justice." *The Journal of Asian Studies*, Vol. 29, No. 2, pp. 255–66.

exploit the cracks in the structures of domination. Such resistance was possible in large part because of the uneven geography of British surveillance and control. There were large areas throughout the island where a subculture of complicity and a conspiracy of silence could thrive, which made it extremely difficult for outsiders, whether they were British or non-local Ceylonese officials, to collect information on village affairs.[51] This was compounded by the normalisation of dissimulation when dealing with outsiders, which meant that even when the conspiracy of silence was broken, the authorities found it extremely difficult to discover who was telling the truth.

In this book I trace the contours of resistance through three bureaucracies: the police, the courts and the prisons. In all three cases, the peasants, and just as importantly the civil servants who formed the lower echelons of the bureaucracies, systematically disregarded and undermined the British laws and regulations that those bureaucracies were set up to enforce. Lacking loyalty to the colonial state, they saw the goals of the bureaucracies as incongruent with their own interests. And so villagers cooperated with the police only when they considered it to their advantage to do so. And the village police headmen (*vidanes*) opposed where possible the central government police. The latter in turn often extorted and brutalised the peasantry when they thought they could operate unseen by higher level administrators. The peasants embraced the courts as this was the most effective way to use the power of the state. They flooded the courts with false cases against their neighbours. Perjured testimony was the norm, more often than not practiced in collusion with subordinate police and court officials. Consequently, the whole judicial system from the police through to the courts and prisons was rife with bribery and other forms of resistance to the system.[52] The prisons, which in theory constituted spatially enclosed, total institutions, where every activity was designed to be visible and regulated, in fact became sites of artful resistance to disciplinary control by prisoners and guards alike.

Such was the scale of resistance throughout the three bureaucracies that the British wondered at times where, if anywhere, and to what extent rule of law could be said to exist on the island. And precisely because resistance was spatially dispersed and individual rather than coordinated at a few centralised sites, and because it was very often regarded as a product of the racial failure of a whole population rather than systematic political opposition, the British despaired at how to confront and combat it. The Ceylonese in turn were clearly aware that the British were unwilling to pay the high financial price necessary

---

51 Hay found a similar conspiracy of silence among villagers in eighteenth-century England in regard to poaching. Hay, D. 1975. "Poaching and the game laws on Cannock Chase," in Hay, D., P. Linebaugh, J.G. Rule, E.P. Thompson and C. Winslow (eds.), *Albion's Fatal Tree: Crime and Society in Eighteenth-Century England.* New York: Pantheon, pp. 189–254.

52 Saha, *Law, Disorder and the Colonial State*, argues that in early twentieth-century Burma such corruption, rather than being an aberration, came to define the nature of the colonial state.

to bring them into line and so, as we shall see, continual everyday resistance through small individual acts had, over time, a massive aggregate effect.[53]

While I draw on an expansive notion of resistance developed by Scott and others,[54] I wish to push the notion a bit further by arguing that subordinate groups exploiting each other also constitutes resistance. What I have found in trying to understand the criminal justice system is that the Ceylonese peasants subverted the system principally by exploiting it in order to take revenge on fellow villagers. Their goal was usually to achieve justice where the ends as they saw them would justify the means. Thus it was peasants flooding the courts with false charges against other peasants and low-level bureaucrats extorting poor peasants that put the system under severe strain. I have followed Courpasson and Vallas' advice to avoid sanitising the internal politics of the dominated, for what I have found is that when subalterns defied the powerful, they often did so in ways that exploited those who were weaker than themselves.[55] It seems that their resistance to the criminal justice system was not always consciously intended as such, but was effective in subverting it nevertheless. They used the system as a technology "to hand" in the sense of bricolage, for as Judith Butler argues, the weak can only use the structures available to them. They do not have the power to create anew, but become agents of the systems that oppress them even as they undermine them and shape them to their own ends.[56] Similarly, in *Law, Disorder and the Colonial State*, Saha discusses the subversion by lower level officials, not as the corruption of an ideal but as constitutive of the "everyday state."[57]

Steve Legg writing on Foucault's use of the concept of *parrhesia* as courageous speech or "speaking truth to power," argues that resistance should "be viewed as 'productive,' rather than romantically and uniformly 'positive.'"[58] In my discussion of the Ceylonese use of the courts, I will give some examples of

53 Scott, *Decoding Subaltern Politics*, p. 71.
54 Scott, J.C. 1985. *Weapons of the Weak: Everyday Forms of Peasant Resistance*. New Haven: Yale University Press; Scott, *Domination and the Arts of Resistance*; Scott, *Decoding Subaltern Politics;* De Certeau, M. 1984. *The Practice of Everyday Life*. Berkeley: University of California Press. For excellent reviews of conceptions of resistance, see Vinthagen, S. and A. Johansson. 2013. "Everyday resistance: exploration of a concept and its theories." *Resistance Studies Magazine*, Vol. 1, No. 1 pp. 1–46; Courpasson, D. and S. Vallas, eds. 2016. *The Sage Handbook of Resistance*. Los Angeles: Sage.
55 Courpasson, D. and S. Vallas. 2016. "Resistance studies: a critical introduction," in Courpasson, D and S. Vallas (eds.), *The Sage Handbook of Resistance*. Los Angeles: Sage, p. 24.
56 Butler, J. 1997. *The Psychic Life of Power: Theories in Subjection*. Stanford: Stanford University Press.
57 On the everyday state see, Véron, R., S. Corbridge, G. Williams and M. Srivastava. 2003. "The everyday state and political society in Eastern India: structuring access to the employment assurance scheme." *The Journal of Development Studies*, Vol. 39, No. 5, pp. 1–28; Fuller and Benei, *The Everyday State*.
58 Legg, S. 2019. "Subjects of truth: resisting governmentality in Foucault's 1980s." *Environment and Planning D.: Society and Space*, Vol. 37, No. 1, p. 29.

more common, less heroic agency than *parrhesia*. In court "speaking lies to power" was a form of resistance which, although perhaps not "transformative," was productive of change and adaption in the criminal justice system.[59] I will argue that "speaking lies to power" was effective in achieving customary notions of justice, what Legg calls resistant "truths" in the sense that Foucault speaks of competing "regimes of truth."[60] It appears that often in court the Ceylonese believed that the ends justified the means, with the caveat that of course not all such deception was for noble causes. Alternative views of justice or "truth" refers not so much to facts, but what "should be" in the normative sense.[61] As we will see, false cases were sometimes brought into court by villagers with the intention of getting retribution for perceived injustices done by individuals or factions in the past or over time. The actual facts of the case being tried were for the villagers often of little importance.

In this regard, I use the term corruption throughout this study, but given that it had a negative valence in the nineteenth century as it does at present, I should make clear at the outset that I am using it in an ambivalent manner. First, there is little question that corruption, understood as the use of public office for private gain, destabilised the criminal justice bureaucracy in Ceylon and greatly troubled the British. In this sense it was clearly a form of resistance to the forced importation of a foreign institution.[62] Second, the notion of corruption is intimately linked to the western cultural model of bureaucracy and the moral role of the individual within it. Writers like Das, Jeffrey and Sneath have suggested that in some cultures the line between gift giving and bribery is ambiguous.[63] Dwivedi goes further in challenging the bureaucratic-centric view of corruption. He argues that the charge that corruption is unethical can be called into question if one holds in mind that at many times and in many places, loyalty to family and faction are considered more important than loyalty

59 Similarly in nineteenth-century Ireland dissimulation was used as a substitute for weapons. Personal communication from David Nally.

60 Legg, S. 2019. "Colonial and nationalist truth regimes: empire, Europe and the later Foucault," in Legg, S. and D. Heath (eds.), *South Asian Governmentalities: Michel Foucault and the Question of Post-colonial Orderings*. Cambridge: Cambridge University Press, p. 115.

61 See Legg on truth telling as the revealing of inequities that can occur through actions as well as words. One might say resistance as the undermining of the court processes could be seen in this light. *Ibid.*, p. 115.

62 For African examples of this see, Felices-Luna, M. 2012. "Justice in the Democratic Republic of Congo: practicing corruption, practicing resistance." *Critical Criminology*, Vol. 20, pp. 197–209; Smith, D.J. 2007. *A Culture of Corruption: Everyday Deception and Popular Discontent in Nigeria*. Princeton: Princeton University Press; Pierce, S. 2016. "The invention of corruption: political malpractice and selective prosecution in colonial northern Nigeria." *Journal of West African History*, Vol. 2, No. 2, pp. 1–28.

63 Das, V. 2015. "Corruption and the possibility of life." *Contributions to Indian Sociology*, Vol. 49, No. 3, pp. 322–43; Jeffrey, C. 2002. "Caste, class, and clientelism: a political economy of everyday corruption in rural north India." *Economic Geography*, Vol. 78, No. 1, pp. 21–41; Sneath, D. 2006. "Transacting and enacting: corruption, obligation and the use of monies in Mongolia." *Ethnos: Journal of Anthropology*, Vol. 71, No. 1, pp. 89–112.

to a bureaucracy or the state. In fact "to resist bribery and not resort to nepotism may very well constitute avoidance of the responsibilities of customary citizenship."[64] Such a view fundamentally contradicts the bureaucratic culture of western modernity that the British sought to impose on Ceylon. It suggests that the British view of the immorality of corruption was not always shared by a majority of the Ceylonese, who tended to theorise morality in terms of traditional obligations.[65] Pierce, "The invention of corruption," pp. 8, 21 argues that the system of patron-clientage that characterised many pre-colonial societies appeared to the British as corruption. This attitude pathologised indigenous political cultures. He further argues that while the British deplored corruption, they found it politically useful as it justified their disciplinary presence.

Holmes' analysis of contemporary corruption helps us understand corruption in nineteenth-century Ceylon. He argues that what is called corruption is more likely to be found under the following conditions: where officials have a tradition of accepting gifts as payment for service, where the state lacks legitimacy and therefore people have little public trust in the state, where officials themselves are widely seen to evade the rule of law, where people have little stake in the system and therefore feel little incentive to obey laws they do not recognise, and where there are low levels of social trust and people believe that others are not following the rules and therefore they will get taken advantage of if they do.[66] Nineteenth-century Ceylon, as we shall see, was a perfect storm in terms of creating conditions that encouraged corruption both in the negative sense of failure of civil servants to abide by the rule of law and in the potentially productive sense of resistance that undermines unjust foreign rule. Muir and Gupta acknowledge the ambivalence of corruption when there is a "lack of alignment between the ethical and the legal that makes the corruption/anti-corruption complex such a site for debate, mobilization and contestation."[67] However, Gupta forcefully argues that corruption can be a form of structural violence in that it often serves to further oppress the poorest members of society.[68] I agree with Gupta in that my study provides many clear examples of resistance to bureaucratic rules that take the form of oppression of the weak.

64 Dwivedi, O.P. 1967. "Bureaucratic corruption in developing countries." *Asian Survey*, Vol. 7, No. 4, p. 248.
65 Similarly, Jeffrey, "Caste, class and clientelism," p. 37, found that whether an act was considered a "bribe" or a "gift" depended on who was doing it. All of this suggests that, in practice, the notion of bribe is ambiguous.
66 Holmes, L. 2015. *Corruption: A Very Short Introduction*. Oxford: Oxford University Press.
67 Muir, S. and A. Gupta. 2018. "Rethinking the anthropology of corruption: an introduction to supplement 18." *Current Anthropology*, Vol. 59, Supplement. 18, p. 57. Also see, Pardo, I. 2004. "Introduction: corruption, morality and the law," in *Between Morality and the Law: Corruption, Anthropology and Comparative Society*. New York: Routledge.
68 Gupta, A. 2012. *Red Tape: Bureaucracy, Structural Violence and Poverty in India*. Durham: Duke University Press.

Colonised Ceylon

The island of Ceylon lies forty miles off the southern tip of India, and has experienced a continual flow of peoples in both directions over the millennia.

From 250 BCE the island came under the influence of Buddhism, but the Buddhist kingdoms were usually in competition with competing Tamil kingdoms in the northern parts of the island. In the early sixteenth century, as the Portuguese were moving to control trade in the Indian Ocean, the island was divided among the Tamil Kingdom of Jaffna in the north, the Kandyan Kingdom in the central highlands, and the Kingdom of Kotte in the south. The Portuguese began trading with Kotte in 1505 and a century later had conquered the island with the exception of the Kandyan Kingdom in the inaccessible central highlands. Although the Portuguese were unsuccessful in their attempts to conquer the Kandyan Kingdom, they nevertheless were able to reduce it to penury by largely blocking external trade. At the beginning of the seventeenth century the Dutch East India Company (VOC) entered the Indian Ocean to challenge Portuguese control over the lucrative trade routes and four decades later in 1640 a combined force of Dutch and Kandyans expelled the Portuguese from the island. But the Kandyans discovered that they had merely exchanged one adversary for another, as the Dutch took over the Portuguese territories along the coasts and squeezed Kandyan trade even harder. At the end of the eighteenth century the Kandyans enlisted the British to drive out the Dutch, only to find once again that their new allies refused to leave. The British not only took over the Dutch possessions along the coast but in 1815 managed to accomplish what the Portuguese and Dutch had been unable to do: to conquer the Kandyan Kingdom and take possession of the whole island. Although by the early nineteenth century the island was administered by the British, the imprint of earlier colonial socio-political patterns remained, with a Tamil population concentrated in the dry north and northeast of the island, a maritime Sinhalese population concentrated in the south and a Kandyan Sinhalese population in the central highlands. Muslim traders, known as "Moors," had migrated to Ceylon beginning in the seventh century forming a sizeable community of perhaps a tenth of the population and were scattered across the island. There were also small communities of Malay Muslims brought as soldiers first by the Dutch and later by the British, and a few very small groups of indigenous *Veddahs*. To this mix the British added, beginning before mid-century, a large number of poor Tamil seasonal workers from India working on the British-owned plantations, as most Ceylonese refused plantation labour due to low wages. Unlike in the south of India where the peasantry were land poor, in Ceylon peasant agriculturalists usually had enough land to sustain themselves. Having said that, the system of inheritance that divided land into increasingly small portions generated a great deal of conflict throughout the nineteenth century and left some Ceylonese landless and thus willing to work on the plantations.[69]

---

69 For a discussion of this, see Duncan, *In the Shadows of the Tropics.*

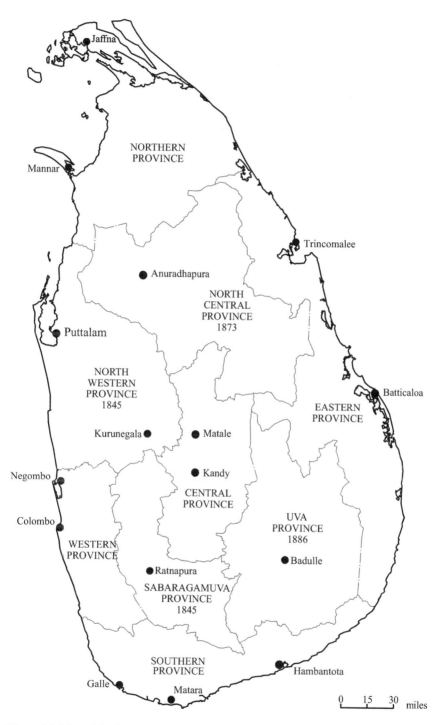

Figure 1.1 Map of Ceylon.

While the British initially seized the island to deny the Dutch and their French allies the strategically important port of Trincomalee, by the late 1830s they realised that the island could be made into an exploitation colony largely through the development of export crop plantations in the central highlands. To this end, beginning in 1840, extensive common (*chena*) lands that had been used to supplement the peasantry's rice lands were expropriated by the state and sold to British and Ceylonese capitalists for coffee plantations.[70] Increasingly throughout the nineteenth century, socio-cultural changes to Ceylonese society followed the geographically uneven penetration of capitalism across the island. The areas in the maritime south that had been longest exposed to a European presence were affected the most along with the areas around the plantations in the highlands, while the more remote areas that attracted little economic change experienced much less social change.

The colonial bureaucracies

At the apex of the administration of the island stood the governor and below him the Colonial Secretary who served as his assistant and in many respects the principal administrator on the island. In 1833 a Legislative Council was created to advise the governor and vote on legislation. During the nineteenth century it evolved to consist of eighteen members: ten government officials and eight non-official members (three Europeans, one lowland Sinhalese, one Kandyan Sinhalese, one Tamil, one Muslim and one Burgher) representing the different sub-populations of the island. While every member had a vote, the government officials held a majority. Furthermore, all of the non-official members were appointed by the Governor and consequently were carefully chosen people whom the administration thought would be seen to represent their ethnic, and in the case of the Europeans, commercial constituencies without seriously rocking the boat.

The administration of the island was divided into separate departments and while it was held as a tenet of good government that throughout the colonies there should be a separation between the executive and the judiciary, in fact, as we shall see, there was considerable blurring, as most judicial positions were appointed from the civil service rather than from the bar. The structure of the government was as follows: next in importance to the Governor and Colonial Secretary were the government agents (GAs) of the provinces. Each was the chief civil authority in his province and was based in the *kachcheri* (government office) of the principal town. Below him was the assistant government agent (AGA) based in an "outstation" in one of the more important districts of the province. Because of the size of provinces and frequent absence of the GA, the AGAs wielded considerable power. Below the AGA were junior civil servants

---

70 The economic, political, social and ecological consequences of this are the subject of Duncan, *In the Shadows of the Tropics*.

Table 1.1 The population of the island 1881–1911

	1881	%	1891	1902	1911
Sinhalese	1,848,842	67.0	2,041,853	2,381,013	2,715,456
Tamils	687,248	25.0	723,853	963,372	1,059,007
Moors	179,542	6.5	197,166	230,150	233,901
Burgher	17,886	0.65	21,213	23,732	26,663
Malays	8,895	0.3	10,133	11,939	12,990
Others	8,506	0.3	9,570	13,444	18,017
Europeans	4,639	0.17	6,068	6,336	7,592
Total	2,755,558		3,008,466	3,629,98	4,106,350

who were "learning the ropes."[71] All of these positions were filled by Europeans into the second decade of the twentieth century. Below these European officials in the *kachcheris* were a battery of minor clerks, interpreters and peons who were overwhelmingly Ceylonese. The more responsible of these positions tended to be given to Burghers, especially before mid-century. Beyond the *kachcheris* was an array of Ceylonese officials. In 1881 there were 208 Ceylonese officials serving under the provincial GAs and AGAs, including thirty-four presidents of the *gansabhavas* (village courts) and various other important officials.[72]

The island was divided into three main divisions: the maritime districts inhabited mainly by the lowland Sinhalese; the Kandyan districts; and the Tamil districts in the north and east inhabited overwhelmingly by Tamils and Muslims. Each of these divisions was composed of districts under an AGA. Under the AGA were various Ceylonese officials. Each district had a principal headman, called a *mudaliyar* in the maritime districts, a *ratemahatmaya* in the Kandyan districts and a *maniakar* or *mudaliya* in the Tamil districts. Under them were a number of minor officials including *vidane archchis* and under them village headmen called *vidanes*, some of whom had police responsibilities and some of whom had revenue duties. The titles of these positions differed between Tamil-speaking areas, the Kandyan highlands and the maritime districts, but to avoid confusion I use the terms headman or *vidane*, and police headman or police *vidane*.[73]

Up until 1845, almost all senior administrative positions were in the hands of resident British families. But as this pool of talent was small, young, single

71 Samaraweera, V. 1981. "British justice and the 'oriental peasantry'; the working of the colonial legal system in nineteenth century Sri Lanka," in Crane, R.I. and G. Barrier (eds.), *British Imperial Policy in India and Sri Lanka, 1858–1912*, Columbia: South Asia Books, p. 112.
72 Mills, L.A. 1964. *Ceylon under British Rule, 1795–1932*. London: Frank Cass, p. 97.
73 Giles, A.H. 1889. *Report on the Administration of Police, Including the Actions of the Courts and the Punishment of Criminals in Ceylon*. Colombo: J.A. Skeen, Government Printer, pp. 392–95.

men from Britain were actively recruited after this date.[74] Henceforth there were two separate streams of recruitment, with the one from Britain thought to have the most promising candidates for eventual promotion to senior positions. Up until 1854, their recruitment was based upon the Haileybury Entrance Examination, taken after two years' attendance at Haileybury, the old East India College founded in 1808 to train administrators for South Asia. Candidates were drawn from Haileybury, rather than from universities, in part because the latter were less likely to consider a position in a small colony like Ceylon, but also because it was argued, that "what was needed in a service which had to deal with backward peoples was not so much brains as personality and character."[75] Leonard Woolf, the future husband of Virginia, who joined the Ceylon Civil Service in 1904 after completing his degree at Cambridge, concurred with this low assessment of the intellect of his fellow civil servants. He claimed he had to hide his intelligence to fit in with his fellow civil servants and planters and that the most important quality for getting ahead were those old public school "virtues" of being a good fellow and a gentleman.[76] Young civil servants coming out to Ceylon were expected to learn enough of either Sinhalese or Tamil after they arrived to allow them to communicate with the people, but the enforcement of this rule was lax.[77]

A second stream of candidates was drawn from Ceylon itself, principally from the European community, and the Burghers. They were appointed by the Governor after taking a non-competitive examination to test their knowledge of

74 De Silva, K.M. 1973. "The development of the administrative system, 1833 to c. 1910," in De Silva, K.M. (ed.), *History of Ceylon, Vol. 3*. Colombo: University of Ceylon Press Board, p. 214.

75 Warnapala, W.A.W. 1974. *Civil Service Administration in Ceylon*. Colombo: Department of Cultural Affairs, p. 49. There was a distinct hierarchy within Britain's civil service. Those candidates with the highest marks in the civil service exams normally chose the Home Civil Service. The next group chose to go to India, and those with the lowest marks had to settle for an Eastern Cadetship. Within this latter category, Ceylon would normally be the first choice, followed by Hong Kong and Malaya. Some governors, such as Arthur Gordon in the late nineteenth century, were highly critical of the Ceylon Civil Service, as in his opinion, the British candidates were often not strong and the local-born European pool was too small to provide a strong set of candidates. Peebles, P. 1981. "Governor Arthur Gordon and the administration of Sri Lanka, 1883–1890," in Crane, R. T. and N.G. Barrier (eds.), *British Imperial Policy in India and Sri Lanka, 1858–1912*. Columbia: South Asia Books, pp. 86, 97.

76 Woolf, L. 1961. *Growing: An Autobiography of the Years 1904–1911*. New York: Harcourt Brace Jovanovich, pp. 36–37. While Woolf is dismissive of his fellow servants, they were at the top of the social heap of European society in Ceylon. Civil servants tended to look down on European businessmen and even more so on planters whom they found rough and aggressive, even violent, especially towards the migrant labourers from India. Woolf notes that in his seven years in Ceylon he never once dined with a European businessman. While on issues of political and commercial importance they tended to support each other, socially, the European community was both fractured and hierarchically ordered.

77 Coperehewa, S. 2011. "Colonialism and problems of language policy: formulation of a colonial language policy in Sri Lanka." *Sri Lanka Journal of Advanced Social Studies*, Vol. 1, No. 1, pp. 27–52.

English composition, accounts and book keeping, algebra and geometry and geography. When non-Europeans were admitted to the higher levels of the civil service they were placed in the judicial, rather than the better paid, revenue branch of the administration. It was telling of British attitudes toward the criminal justice system that the judicial was considered less important. In the early 1870s every single non-European in the civil service was in the judicial branch, along with the less successful British civil servants. The particularly successful Ceylonese could progress up to the relatively powerless but prestigious post of solicitor general.[78] There were, however, large numbers of Ceylonese holding lower level positions. In 1868, out of 1,084 appointments, 894 were held by Ceylonese including many Burghers and only 190 by Europeans. However, all of the high positions were held by Europeans with the Ceylonese filling out the ranks as interpreters, clerks, surveyors, etc.[79] By the 1880s the number of non-European clerks, interpreters and other minor officials had grown to 1,300.[80] But there were still no non-Europeans occupying senior positions in the revenue department. In fact, it was not until after the first decade of the twentieth century that a Ceylonese was appointed to the position of AGA.[81]

In 1883 a Committee of the Legislative Council recommended that "if Ceylon is to be administered on a more economical basis, it can only be by more extended employment of natives."[82] There was also pressure brought to bear in the Legislative Council by Ceylonese members and one of the members for the local European community to open up the revenue department to the Ceylonese. In both cases vague assurances were given that change would take place.[83] The policy became to appoint Ceylonese as office assistants and magistrates at the larger stations "where they would receive the guidance of higher European officers."[84] In 1891 a Lower Division of the Ceylon Civil Service was created that was reserved for locals; but it was mainly Burghers and members of the local European community who filled these lower posts.

Having said this, it is important to emphasise that throughout the nineteenth century the colonial bureaucracies were run by a small handful of Europeans located mainly in the capital, Colombo with a few district officers spread out across the country in the major towns. As was the case with British colonies elsewhere in India and Africa, the government was dependent upon local people to fill over 90% of the positions within the government bureaucracies. As such, although British officials made the policies, it was overwhelmingly Ceylonese

78 Warnapala, *Civil Service Administration*, p. 53.
79 Mills, *Ceylon under British Rule*, p. 91.
80 Governor to Secretary of State, CO 54/531.
81 De Silva, "The development," p. 223.
82 *Sessional Papers* (*SP*)XXVII, 1897.
83 "The Ceylonese in the civil service," *Ceylon Hansard CH* 1882–83, pp. 47–48; "The admission of native and colonial-born Europeans into the Ceylon Civil Service." *CH* 1888–89, pp. 93–94.
84 *SP* XXVII, 1897.

who were supposed to implement them.[85] In effect, the British were forced to govern at a distance not only culturally, but physically as well, due to the existence of thousands of villages that the government in Colombo could not effectively penetrate.[86] Saha's claim about colonial Burma applies equally to Ceylon. It was subordinate "native" officials who overwhelmingly made the colonial state through their quotidian practices.[87]

On terminology

I use the names Ceylon and Ceylonese throughout because those were the names used by the British at the time. The name of the country was changed to Sri Lanka in 1972. The British reports divided the island's peoples into various ethnic groups. The terms British and European were often used interchangeably as European was considered a racial category at the time. Europeans born in Ceylon were often considered inferior to those born in Europe, due to the prevailing Lamarkian belief in tropical degeneration.[88] The term Burgher included people of Dutch ancestry and of mixed Dutch and other non-European ancestry. The largest group were called "natives" by the British and these included the Sinhalese, Tamils, Ceylon Muslims who were called "Moors" and Malayan Muslims called "Malays." The Tamils included Ceylon Tamils and the recently arrived Indian Tamils who were migrant workers. The British term "native" referred to the non-European Ceylonese, originally not including Burghers. However, with the popularity of racial "science" towards the end of the century, being of mixed race was increasingly seen as negative and so Burghers began to be included officially in the category "native."[89] Throughout the book I have made reference to some offensive terminology current at the time and present

85 Crooks, P. and T.H. Parsons. 2016. "Empires, bureaucracy and the paradox of power," in Crooks, P. and T.H. Parsons (eds.), *Empires and Bureaucracy in World History: From Late Antiquity to the Twentieth Century.* Cambridge: Cambridge University Press, pp. 1–20; Parsons, T.H. "The unintended consequences of bureaucratic 'modernization' in post-World War II British Africa," in Crooks, P. and T.H. Parsons (eds.), *Empires and Bureaucracy in World History: From Late Antiquity to the Twentieth Century.* Cambridge: Cambridge University Press, pp. 412–33; Heath, D. 2016. "Bureaucracy, power and violence in colonial India: the role of Indian subalterns," in Crooks, P. and T.H. Parsons (eds.), *Empires and Bureaucracy in World History: From Late Antiquity to the Twentieth Century.* Cambridge: Cambridge University Press, pp. 364–90.

86 On governing at a distance see Berry, S. 1992. "Hegemony on a shoestring: indirect rule and access to agricultural land." *Africa*, Vol. 62, No. 3, pp. 327–55; Rose, N. and P. Miller. 2010. "Political power beyond the state: problematics of government." *British Journal of Sociology*, Vol. 61, No. 1, pp. 271–303, and in regard to India see Cohn, B. 1987. *An Anthropologist among the Historians.* Oxford: Oxford University Press, p. 512.

87 Saha, *Law, Disorder and the Colonial State*, p. 49.

88 On Lamarkianism and degeneration see Schuller, K. 2018. *The Biopolitics of Feeling: Race, Sex and Science in the Nineteenth Century.* Durham: Duke University Press.

89 For British ambivalence about mixed-race people see Sivasundaram, *Islanded*, pp. 294–99.

some astonishingly racist quotations in order to authentically convey the attitudes of many of the British about race.

I use the term subaltern to describe the most powerless and oppressed of the Ceylonese populations. However, many of those who committed crimes cannot be considered subaltern. As Rogers points out much of the crime in Ceylon was not committed by the poorest classes.[90] Hunger or famine crime existed, but was not as prevalent as other types of crime. Many criminals, he points out, were from the so-called "respectable" classes and castes. For example, headmen, many of whom were rich land owners, were often involved in crime. On the other hand, the lower level bureaucrats, including prison guards, court peons and policemen, who were unskilled, landless and poorly paid, can properly be considered subaltern. As we shall see, all classes were actively and productively engaged in the criminal justice system including subalterns who had what Legg describes in his essay on subalternity, as attenuated or "partial agency."[91]

On method

With some prominent exceptions among the English speaking elite, such as the Legislative Council members and the judges whom I quote, the Ceylonese have no significant voice in the historical records. As Amir points out, "Peasants do not write; they are written about."[92] The voices of the subaltern are rarely found in the archives, not only because of illiteracy, but because those in power seldom recorded their voices except under unusual circumstances; circumstances that usually boded ill for them. Also included among the most invisible are the majority of women of all classes as their presence in the records is virtually absent. Women rarely participated in the criminal justice system either as litigants or officials and so there is virtually nothing I can reliably say about them. The British were astonished at how few women committed crimes and women only constituted between 1 and 2% of all prisoners. Their influence is not to be doubted, but it would have been in unofficial and private, unrecorded capacities.[93]

90 Rogers, *Crime, Justice and Society*, p. 2. "Unlike most societies, the social profile of persons treated as criminal was not weighted towards the poor and otherwise disadvantaged. Crime was committed largely by Sinhalese adult men of respectable caste."

91 I use the term subaltern as a relative rather than absolute term. In other words, I am not using it in the more philosophically rigorous sense of radical alterity as exemplified in Spivak's question, "Can the subaltern speak?" Spivak, G.C. 1994. "Can the subaltern speak?" in Williams, P. and L. Crisman (eds.), *Colonial Discourse and Postcolonial Theory*. New York: Columbia University Press, pp. 66–111. Legg, "Empirical and analytical subaltern space?" Also see Legg and Jazeel, *Subaltern Geographies*.

92 Amin, S. 1995. *Event, Metaphor, Memory: Chauri Chaura 1922–1992*. Delhi: Oxford University Press, p. 1.

93 They were, however, victims of male violence and were held by some European officials to be a cause of male on male violence. For a discussion of the causes of crime see Rogers, *Crime, Justice and Society*.

Even when one finds the subaltern voice in official writings, as Scott argues, it must be problematised for speaking truth to power is extremely dangerous.[94] What we are likely to find, he tells us, is "the dissembling of the weak in the face of power."[95] As the public transcript is controlled by the powerful, it is unlikely to reveal a true subaltern voice. It tends to display "performances of deference" from subordinates, as they carefully tailor their views so as not to suffer the consequences of holding views subversive to the system. When the opinions of villagers, prisoners, police constables, police *vidanes* and court clerks appear in the public transcript, they are normally quoted to support an administrator's view.[96] The voices of a small number of Ceylonese elites appear occasionally in the official records and newspapers, especially those of judges and members of the elite who represented their ethnic communities on the Legislative Council. However, as these were government appointments, they cannot be considered truly representative. They did often claim to present local views, but there is no way to know the extent to which these elites were really knowledgeable or concerned about the peasantry and urban lower classes. Certainly the Ceylonese judges' views of their fellow countrymen were not dissimilar from those held by British judges, although unsurprisingly the former tended to qualify their negative opinions by suggesting that they only applied to non-westernised Ceylonese.

While the voices of the non-elite Ceylonese were rarely heard, there was a great deal of commentary by officials on their behaviour and imputed motives. While I am greatly suspicious of administrators' descriptions of the motives behind or causes of the behaviour of Ceylonese, for they are often self-justificatory and riven by racism, I am much more ready to accept that much of the behaviour described is accurate. For example, when an AGA wrote that the people of his district are too lazy even to steal, I accept that behind this slur, there is little reported theft in that district. Or when a judge wrote that the Ceylonese have a natural talent for lying, I accept that people lied in the courts of law, while suspecting that the reason for this behaviour might have to do with the lack of moral authority of the courts and alternative conceptions of justice. My approach is somewhat similar to Guha's in using elite sources. However, while he argues that one can begin to reconstruct subaltern consciousness from

94 Scott, J.C. 1992. "Domination, acting and fantasy," in Nordstrom, C. and A. A. Martin (eds.), *The Paths to Domination, Resistance and Terror.* Berkeley: University of California Press, pp. 55–84.

95 *Ibid.*, p. 56.

96 Two notable exceptions to this methodological constraint within the south Asian historical literature are found in the work of Anderson and Pieris. They were able to draw upon letters of appeal to officials as well as the very rare diaries of elite prisoners. Both authors have been able to skilfully use these to reconstruct a number of biographies of transported prisoners. Anderson, C. 2012. *Subaltern Lives: Biographies of Colonialism in the Indian Ocean World, 1790–1920.* Cambridge: Cambridge University Press; Pieris, A. 2009. *Hidden Hands and Divided Landscapes: A Penal History of Singapore's Plural Society.* Honolulu: University of Hawai'i Press.

these sources by, as he puts it, reading these sources as "writing in reverse."[97] I sincerely doubt that one can actually recover such consciousness. On the other hand, I do believe that one can reconstruct specific actions and, to an extent, generalised behaviour, while trying to be acutely aware of exaggerations and crude stereotypes. This can be done by focusing on what the authorities considered to be the problems they had in dealing with the local population. Their reports reveal the strategies they used to counter what were seen as the continual violations of the rule of law.[98] The reports also open a window on the relative failure of these strategies as they were full of recriminations against not only the general public who filled the courts, but also against subordinate officials whom they believed to be corrupt, and at times their fellow senior officials who they judged to be wrong-headed or simply incompetent. By using behavioural data, I can find out what people were actually willing to do in order to appropriate the British administrative structures in an attempt to make the best of a bad situation.

Organisation of the book

In the following chapter I first discuss the theories of criminality that informed European views of crime and its deterrence in the nineteenth century. Particular attention is paid to utilitarian theories, as these were most influential on senior administrators in Ceylon. Having said that, most administrators were lightly educated pragmatists who understood their utilitarian ideas to be empirically derived common sense. In the second part of the chapter, I survey administrators' attitudes towards crime as expressed in annual reports and commissions, showing how their ideas were in practice cross-cut by evolving and in many respects hardening ideas of racial difference, degeneration and tropicality. In the third part of the chapter I briefly discuss administrators' views of the types and amount of crime on the island, noting their doubts that the figures they were working with were reliable.

Chapter 3 examines the difficulties of administering the police bureaucracy, which included policing the police themselves. The government expected Ceylonese police officers and police headmen to adopt British bureaucratic values, which required them to place their official duties above family obligations, something poorly paid officers were reluctant to do. Senior officials feared that the police on the beats loitered and exploited the citizens and so they organised the beats into elaborate space-time grids in order to supervise the policemen.

---

97 Guha, *Elementary Aspects*, p. 333. On the use of crime reports for information on the subaltern, see Arnold, D. 1985. "Crime and crime control in Madras, 1858–1947," in Yang, A.A. (ed.), *Crime and Criminality in British India*. Tucson: University of Arizona, p. 62.

98 Guha, R. 1989. *The Unquiet Woods: Ecological Change and Peasant Resistance in the Himalayas.* Delhi: Oxford University Press, p. 126 uses the same methods to get at resistance to ecological change in colonial India.

The government was never able to effectively survey the police no matter which configurations were put in place. I then examine the even more intractable problem of monitoring detectives who needed freedom of movement and anonymity and so were of necessity off the grid.

Chapter 4 examines the policing of the countryside where governing at a distance was the principal issue. As it was impossible to ensure law and order throughout the many villages and remote spaces in the countryside, the British concentrated their efforts by setting up police posts in the plantation districts and the places where the village social structure had been most disrupted by capitalism and the mobility of populations. The countryside beyond was left to the delegated authority of the village headmen. This policy left 99% of the area of the island under the control of village headmen. The British then imposed new boundaries, which caused irresolvable confusions and conflicts as police activity in rural areas was seen by village headmen as an attempt to undermine their traditional authority.

Chapter 5 first examines the organisation of the island's courts as a hybrid mix of customary, Dutch-Roman, English law and later the Indian Penal Code. It then goes on to discuss the practices of officials in the court bureaucracy from the judges to the court clerks showing how members of each group used their power tactically in ways that undermined the functioning of the courts. I make the argument that while lip service was paid to rule of law, it was under assault not only from low-level court officials, but also from some judges and senior revenue officers who preferred that judges make pragmatic rulings rather than hewing too closely to the letter of the law.

Chapter 6 turns to a consideration of how a significant percentage of the general Ceylonese population embraced the court system, overwhelming it with both legitimate and false cases against their neighbours. I explore how government strategies to counter this appropriation were constrained by rule of law principles and also hampered by the incompetence of many of the judges. I then show how the lack of a trustworthy police force and inadequate detective work in the countryside impeded the judges who felt greatly handicapped by lack of information and their inability to discover when the accusers, defendants or witnesses were committing perjury.

Chapter 7 examines the introduction of the reformed prison system in Ceylon during the 1860s and the way in which government officials attempted to resist the English model on racial, climatic and financial grounds. They argued that in a tropical colony it was very difficult to create a prison environment harsh enough to deter criminals and yet stay within the guidelines set by the reformed English model based on the more humane ideas of Bentham and Beccaria. I describe the space-time organisational grid of the prison which was set up as the framework for strict biopolitical control.

Chapter 8 focuses upon the informal experiments that were conducted to calculate the maximum amount of hard labour that could be demanded in order to push the bodies of prisoners to their absolute physical limits. Drawing upon European studies of motion and human energy, jailers tried multiple types of deterrent technologies, each carefully calculated in foot/tons and matched to the average weight and strength of a Ceylonese male prisoner. I then show the many ways in which prisoners attempted to resist the full impact of hard labour on their bodies and the role that guards played in these struggles, at times enforcing prison rules and at others aiding prisoners in evading them. Even in highly controlled prison environments, surveillance was a problem because the guards who were charged with supervising the prisoners were inadequately supervised themselves. Finally, I show how the treatment of European prisoners demonstrated that for the British the rule of racial difference trumped the liberal rule of law.[99]

Chapter 9 examines the numerous biopolitical techniques developed through experimentation using prisoner bodies. Officials tried to discover the minimum amount of food prisoners could survive on and the maximum amount of hard labour possible in order to punish prisoners to the absolute lawful limit, that is the limit of "bare life." In other words, prison administrators sought to calculate the threshold between health and long-term bodily damage to ensure the rates of sickness and death fell within an acceptable range.

Prisoner resistance to these dietary regimes took a number of forms. Particular attention is paid to the way in which prisoners enrolled doctors in their struggles against the harshest penal diets and cruel punishments. I discuss the ambivalent relationship between prisoners and their guards, whose lack of commitment to the prison system, due in part to their very low salaries, meant that they sometimes colluded with prisoners and felt entitled to their bribes as a supplement to their meagre earnings.

---

99 See Chatterjee on the rule of colonial difference. Chatterjee, P. 1993. *The Nation and Its Fragments: Colonial and Postcolonial Histories*. Princeton: Princeton University Press.

2 Criminological theories and the "men on the spot"

The annual administration reports, government reports on crime and local newspaper articles during the second half of the nineteenth century in Ceylon contain a mélange of common sense theories of crime, its causes and prevention. Most of the opinions were the thoughts of "men on the spot" attempting to administer punitive discipline in the interest of deterring crime in Ceylon. These people, principally British, Burghers and some members of the Ceylonese elite, had no formal training in the new field of criminology nor in the philosophical thought on criminal justice.[1] And so they tended to draw on utilitarian theories of crime and deterrence that had become popularised in British society. These ideas, however, were tempered by earlier religious views of crime as sin or vice and shifting theories of racial difference and tropical degeneration which questioned whether enlightenment ideas of innate reason were universally applicable. For example, some wondered if tropical peoples had different ways of calculating pain and pleasure and thus might not be deterred from committing crime in ways they could predict.

It is therefore useful to trace this mélange of ideas on crime back to earlier sources. For it has been said that much of what is called "common sense" in any time and place is based on the ideas of long dead theorists. What may have seemed to these "men on the spot" to be empirical evidence gathered from their personal experience working in Ceylon was in fact theory-laden.

Theories of criminality

Eighteenth and nineteenth-century thinkers writing about criminality, its definition and causes, derived their ideas from a diverse range of religious and secular sources. Religious beliefs concerning the salvation of criminals informed the late eighteenth and early nineteenth-century prison reform movement perhaps even more than utilitarian principles.[2] For example, the Penitentiary Act of

---

1 The term criminology was coined in 1885 by Raffaele Garofalo, a student of Cesare Lombroso, known as the father of modern criminological theory.
2 Cooper, R. 1981. "Jeremy Bentham, Elizabeth Fry, and English prison reform." *Journal of the History of Ideas*, Vol. 42, No. 4, pp. 675–90.

1779 connected punishment to penitence and influential prison reformers such as John Howard believed in the role of original sin as the principal cause of crime. Punishment through imprisonment was intended to promote a sense of religious guilt and offer a place for moral reflection.[3]

By contrast, secular streams of thought on criminality embraced enlightenment rationality. The contractualist theorist Thomas Hobbes and his followers viewed the origins of criminal behaviour in free will and rational assessments of self-interest, as did Beccaria and utilitarians such as Bentham. Beccaria reasoned that the harm of punishment should exceed the good derived from the crime. Thus, harsh punishments were calculated to appeal to the perceived self-interest of the criminal. Despite the obvious differences, the penal solutions offered by the utilitarians converged with those of the religious thinkers.[4] While religious thinkers viewed hard penal labour as a tool for reforming the criminal through penitence, the secular view of labour as gratifying assumed that it was idleness that led to crime and so penal labour might inculcate an honest work ethic.[5]

Garland points out that the classical view of criminality as rational was based on imported liberal economic concepts.[6] According to classical utilitarian theory, criminal behaviour results from a failure to develop the habit of long-term consequential thinking, instead seeking the immediate gratification of sensual desires. The solution was to demonstrate to the criminal that loss of freedom rather than the gratification of desire was the likely result of criminal behaviour. The phrase "crime doesn't pay," with its overtones of economic liberalism, captures the utilitarian view of criminal deterrence. Ideally external social constraints would eventually be replaced by self-discipline, or what Foucault refers to as the "conduct of conduct," the coming together of the governor and the governed in one body.[7] The ideal is a form of hegemony such that individuals and populations have been persuaded to want what the government intends them to want. Those unable or unwilling to enter into this Faustian bargain, the poor, criminals, and subject peoples, were to be controlled directly through supervision and other forms of guided discipline and moral stewardship.

3 McLynn, F. 1991. *Crime and Punishment in Eighteenth-Century England*. Oxford: Oxford University Press, pp. 243, 254.

4 Vold, G.B, P.J. Bernard and J.B. Snipes. 2002. *Theoretical Criminology*. New York: Oxford University Press, p. 15.

5 Darby, N. 2012. "A Protestant purgatory: theological origins of the penitentiary act, 1779." *Journal for Eighteenth Century Studies*, Vol. 35, No. 4, p. 617.

6 Garland, D. 1985. *Punishment and Welfare: A History of Penal Strategies*. Aldershot: Gower, p. 17.

7 Foucault, M. 1991. "Governmentality," in Burchell, G., C. Gordon and P. Miller (eds.), *The Foucault Effect: Studies in Governmentality*. Chicago: University of Chicago Press, pp. 87–104; Gordon, C. 1991. "Governmental rationality: an introduction," in Burchell, G., C. Gordon and P. Miller (eds.), *The Foucault Effect: Studies in Governmentality*. Chicago: University of Chicago Press, pp. 1–52.

During the eighteenth and early nineteenth centuries, the causes of crime were increasingly seen as environmental rather than innate. Although environmental determinism can be traced back to ancient Greek ideas, eighteenth-century environmentalism allowed for progress through the application of rationality. Thus, although physically noxious environments were held to foster crime, it was believed that through enlightened efforts a more beneficial physical and social environment might be created that would discourage crime and reform those who had been led astray.[8] This view of the causation of crime opened up vast bio-technocratic possibilities for the psychological transformation of individuals and the social transformation of society and there was the strong feeling the one begat the other. The notion of environment was broad; it included climate, the physical condition and arrangement of communities, as well as the social and economic organisation of society.

Another secular approach to crime was the French statistician Quetelet's social mechanics derived from Comtean positivism, which was less progressive in orientation in that it did not hold out much hope for reform either through manipulating physical environments or through penal practices. Quetelet believed in the statistical constancy of crime and the need for controlling "dangerous classes."[9] Tensions arose between those who believed in the possibility of the improvement of character and moral choices of offenders and positivists who looked more towards surveillance and crime prevention through enhanced policing and incarceration. The latter largely rejected the voluntarism of the classical perspective, believing that criminal behaviour was produced by factors beyond an individual's control. An individual criminal's reasoning they argued, served only to rationalise predetermined behaviour. As such, the positivists looked for what they considered to be deeper, more fundamental causes of crime, those rooted in biology, psychology, evolution, mental heredity, the physical environment and the lack of exposure to civilising forces.[10]

The roots of the positivist approach to the study of populations can be traced back to the beginnings of the collection of statistics by the state. By the 1500s systematic records of births and deaths were increasingly collected in Europe. In the following centuries economic and social data were collected as well and assembled in what Latour has usefully referred to as "centres of calculation."[11] Statistical analysis was developed further in the early eighteenth century when Adam Smith and Thomas Malthus used numerical data to bolster their argu-

---

8 Wiener, M.J. 1990. *Reconstructing the Criminal: Culture, Law and Policy in England*, 1830–1914. Cambridge: Cambridge University Press, pp. 39–43.

9 Beirne, P. 1987. "Adolphe Quetelet and the origins of positivist criminology." *AUS*, Vol. 92, No. 5, pp. 1140–69.

10 Vold, et al., *Theoretical Criminology*, p. 9.

11 Latour, B. 1987. *Science in Action: How to Follow Scientists and Engineers through Society*. Cambridge: Harvard University Press.

ments concerning population growth and social control.[12] In England in 1810, the Clerk of Courts recorded annually all committals for trial, serious crimes and executions.

For Ceylon, comprehensive trade statistics were collected by Anthony Bertolacci, the first economic historian on the island during the British period.[13] At the time there was great confidence in statistics as a tool for understanding the amount, if not the causes of crime. There was a great outpouring of statistically based criminological studies in Britain and the colonies, which lasted at least until the late nineteenth century when confidence was somewhat undermined by distrust in both the reporting of crime and the collection of statistics.[14]

In 1827 France became the first nation to begin to collect national crime statistics annually. It appeared from these data that crime exhibited regularity over time and place. In 1835 Quetelet offered the classic statement on these findings, "the crimes which are annually committed seemed to be a necessary result of our social organisation ... Society prepares a crime, the guilty are only the instruments by which it is executed." He continued,

> The share of prisons, chains and the scaffold appears fixed with as much probability as the revenues of the state. We are able to enumerate in advance, how many individuals will stain their hands with the blood of their fellow creatures, how many will be forgers, how many poisoners, pretty nearly as one can numerate in advance the births and deaths, which must take place.[15]

Liberal thinkers took statistical regularity as evidence that the organisation of society and particularly the economy was self-regulating and should not be interfered with by government bureaucracies. Others, such as Edwin Chadwick, took the opposite view, seeing the regularity as a sign that scientific management and policing of populations was possible. If human behaviour was lawful, like other natural phenomena, it could be managed by technical bureaucratic means.[16] Foucault claims that the rise of statistics made possible the concept of a population as a statistically derived artefact that could be acted upon by government. The idea that criminals constitute a distinct subgroup, a statistically understood population, facilitated their management by the state.

12 Vold et al., *Theoretical Criminology*, pp. 21–22. On Malthus, see Mayhew, R.J. 2014. *Malthus: Life and Legacies of an Untimely Prophet*. Cambridge: Harvard University Press; Mayhew, R. J., ed. 2016. *New Perspectives on Malthus*. Cambridge: Cambridge University Press; Bashford, A. and J.E. Chaplin. 2016. *The New Worlds of Thomas Robert Malthus: Rereading the Principle of Population*. Princeton: Princeton University Press.
13 Bertolacci, A. 1817. *A View of the Agricultural, Commercial and Financial Interests of Ceylon*. London: Black, Parbury and Allen.
14 Godfrey, B. 2014. *Crime in England, 1880–1945*. London: Routledge, pp. 18–20.
15 Quetelet, cited in Vold et al., *Theoretical Criminology*, p. 25.
16 Wiener, *Reconstructing the Criminal*, pp. 163–64.

Another major focus of the positivists was on biological explanations of crime. In 1775, Lavater published a four-volume work on physiognomy, the study of the face and phrenology, the study of the shape of the head. His work became a key resource for positivist criminologists. In 1791, Franz Gall posited a relationship between the shape of the head and personality. Such ideas were further developed by a number of thinkers interested in criminology and race. Among the best known was Cesare Lomboso, who in *Criminal Man* developed a theory that criminals were atavistic, biological throwbacks to an earlier evolutionary stage.[17] Although there was much criticism of his theory, his work was nevertheless influential during the latter part of the nineteenth century. The more general view that criminals were biological degenerates was reinforced and popularised by the writings of Charles Darwin and Herbert Spencer, although Darwin never accepted the application of evolutionary principles in the social sciences and rejected social Darwinism.

Although, during the nineteenth century, theories of criminality were based on competing models of why certain individuals committed crimes, virtually all agreed that crime was connected in some manner to the growth of populations, urbanisation, the increase of wealth and broader access to consumer goods.[18] Capitalism, as the great engine of economic growth, was thought to promote moral dissolution and social disorder by disrupting traditional forms of deference and authority. The rise of individualism was thought to be productive of crime, by weakening limits on freedom of action and thus allowing lower classes to commit crimes without traditional social constraints. In this view, exposure to luxuries, although thought harmless for the upper classes, was seen as tempting the poor into dangerous dissatisfaction with their own social position, and could promote idleness and ultimately encourage crime.[19]

While there was a common belief in the interconnection of poverty and crime, the exact nature of that link was disputed.[20] One view was that it was the mobile poor who were most likely to be criminal.[21] Mayhew, in *London Labour and the London Poor*, wrote,

> Of the thousand millions of human beings that are said to constitute the population of the entire globe, there are—socially, morally, and perhaps even physically considered—but two distinct and broadly marked races,

17 Lombroso, C. 2006. (1876). *Criminal Man*. Trans. M. Gibson and N.H. Rafter. Durham: Duke University Press.

18 McLynn, *Crime and Punishment*, p. 247.

19 *Ibid.*; Wiener, *Reconstructing the Criminal*, pp. 11, 16, 18.

20 Yang, A.A. 1985. "Dangerous castes and tribes: the Criminal Tribes Act and the Magahiya Doms of northeast India," in Yang, A.A. (ed.), *Crime and Criminality in British India*. Tucson: University of Arizona, p. 114.

21 Radhakrishna, M. 2008. "Laws of metamorphosis: from nomad to offender," in Kannabiran, K. and R. Singh (eds.), *Challenging the Rule(s) of Law: Colonialism, Criminology and Human Rights in India*. New Delhi: Sage, p. 3.

viz., the wanderers and the settlers—the vagabond and the citizen—the nomadic and the civilized tribes ... The nomadic or vagrant class are all an universal type, whether they be the Bushmen of Africa or the 'tramps' of our own country.[22]

No less a respected authority than Darwin concurred when he wrote, "nomadic habits, whether over wide plains, or through the dense forests of the tropics, or along the shores of the sea, have in every case been highly detrimental to civilization."[23] Such was the fear of mobility brought on by the industrial revolution that the category of the mobile poor seems to have been one of the main reasons for the professionalisation of the police and the widespread introduction of new vagrancy laws into Western Europe, especially into England.[24]

An important qualification to the premise that landlessness and poverty leads to crime is the notion of indigence. John Wesley in the eighteenth century held that it was indigence that leads to crime. This distinction, which persisted throughout the nineteenth century, was that the labouring poor who could sustain themselves, albeit in a hand-to mouth existence, were morally superior to the sub-population of the indigent or able-bodied paupers who would not, or could not, support themselves.[25] The latter condition Wesley linked to drinking, gaming and amusements. During the nineteenth century, drunkenness was seen as a key component of indigence and by mid-century it was held to be a leading cause of crime.[26] Drinking and gambling, and luxuries more generally, again were held to be relatively harmless to the upper classes due to their greater powers of self-control, but ruinous for the poor.[27] The *Report of the Royal Commission on the Rural Constabulary* of 1839 echoed the view that it was not poverty per se that led to crime. Rather the criminal poor suffered from the vices of "indolence or the pursuit of easy excitement and were drawn to commit crimes by the temptation of profit of a career of depredation, as compared with the profits of honest and even well-paid industry."[28] The Christian

22 Mayhew, H. 1862. *London Labour and the London Poor*, vol. 1. London: Griffin, Bohn and Company, p. 341.
23 Darwin, C. 2003. (1871). *The Descent of Man*. London: Gibson Square Books, p. 133.
24 Lucassen, L., W. Willems and A. Cottar. 1998. *Gypsies and Other Itinerant Groups: A Socio-Historical Approach*. New York: Palgrave, pp. 66–67.
25 Nally, D. 2011. *Human Encumbrances: Political Violence and the Great Irish Famine*. South Bend: University of Notre Dame Press, p. 103.
26 Wiener, *Reconstructing the Criminal*, p. 79. Beckingham, D. 2017. *The Licensed City: Regulating Drink in Liverpool, 1830–1920*. Liverpool: Liverpool University Press.
27 McLynn, *Crime and Punishment*, pp. 244, 248.
28 Nijhar, P. 2009. *Law and Imperialism: Criminality and Constitution in Colonial India and Victorian England*. London: Pickering and Chatto, p. 44. Likewise Mayhew and Binny, in their influential mid-nineteenth century book on London's prisons, wrote of the criminal's "innate love of a life of ease, and aversion to hard work, which is common to all natures." Mayhew, H. and J. Binny. 1862. *The Criminal Prisons of London and Scenes of Prison Life*. London: Griffin, Bohn, and Company.

proscription against idleness was expanded at the end of the eighteenth century in Britain into a moral condemnation of the persistence of pre-industrial forms of behaviour, which were seen as antithetical to the new industrial spirit of time-keeping and disciplined labour. It was suggested that such premodern behaviour could be reformed by work discipline. All such assumptions about the indiscipline of the poor in Britain were seen as especially true of the poor in the "less civilised" tropical colonies.

Although the middle of the nineteenth century saw a general decrease in crime in Britain, people became concerned about "habitual" criminals who were thought to be addicted to a life of crime and virtually impossible to reform. Habitual criminals were considered a population, a statistically derived grouping of individuals, who shared a common deviation from the social norm. Recidivism was targeted by government in the 1869 Habitual Criminals Act which called for extended prison terms for previously convicted criminals who were by definition considered incapable of conforming to societal norms. Increasingly throughout the rest of the century habitual criminals became a major focus of the criminal justice system in Britain.[29] While the notion that there was a population of habitual criminals was common in late nineteenth-century criminology, not all theorists believed that the criminal class was a biologically determined sub-population.

As the century drew to a close, criminologists began to think that recidivism was rooted in social conditions and that it was the result of poor upbringing.[30] Such a perspective incorporated elements of the classical approach which had emphasised that over time choices could become embodied as unconscious habits; these habits were sometimes seen in Lamarkian terms as acquired characteristics that could be passed down through the generations. Social influences, however, could modify these acquired patterns of criminal behaviour.[31] As Wiener points out, "the concept of habit as congealed will was to serve as a bridge between the voluntarism of the first-half of the century and the increasing determinism of later Victorian naturalism."[32]

Others argued that habitual criminals suffered from a type of insanity and by 1899 there was consensus in the Home Office that habitual criminals should be treated as lunatics and when possible confined, albeit for an indeterminate period, due to the inability of prison officials to determine when such an insane criminal could be safely freed.[33] Such harsh measures were based on the

29 Nijhar, *Law and Imperialism*, pp. 10–11; Radzinowicz, L. and R. Hood. 1990. *A History of English Criminal Law and Its Administration from 1750, Vol. 5: The Emergence of Penal Policy in Victorian and Edwardian England*. Oxford: Clarendon Press.
30 Yang, "Dangerous castes," p. 111; Wiener, *Reconstructing the Criminal*, p. 359.
31 Bennett, T. 2011. "Habit, instinct, survivals: repetition, history, biopower," in Gunn, S. and J. Vernon (eds.), *The Peculiarities of Liberal Modernity in Imperial Britain*. Berkeley: University of California Press, p. 104.
32 Wiener, *Reconstructing the Criminal*, p. 43.
33 *Ibid.*, pp. 348–49.

differentiation between occasional and habitual or recidivist criminals who were considered fairly irredeemable unless placed for an extended period of time in an environment where they could learn new habits.[34]

The "nature" of crime in Ceylon

The men in Ceylon who ran the criminal justice system filtered their observations of criminal behaviour in Ceylon through popular interpretations derived from the more formal theories outlined above. As Chanock put it, the British understanding of crime in colonial societies was "dependent upon frameworks from elsewhere."[35] The "elsewhere" in the case of Ceylon was Britain, other colonies, especially India, and other European countries, and the United States. These popularised criminological theories were in turn cross-cut by ideas of racial and climatic tropicality as it had been experienced in India and other colonies.[36] Whereas very little ethnological work on criminals was done in Ceylon in the nineteenth century, there were many studies of Indian organised crime and criminal tribes.[37] As many of the senior officials in the Ceylon criminal justice system had prior experience in India, the situation there was often projected onto Ceylon, with the caveat that Ceylon appeared to have no criminal tribes as were found in India.

The tropical environment was thought to have a deep structural impact on the psychology of the Ceylonese. The tropics were considered a crime-producing environment, where it was possible to survive without doing much work due to the lush natural growth of fruit and vegetables. These beliefs, combined with a Lamarckian or neo-Lamarckian theory of evolution, were thought to produce over the generations people who were apathetic, lazy, given to the pursuit of pleasure and the avoidance of hard work.[38] The heat of the tropics was also thought to produce people with hot, excitable temperaments, characterised by uncontrollable rages and general lack of self-control. The tropics, in other words, fostered what European criminologists thought to be two of the principal causes of crime; a love of ease and pleasure and a lack of self-control. In addition to this it was assumed that the tyrannical rule of the kings in the Kandyan

34 Nijhar, *Law and Imperialism*, p. 50.
35 Chanock, M. 1995. "Criminological science and the criminal law on the colonial periphery: perception, fantasy, and realities in South Africa, 1900–1930." *Law & Social Inquiry*, Vol. 20, No. 4, p. 914.
36 On nineteenth-century beliefs about the impact of climate on the self, see Livingstone, D.N. 1991. "The moral discourse of climate: historical considerations on race, place and virtue." *Journal of Historical Geography*, Vol. 17, pp. 413–34; Livingstone, D.N. 1999. "Tropical climate and moral hygiene: the anatomy of a Victorian debate." *British Journal for the History of Science*, Vol. 32, pp. 93–110.
37 Brown, M. 2003. "Ethnology and colonial administration in nineteenth century British India: the question of native crime and criminality." *The British Journal for the History of Science*, Vol. 36, No. 02, pp. 201–19.
38 Alatas, S.H. 1977. *The Myth of the Lazy Native*. London: Frank Cass.

highlands before the arrival of the British had produced subjects with little moral compass of their own.

Along with the turn to psychological explanations of crime came sociological and cultural explanations. One was that traditional patterns of land inheritance, whereby all members of a family inherited an equal share, were thought to be socially dysfunctional, producing endless quarrels over fractional shares of the fruits of small slivers of land.[39] Another idea was that crime naturally rose with the social change ushered in by European capitalism, with its attendant breakdown of the subsistence-oriented peasant society and weakening of local authority and communal cooperation. Such disruption was thought to have removed an important check on crime as it had in Britain.

The opinion expressed in some quarters was that the collapse of coffee, the dominant cash crop, in the 1880s and the ensuing loss of peasant land for failure to pay the grain tax led to hunger crime. The latter theory was controversial with the British as it held colonial government taxation policy responsible for peasant poverty and resulting crime. In 1889 a Government Agent reported in the administration reports that rice cultivation was not a commercially successful product and so it was important not to over-tax the farmers. He explained this in simplistic, racialised terms:

> to hamper the rice cultivator is to make him unsettled and discontented, and to endanger the village ... so long as a villager has a paddy field, and can spend his time and energies upon and get a living from it, so long he will remain happy and contented. Take it away or divert his energies into a different groove, and you change his nature and make a wandering, discontented, money-grubbing and dangerous character of him.[40]

While some administrators acknowledged the impact of government policy on criminal behaviour, most remained more drawn to utilitarian free-choice models. But as I have argued, popular criminology in Ceylon was an unsystematic, interdisciplinary mix. For example, while the ideas of biological determinists such as Lombroso were not mentioned by name in the British records, many officials expressed their belief in inherited criminological tendencies and biological degeneration caused by the tropical environment.

It is important to note that while the British often relied on broad generalisations about the Ceylonese character, they frequently acknowledged these as roughly drawn stereotypes. Some Ceylonese were thought to resist the cultural influences and environmental press to a greater degree than others, through force of character. They were described as "hard working," "honourable," "responsible," or "self-disciplined" and were seen as exceptions to the rule.

---

39 This was common practice in Ireland too, and much frowned on by the English who took the law of primogeniture as the norm. Personal communication from David Nally.

40 *Administration Reports, Ceylon (ARC)*, 1889, p. C34.

Likewise, the British espoused stereotypes of European character, but these were cross-cut with beliefs about class difference. Class stereotypes were then projected onto Ceylonese society, as they distinguished Kandyan and lowland elite families from the lower castes and classes. Similarly, when Ceylonese officials were critical of the character of their own countrymen, they tended to qualify their generalisations as referring to the uneducated poor.

Questioning crime statistics

Throughout the period, officials expressed astonishment at the number of cases that were brought to court in Ceylon. However, due to a very large number of false cases and the presumption that there was a great deal of unreported crime, they were in the dark about how much crime actually existed. They were aware that their methods of collecting statistics were unsatisfactory. The lack of statistical information and their inability to recognise false cases or identify criminals did not sit well with colonial administrators as it undermined their faith, typical of the period, that expertise and statistical analysis could facilitate administration. In a government-commissioned report on crime in 1889, Giles, who was an inspector of police from Bengal, expressed his frustration thus: "Coming now to the main object of my inquiry, the prevalence of crime in Ceylon, I must premise that I have met with great difficulty owing to the absence of complete and accurate statistics." He suggested a new system of collecting statistics on crime and concluded hopefully,

> In the course of a few months it will be possible for the first time to say, with some approach to accuracy, what proportion of each description of cognizable offences instituted is true and what proportion false; and unless this important distinction can be made, it is vain to attempt to judge with any degree of certainty, whether or not serious crimes are unduly prevalent. Inferences drawn from existing data can be at best but shrewd guesses.[41]

And yet in spite of Giles' optimism, by the end of the century the statistics remained unreliable. The Solicitor General in his Special Report on Crime in 1897 wrote that "the reported cases of crime really include both true and false cases, and who can say how much of it is true and how much false?"[42]

Although Giles had reservations about the accuracy of the statistics on crime, he claimed that he could read the broad picture well enough to make an assessment of crime on the island. Having served as Deputy Inspector General of Police in Bengal before his arrival in Ceylon, he used Bengal as a benchmark

---

41 Giles, A.H. 1889. *Report on the Administration of Police, Including the Actions of the Courts and the Punishment of Criminals in Ceylon*. Colombo: J.A. Skeen, Government Printer, p. 357.
42 R. Ramanathan, Solicitor General, Special Report on Crime in Ceylon, *Ceylon Sessional Papers (SP)*, 1897, p. 333.

of crime. He was particularly struck by the extent of violent crime in Ceylon and was shocked that the ratio of murders to population in Ceylon was presumed to be 1 case in 74,587 persons, while in Bengal it was thought to be 1 case in 260,000. He explained this difference as based on "the difficulty of obtaining redress from the courts, under the present system, so favorable to false witnesses, [which] may in some cases lead to the people taking the law into their own hands and administering it with ferocity." He added that,

> the fact that the execution of from four to twelve persons annually in every million of population has no marked deterrent effect, looks as if the causes are deep-seated and of long standing. In Bengal and England the executions hardly exceed one person per million.[43]

The ratio of thefts and robberies in Ceylon to population was one to 219 while in Bengal it was one to 2,565 and in Madras one to 1,407.[44] In the countryside petty thefts of crops were most common while in cities like Colombo burglary and theft were the most common crimes.[45] Homicide and other types of violent crime were more common in the countryside, and the overall greatest offence in Ceylon was cattle theft. Giles found that in 1887 there was one case of theft for every 1,152 persons, while in Bengal the ratio was one in 361,603 people. He argued that even given that a very large number of these cases were probably false, cattle theft was still extraordinarily common.[46]

Eight years later, in his special report on crime, the Solicitor General compared serious crime in Ceylon and England and concluded that England had far more rape, forging of coin and theft, while Ceylon had twelve times more murder per capita, over double the manslaughter, and nineteen times the grievous bodily harm. He added,

> Lust and greed for money are the most prominent motives to serious crime in England, while Ceylon seems to labour under uncontrollable anger and utter disregard of the sanctity of human life. These remarks apply of course only to that stratum of society which, for want of education and reflection, is unable to regulate itself under temptation.[47]

Another difference cited was in the percentage of recidivism. Between 1892 and 1896, 20% of criminals in Ceylon were considered "habituals," while in England 54.5% were.[48] And, finally, between 1887 and 1896 only around 1%

---

43 Giles, *Report on the Administration*, p. 359.
44 *Ibid.*, p. 360.
45 *Ibid.*, p. 363.
46 *Ibid.*, p. 360.
47 R. Ramanathan, Solicitor General, Special Report on Crime in Ceylon, *SP*, 1897, p. 344.
48 *Ibid.*, p. 348.

of convicts were women, compared to Britain where the percentage was considerably higher. This was explained by the claim that "in Ceylon and India, where women live under greater restraints than in England, their appearance in court as offenders is considered particularly disgraceful."[49] Furthermore, in Ceylon as in Britain, where alcohol was a major cause of crime, Ceylonese women did not drink as British women did, nor did they carry knives as did most peasant Sinhalese men.

British officials tended to understand human differences in terms of race and their own empirical observations led them to classify the different races in Ceylon according to their propensity to commit crime and to assume that the different races committed different types of crimes.[50] In his comparative history of criminal identification, Simon Cole notes that "the most acute problem facing the nineteenth century police and penal bureaucracies was not in the recording of information, but in ordering it."[51] Criminal justice administrators in Ceylon did in fact try to make criminal populations legible through classification. So the report of the District Court in Badulla in 1869 produced the following table of offences by race in the plantation districts:[52]

High-Country Sinhalese	Petty assaults and quarrels
Low country Sinhalese	Petty theft and vagrancy
Immigrant coolies	Desertion of service, petty theft of coffee
Moors	receiving stolen coffee
Chetties and settled Tamil labourers	nothing in particular.

Twenty years later in 1889, Giles, in the section of his report entitled "Criminality of the Various Races," wrote that "the Sinhalese are not only the most homicidal, but the most criminal in every respect."[53] To this he adds, "the penchant for gambling, drinking and debauchery of all kinds ... debases the lower orders of the Sinhalese. Evils of this sort cannot be eradicated until the people themselves reform."[54] In his opening address to the Legislative Council in 1897, Governor Ridgeway stated that the Sinhalese are "notorious for the

---

49 *Ibid.*, p. 348.
50 The term race was often employed when talking about the different ethnic and religious groups in Ceylon. Race was a chaotic concept in the nineteenth century and was used in a loose way to describe a wide variety of human groupings. On the concept of criminal types in India see Brown, M. 2001. "Race, science and the construction of native criminality in colonial India." *Theoretical Criminology*, Vol. 5, No. 3, pp. 345–68.
51 Cole, S. 2002. *A History of Fingerprinting and Criminal Identification*. Cambridge, MA: Harvard University Press, pp. 2–3.
52 *ARC*, 1869, p. 194.
53 Giles, *Report on the Administration*, p. 364.
54 *Ibid.*, p. 339.

reckless use of the knife, partly due to an absence of self-control under provocation and partly to the fatal Sinhalese practice of habitually carrying knives." He went on to add that "the Moors and the Tamils, although the least criminal are the most litigious."[55] Giles and later Ridgeway supported their personal observations by statistics assembled from the 1870s onward which showed the race of prisoners.

Although there were a few inter-ethnic riots in the second half of the nineteenth century, officials noted approvingly that violent crime was overwhelmingly intra-group. Inspector General of Police, Campbell wrote,

> It may be stated that almost every murdered person was killed by one of his own race, Sinhalese by Sinhalese, Tamils by Tamils, Malays by Malays. This is probably owing, among other causes, to the fact that Ceylon is singularly free from race or caste antipathies.[56]

All of this lends support to Rao and Dube's assertion that "the notion of culture had a central place in justifying colonial discipline by devolving onto culture and religion an agency that natives" (as individuals) "were seen to lack."[57] As I will show, the classification of sub-populations, each with their own slightly different collective psychologies, pathologies and usefulness to the colonial bureaucracy, were used in ordering and targeting the different populations in the organisation of policing and penal policy.

It is important to recognise that cultural understandings of criminality and legal consciousness differed markedly between the British and the Ceylonese who had competing social rules and norms.[58] Ceylonese extra-legal sanctions at times contradicted British law and at times became colonised by law, as law became normative and disciplinary.[59] In other words, when crime is viewed

---

55 Sir J. West Ridgeway on opening the session of the Legislative Council on Friday November 5, 1897, *Ceylon Hansard (CH)*, 1897–98, p. x.

56 *ARC*, 1879, p. B29. For a discussion of Sinhalese attacks on estate Tamils in the Kandyan highlands, see Duncan, J.S. 2016. *In the Shadows of the Tropics: Climate, Race and Biopower in Nineteenth Century Ceylon*. London: Routledge.

57 Rao, A. and S. Dube. 2013. "Questions of crime," in Dube, S. and A. Rao (eds.), *Crime through Time*. Oxford: Oxford University Press, p. xxxiii.

58 Legal consciousness is a social practice through which law and justice are understood, sustained and contested. See Sarat, A. 1990. "Law is all over: power, resistance and the legal consciousness of the welfare poor." *Yale Journal of Law and the Humanities*, Vol. 2, p. 343. Shaw argues that although a shared ontology may be a precondition for legitimate authority, it provides a terrain of resistance and the possibility of reconfiguring that ontology. However, it remains limited by the grammar of sovereign power. Shaw, K. 2004. "Creating/negotiating interstices: indigenous sovereignties," in Edkins, J., Shapiro, M. and Pin-Fat, V. *Sovereign Lives: Power in Global Politics* London: Routledge, pp. 34–56.

59 See Legg, S. 2007. *Spaces of Colonialism: Delhi's Urban Governmentalities*. Oxford: Blackwell, p. 97. Also see Migdal, J.S. 1994. "The state in society: an approach to struggles for domination," in Kohl, A., V. Shue and J.S. Migdal (eds.), *State Power and Social Forces: Domination and Transformation in the Third World*. Cambridge: Cambridge University Press.

from a purely legalistic perspective, it may sometimes clash with normative perspectives which derive from local cultural traditions. From a normative perspective, crime is behaviour that violates cultural expectations of how members of a society ought to act. Whether behaviour is seen by one's community as acceptable has more to do with whether it deviates from a culturally specific moral code than whether it is deemed criminal by a ruling power that is considered illegitimate. As I shall show in the following chapters, the problems that caused the British the most anxiety and concern were the litigiousness, perjury and attitudes towards incarceration of the Ceylonese. The British tended to depoliticise Ceylonese attitudes by interpreting them as innate racial differences and racial failings and, as I have said, this was sometimes advantageous to the Ceylonese from whom less was expected. As we shall see, the Ceylonese played these British officials' prejudices to their own advantage whenever possible.

Part II

The police and the arts of subterfuge

3 Struggles in space and time

Policing the towns

Policing in nineteenth-century Ceylon, as in other parts of colonial Asia, was a hybrid enterprise blending recently developed British methods of policing with a Ceylonese system that had evolved in the context of Portuguese and Dutch rule in Ceylon before the British arrival.[1] Although this hybridity was found to be cumbersome and difficult to administer, a fully British-staffed police force was not financially feasible. While devolution of power to local authorities had been promised in the Kandyan Convention of 1815, after the Kandyan Revolt of 1818 the British sought to restrict the power of the Kandyan nobles. The Colebrooke-Cameron Commission which led to the Constitution of 1833 then aimed to further re-centralise and governmentalise power. The Commission recommended the training of the Ceylonese for government bureaucracies; however, the paucity of the resources allocated to the administration of the police force made attempts to train police and inculcate bureaucratic values largely ineffective. A constant tension between modernisation of the police bureaucracy and financial constraints forced the British into relying on a largely untrained force that was only lightly supervised, especially away from the central towns. And so the devolution of power was geographically uneven.[2]

There was a policy that Ceylon and other colonies should be largely self-funded. Revenue from taxation in Ceylon was allocated primarily to develop the infrastructure for the benefit of the plantation economy and to encourage

1 On the hybrid nature of policing in colonial India see Arnold, D. 1985. "Bureaucratic recruitment and subordination in colonial India: The Madras Constabulary, 1859–1947," in R. Guha (ed.), *Subaltern Studies IV*. Delhi: Oxford University Press, pp. 1–53; Arnold, D. 1985. "Crime and crime control in Madras, 1858–1947," in A.A. Yang (ed.), Crime and Criminality in British India. Tucson: University of Arizona Press, pp. 62–88; Bayly, C.A. 1996. *Empire and Information: Intelligence Gathering and Social Communication in India, 1780–1870*. Cambridge: Cambridge University Press; Freitag, S.B. 1991. "Crime in the social order of colonial north India." *Modern Asian Studies*, Vol. 25, No. 2, pp. 227–61. On indirect rule as a way to reduce government expenditures see Berry, S. 1992. "Hegemony on a shoestring: indirect rule and access to agricultural land." *Africa*, Vol. 62, No. 3, pp. 327–55.
2 Casinader, I., R.D. Wijeyaratne and L. Godden. 2018. "From sovereignty to modernity: revisiting the Colebrooke-Cameron Reforms—transforming the Buddhist and colonial imaginary in nineteenth-century Ceylon." *Comparative Legal History*, Vol. 6, No. 1, pp. 34–64.

peasants to participate in cash crop production that provided not only valuable exports for Britain, but a tax base as well. While law and order was seen as necessary to the colonial economy and Britain's greatest "gift" to the colonies,[3] the development of a criminal justice system modelled on the English system was in fact given a lower priority than infrastructural development. Consequently, there was a persistent reluctance to adequately fund law enforcement initiatives. The police were paid what were known as "coolie" wages for long hours of work. As a result, only those desperate for any sort of work or those intending to prey on their fellow peasants could be recruited to the force. The Ceylonese police occupied a liminal position; they were subaltern, as they were drawn from the lower levels of society, but also served as the chief enforcement arm of the colonial state.[4]

In one police division after another, in one region after another, and in one decade after another, there was an unending struggle among senior police officers, the lower ranks of the police, and local headmen over the policing of the island. The bureaucratic strategies and the resistance to these strategies played out repeatedly throughout the decades with only slight variations depending upon geographical location and varying degrees of budgetary constraint. While European supervision was thought to be key to the success of policing, due to budgetary priorities there would never be enough surveillance of personnel and so the whole policing enterprise was in a sense undermined from the beginning.[5] As Tania Murray Li, in a somewhat different context, claims: "questions that are rendered technical are simultaneously rendered nonpolitical."[6] In Ceylon issues of policing were consistently rendered technical and financial, but only rarely opened up to a critique of the larger political, economic structures that underpinned them.

In order to cope with their self-imposed budgetary constraints, the British created a tripartite system of policing. In the major towns British officers served as supervisors of Ceylonese recruits who had been given only a modicum of training. The villages, on the other hand, remained under the control of unpaid village headmen (police *vidanes*), with little or no supervision. And, finally, police

3 Olund stresses the negative aspect of rule of law when he writes that colonialism reinvented coercion as a gift. Olund, E.N. 2002. "From savage space to governable space: the extension of United States judicial sovereignty over Indian Country in the nineteenth century." *Cultural Geographies*, Vol. 9, No. 2, pp. 129–57.

4 Arnold uses the term limnality to describe a similar pattern in Madras. Arnold, "Bureaucratic recruitment," p. 3.

5 For a similar situation in Burma see Saha, J. 2013. *Law, Disorder and the Colonial State: Corruption in Burma c. 1900*. London: Palgrave Macmillan, p. 75.

6 Li, T.M. 2007. *The Will to Improve: Governmentality, Development and the Practice of Politics*. Durham: Duke University Press, p. 7. The consequence of this as Ferguson points out is not only to depoliticise but to strengthen bureaucratic control by making political decisions appear to be technical solutions to technical problems. Ferguson, J. 1990. *The Anti-Politics Machine: "Development," Depoliticization and Bureaucratic Power in Lesotho*. Cambridge: Cambridge University Press.

posts were strategically placed in rural areas where social and economic changes brought about by the colonial occupation had eroded the authority of village headmen. The result of this strategy was that less than one percent of the surface area of the island was under the jurisdiction of a British-style police force.[7] The remaining 99%, encompassing approximately 13,000 villages, continued to be policed by police *vidanes*. As such, the British were utterly dependent upon a system of traditional local authority and resigned themselves to ruling at a distance. Throughout the nineteenth century they sought to submit village headmen to their own centralised authority with little success. Unsurprisingly, the headmen pursued their own agendas, paying lip service to bureaucratic norms while continually subverting them. Underpinning this tripartite system was the British decision to focus primarily upon protecting European interests. Consequently, there was a greater police presence in European sections of towns and around European investments in the countryside. Supervision of police was concentrated in these areas and virtually absent elsewhere.

If one considers that all of the village police and 99% of the town police and the regular police in the rural posts were Ceylonese, it is clear that the British "had to share power to wield power."[8] From the point of view of the Ceylonese, the British were a foreign occupying power. Although too powerful to be expelled, they failed to achieve significant moral authority among the Ceylonese, including members of the police force. Given this lack of legitimacy, the Ceylonese persistently undermined the bureaucracy by adapting it to their own ends. They managed this through what Scott terms "the weapons of the weak," a set of tactics of opposition that employs subterfuge, foot-dragging, wilful disregard of rules, failure to cooperate and hiding or fabricating evidence.[9] Because this resistance was leaderless and spatially diffuse, its tactics were communicated discretely by word of mouth from experienced policemen to new recruits, spreading rhizomatically across the country from village to village. Such an organic form of diffusion made it almost impossible to counter.

Before exploring resistance to the police bureaucracy, let us briefly examine the English and Indian models of policing upon which it drew. Rather than simply importing a model that had been tried and tested over time, techniques

---

7　It is important to note that although not all of the island was heavily populated, there were many villages dotted through all the provinces. In 1865 an ordinance specified that the regular police could only legally operate within "police limits," the boundaries of the town or post to which they are assigned. This ordinance was routinely violated by the police at the urging of senior officials. After 1891, the regulations were changed and the police were legally permitted to operate beyond the limits. P. Ramanathan, Solicitor-General. 1892. *Administration Reports, Ceylon (ARC)*, p. A2.

8　Parsons, T.H. "The unintended consequences of bureaucratic 'modernization' in post–World War II British Africa," in Crooks, P. and T.H. Parsons (eds.), *Empires and Bureaucracy in World History: From Late Antiquity to the Twentieth Century*. Cambridge: Cambridge University Press, p. 413; Heath, D. 2016. "Bureaucracy, power and violence in colonial India: the role of Indian subalterns," in Crooks, P. and T.H. Parsons (eds.), *Empires and Bureaucracy in World History: From Late Antiquity to the Twentieth Century*. Cambridge: Cambridge University Press, p. 377.

9　Scott, J.C. 1985. *Weapons of the Weak: Everyday Forms of Peasant Resistance*. New Haven: Yale University Press.

were imported to Ceylon almost simultaneously with their development in England, Ireland and India.

The English and Irish model of policing

In the eighteenth century, law and order in the English countryside was under the control of local magistrates with the assistance of unpaid local constables. Professional policing did not come into being until 1748, when a squad of special constables called the Bow Street Runners was set up in London.[10] By the end of the century this system of policing was extended to other parts of London and thence to other cities around the country.[11] In London in 1829 the first modern, bureaucratically rationalised police force was founded under Sir Robert Peel's Metropolitan Police Act.[12] Although opponents of the Act worried that the formation of such a police force would infringe upon people's freedoms, concerns about rising crime rates convinced Parliament that a police force was necessary. It was organised like an army into divisions with superintendents, inspectors, sergeants and constables. However, in order to convince citizens that the police were not an army of occupation, they wore blue uniforms, top hats and carried truncheons rather than guns. Initially they were hated by nearly everyone, including the judiciary, but within a few years were considered, by the upper and middle classes at least, to be essential in controlling crime.[13] By 1842 all large towns in England had a police force and by 1856 the County and Borough Police Act extended it to all counties. However, in certain parts of England, the police continue to this day to be seen by many members of the working class as alien and hostile.[14]

A second model of policing was the Royal Irish Constabulary, a centralised paramilitary style police force employed by the British principally to crush agrarian, political unrest in Ireland.[15] Although separate from the military, it

---

10 Sauvain, P. 1987. *British Economic and Social History 1700–1870*. Cheltenham: Stanley Thornes, pp. 287–88.

11 Barrie, D.G. 2008. *Police in the Age of Improvement: Police Development and the Civic Tradition in Scotland, 1775–1865*. Cullompton: Willan, p. 63; Taylor, D. 1997. *The New Police in Nineteenth Century England: Crime, Conflict and Control*. Manchester: Manchester University Press, pp. 97–124.

12 Tobias, J.J. 1967. *Crime and Industrial Society in the Nineteenth Century*. London: Batsford, pp. 231–32.

13 Emsley, C. 1996. *The English Police: A Political and Social History*. London: Routledge; Das, D. K. and A. Verma. 2003. *Police Mission: Challenges and Responses*. Oxford: Scarecrow Press, pp. 130–38.

14 Emsley, *The English Police*, pp. 125–49; Dodsworth, F. 2012. "Men on a mission: masculinity, violence and self-presentation of policemen in England, c. 1870–1914," in Barrie, D.G. and S. Broomhall (eds.), *A History of Police and Masculinities, 1700–2010*. London: Routledge, pp. 123–40; Taylor, *The New Police*, pp. 97–124.

15 As Nally, D. 2011. *Human Encumbrances: Political Violence and the Great Irish Famine*. South Bend: University of Notre Dame Press, Chapter 3, points out, Ireland was a testing ground for the British of a paramilitary police in the first half of the nineteenth century. Marquis, G. 1997. "The 'Irish model' and nineteenth-century Canadian policing." *Journal of Imperial and Commonwealth History*, Vol. 25, No. 2, p. 194.

was armed and served as the enforcement branch of the government. It adopted an aggressive, often violent, stance towards resistant, subjugated populations. The Irish model was employed extensively in India and was to some degree the model for the regular armed police in their rural posts in Ceylon.[16] However, the police administration in Ceylon during the nineteenth century was principally concerned with law and order and in preventing what might be called ordinary rather than the overtly political crime that primarily concerned the Royal Irish Constabulary.[17] The major exceptions were when the British military brutally repressed the major rebellion in Uva (1815), the minor one in Matale (1848) and the riots of 1915.[18]

The Indian model of policing

The system of policing in Ceylon also drew upon the Indian model, as the two colonies were considered to be racially and culturally similar. In pre-British Mughal India, towns had chief police officers with their own staff of constables and night watchmen (*chaukidars*). The countryside was mainly policed by low status village watchmen, whose services were paid for by their village. While the *chaukidars'* job was to guard and maintain order in the village, they were also expected to relay information regarding security issues to police constables who were sent periodically to the villages for that purpose.[19]

While the British in India attempted to place a more English style of policing in strategically important cities, in other areas they relied on the traditional policing systems and were consequently never able to effectively control the

---

16 Dhillon, K.S. 1998. *Defenders of the Establishment: Ruler-Supportive Police Forces of South Asia*. Shimla: Indian Institute of Advanced Study, pp. 94–95. It was the Indian adaptation of the RIC that was spread throughout the empire. Das, D.K. and A. Virma. 1998. "The armed police in the British colonial tradition." *Policing: An International Journal of Police Strategies and Management*, Vol. 21, No. 2, pp. 354–67. On the spread of the Irish model throughout the empire see Sinclair, G. 2008. "The 'Irish' policeman and the Empire: influencing the policing of the British Empire/Commonwealth." *Irish Historical Studies*, Vol. 36, No. 142, pp. 173–87.
17 Of course, as Nally (personal communication) points out, the crucial question is who gets to decide which is which. I will demonstrate that the Ceylonese had a real interest in having the British define any violations of the law as non-political.
18 The British misinterpreted the 1915 Buddhist-Muslim riots as a political insurrection. As the above suggests, colonial policing typically drew on a combination of these two models of policing. See Emsley, C. 2014. "Policing the empire, policing the metropole: Some thoughts on models and types." *Crime, Histoire & Sociétés/Crime, History & Societies*, Vol. 18, No. 2, pp. 5–25; Brogden, M. 1987. "The emergence of the police: the colonial dimension." *British Journal of Criminology*, Vol. 27, No. 1, pp. 4–14; Killingray, D. 1986. "The maintenance of law and order in British colonial Africa." *African Affairs*, Vol. 85, No. 340, pp. 411–37.
19 Bayly, *Empire and Information*, p. 16.

countryside.[20] The village police jealously guarded their local power and were usually more loyal to the local headmen and elites who paid them than to the British.[21] Although there were important differences, if one compares policing in Ceylon to that in India during the same period, one can see broad structural similarities in the problems police faced and in their attempts to deal with them.[22]

The town police in Ceylon

Policing in the colonies was inherently problematic for government. It was a dangerous, poorly paid line of work that put a certain amount of local power in the hands of poor, uneducated people. Those who were attracted to the position were often members of a floating population of landless, unemployed men with few employment opportunities. While the use of that power was designed to be law-governed and strictly limited, in fact the temptations to use their positions for personal or familial gain or mere survival was very great. Some of these uses were socially accepted by their communities and others clearly not. There were a limited number of ways the state might have been able to channel the power of the police along bureaucratically approved paths. The first would have been to inculcate in the police an understanding of and appreciation for the rule of law and an acceptance of the ethic whereby personal economic gain was subordinate to the goals of the colonial institution. Typically, policemen in Ceylon in the nineteenth century had little appreciation for the idea of the rule of law, especially law imposed by an alien government, and in many cases even had scant knowledge of the laws that they were expected to enforce. The ideal of loyalty to a bureaucratic agency over familial and other obligations had not yet been well established in Britain, let alone in the colonies. Given the government's unwillingness to pay the police a decent wage and the near impossibility of inculcating loyalty to the colonial bureaucracy among low level officials who, unlike more educated elites, were unlikely to ever personally benefit through such loyalty, there remained but one way to control the police and that was through surveillance.

20 The practice of incorporating the traditional criminal justice system in India followed a century-old British pattern of preserving Mughal systems and avoiding trying to create political society anew. This preservationist policy was to give way increasingly during the nineteenth century, however. See Raman, K. 1994. "Utilitarianism and the criminal law in colonial India: a study of the practical limits of utilitarian jurisprudence." *Modern Asian Studies*, Vol. 28, No. 4, pp. 739–791. On indirect rule through the use of traditional policing systems see Thomas, M. 2012. *Violence and Colonial Order: Police, Workers and Protest in the European Colonial Empires, 1918–1940.* Cambridge: Cambridge University Press.

21 Arnold, "Crime and crime control"; Freitag, "Crime in the social order" Arnold, D. 1976. "The police and colonial control in south India." *Social Scientist*, Vol. 4, No. 12, Jul., p. 4.

22 Arnold, "Bureaucratic recruitment," pp. 1–53.

Figure 3.1 Street scene. Pettah, Colombo, 1852.
(Source: The British Library Board, F. Fiebig Collection)

When the British captured the maritime areas of Ceylon from the Dutch in 1795 they initially followed the Dutch model of employing soldiers to patrol Colombo at night. By the turn of the nineteenth century citizen patrols replaced soldiers, but these patrols were few in number and had little effect in suppressing the modest amount of crime. Colombo was spatially divided into a walled area called the Fort containing British governmental offices and residences, the Pettah area containing the businesses and residences of the Burghers and beyond that a zone of other Ceylonese businesses and residences.[23]

It was not until 1806 that Colombo began to be policed in a more systematic and geographically rationalised manner. The Fort, where the Europeans stayed, continued to be policed by the military and the rest of the town was divided into forty districts with an unpaid Burgher constable assigned to each to oversee night-time citizen patrols. In theory, all residents of Colombo were to take turns

23 Perera, N. 1998. *Society and Space: Colonialism, Nationalism and Postcolonial identity in Sri Lanka.* Oxford: Westview Press, pp. 48–51.

serving in the patrols, but in practice the more prosperous of the Ceylonese often bribed Burgher constables to find paid substitutes. The result was that the members of the patrols were largely drawn from the poorest class and therefore had the least stake in the protection of property.

In 1822, the same year they were established in London, daytime patrols were introduced in Colombo.[24] Although unprofessional, these patrols did manage to help maintain some order and discourage crime. Nevertheless, the British authorities were concerned that residents did not take the unpaid job seriously, and many slept while on patrol. Some, it was claimed, volunteered to replace affluent residents for a fee with the idea of burglarising properties while on their rounds.[25] This system, which was not markedly different from that of England at the time, continued until 1833 when, as part of the Colebrook-Cameron liberal-utilitarian reforms, citizen patrols in Colombo were replaced by a paid police force funded by a city tax and modelled along the lines of the new London Metropolitan Police.[26]

Following the English model, the force was organised like an army, albeit a largely unarmed one, with a constable in charge of each division and under him sergeants and "peons."[27] The superintendent and his second in command were ex-military men hired from Britain. Beneath them were five Burgher constables, whose job it was to supervise the daily operation of the force. Below them were the subordinate police sergeants and peons who were to patrol the beats, 75% of whom were either Malays or Moors (Ceylonese Muslims). Although these two ethnic groups comprised only a small percentage of the population, they were considered fierce, and so particularly useful for supressing criminal activity. It was also thought that they were less likely to fraternise or collude with the local Sinhalese and Tamil populations or to pose a security risk to the British.[28]

As I will show, the policing system as organised in the 1830s stumbled along unsatisfactorily for the next three decades. As the pay was very low, there was

---

24 Pippet, G.K. 1938. *A History of the Ceylon Police, Volume 1, 1795–1870*. Colombo: The Times of Ceylon, pp. 27–38.

25 Pippet, *A History*, p. 42. Ager, A.W. 2014. *Crime and Poverty in 19th Century England: The Economy of Makeshifts*. London: Bloomsbury, p. 122 shows that policing in England suffered from similar problems.

26 Ceylon's police in fact became an amalgamation of the London and Irish models, with the latter more prevalent in the police posts outside of the major towns. Dep, A.C. 1969. *A History of the Ceylon Police, Volume 2, 1866–1913*. Colombo: Police Amenities Fund, p. 1.

27 In 1844 ranks were renamed and peons were divided into three classes of constables with sergeants to supervise them and inspectors to supervise them.

28 Many non-official Europeans refused to recognise the authority of Ceylonese policemen. And so in 1850, twenty British constables were hired despite requiring a higher rate of pay. They were posted to Colombo and Galle primarily in order to deal with European sailors. But by 1861 the Inspector of Police in Galle reported that of the ten British police assigned to his force, only three remained. He had been unable to replace them, as only drunken ex-soldiers and sailors had applied.

a high turnover rate;[29] of the 178 sergeants and constables who joined the Colombo force in 1844, only forty-three were still active three years later, the majority having been dismissed for failure to follow regulations or resigned. The government's choice to offer a bare subsistence wage clearly induced corruption. This pattern of indiscipline, resignation and desertion was also found among the police in England and India and, as Arnold points out, was also similar to the manner in which industrial workers reacted to low pay in early industrial England.[30] Consequently, while the British deplored the behaviour of the police in Ceylon, they found it depressingly familiar.

In 1866, as part of a general reorganisation of the police to bring them in line with current practices in England, G.W.R Campbell was appointed as the first Inspector General of Police. He was chosen as a strict disciplinarian who had served in the Indian Police for nine years. The government set him two principal tasks: to reduce crime and decrease regulatory violations among members of the force. In his first year he dismissed nearly 300 of the 560-man force,[31] seeking replacements based on the reputations of their ethnic groups.[32]

Although, as we shall see, Campbell's tenure was largely unsuccessful, he nevertheless remained in charge until after the 1889 *Report on the Administration of the Police* by A.H. Giles, the Deputy Inspector of Police for Bengal.[33] Giles' damning conclusion was that, although the problems with the force were clear from the outset, the Ceylon Government had been unwilling to make any significant changes in town policing for fifty years. One of the principal reasons that the British did not devote more money to the town police is that there were few attacks over the years on Europeans. In part this was due to a comparatively large police presence in the areas of European commerce and European residential areas in Colombo such as the Fort, parts of the Pettah and, after the 1860s, the new inner suburb of Cinnamon Gardens. In spite of continual pleas by senior police officials for more funding, the government was willing to accept what they took to be a "normal" amount of Ceylonese-on-

---

29 The same pertained in parts of England where the only men willing to become constables were those who could find no other work. Field, J. 1981. "Police, power and community in a provincial English town: Portsmouth 1815–1875," in Bailey, V. (ed.), *Policing and Punishment in Nineteenth Century Britain*. London: Croom Helm, p. 52; Taylor, *The New Police*, pp. 49–50. The British found similar recruitment problems in Madras. See Arnold, "Bureaucratic recruitment."

30 Arnold, "Bureaucratic recruitment," p. 24.

31 Inspector General of Police, *ARC*, 1867, p. 257. Similar problems were found at the time with the police in England. Low pay, long hours and incompetence led to high turnover. In 1857 31% of the force in Kent was either dismissed or resigned. Ager, *Crime and Poverty*, p. 126; Bailey, V. 1981. "Introduction," in Bailey, V. (ed.), *Policing and Punishment in Nineteenth Century Britain*. London: Croom Helm, p. 14.

32 Campbell, G.W.R. Inspector General of Police, *ARC*, 1867, p. 250.

33 Giles, A.H. 1889. *Report on the Administration of Police, Including the Actions of the Courts and the Punishment of Criminals in Ceylon*. Colombo: J.A. Skeen, Government Printer. There were similar problems in Giles' own home turf, which can be seen in a review of police in Bengal two years later. Dhillon, *Defenders*, p. 137.

Ceylonese crime. However, the government responded swiftly to any spike in crime against European property. They also responded, but with fewer resources, if Ceylonese-on-Ceylonese crime rate increased beyond what at any given time was considered normal.[34]

Having provided a brief overview of the organisation of the town police during the nineteenth century, I turn in the remainder of the chapter to an examination of the ways in which the institution was shaped by the specific strategies that senior police officials adopted to force the lower ranks to obey regulations and the lower ranks' resistance to those strategies.

Inspiring loyalty and pride among the town police

There were minor attempts by police administrators to inculcate bureaucratic norms and promote loyalty among the police recruits, but these did not include more adequate pay. Instead they substituted some relatively inexpensive, symbolic sartorial gestures and titles, better training and diffusion of information, as well as some small monetary rewards.

In the early 1830s the police administrators thought that the increase in town crime was due in part to the fact that recently arrived villagers and the lower echelons of the police were unfamiliar with the newly introduced laws. Consequently, a list of felonies and misdemeanours along with the regulations of the newly created police force was posted around the city and published as a supplement in the Colombo newspapers.[35] Familiarising citizens with police regulations was also seen as an important way to stop the police from bullying the residents and engaging in extortion and bribery.[36]

A small reward fund for good police behaviour was instituted in order to supplement the meagre police salaries and keep the police honest. However, these rewards were simply not sufficient.[37] There also existed a reward system for informers from among the general public which was slightly more generous. During the 1860s Inspector General Campbell instituted another, more significant, reward system for police who caught criminals, in the hope that it might curtail what he described as the common police practice of bringing forward "a

---

34 The normal range was a shifting number over time and was periodically reset depending upon economic circumstances.

35 *Provisional Instructions for the Police of Colombo*, 1833 in Pippet, *A History*, Appendix A, p. 289.

36 *Ibid.*, p. 291.

37 *Instructions and Orders for the Regulation of the Police Force.* 1857. Colombo: William Skeen, Government Printer, Ceylon. A similar situation pertained in Bengal and Madras at the time where low salaries and little supervision led constables to extort, lay false charges and accept bribes to hide crimes Chakrabarti, R. 2009. *Terror, Crime and Punishment: Order and Disorder in Early Colonial Bengal, 1800–1860.* Kolkata: Readers Service, pp. 60–101; Arnold, D. 1986. *Police Power and Colonial Rule: Madras, 1859–1957.* Delhi: Oxford University Press. Constables in southern England in the 1850s were also notorious for committing the same sorts of offences. Ager, *Crime and Poverty in 19th Century England*, p. 126.

fictitious informer from among members of the general public. That person would receive the whole reward and then divide it with the policeman who in fact ferreted out the information or effected the capture."[38] Although this move ended this particular scam, rewards were still so small that the police continued to reward themselves in other illegal ways. Some of the methods of self-reward would have been seen as legitimate by the local population, while others fell into a grey zone between moral and immoral all depending upon the circumstances of the individual and his obligations, which may at times have overridden the official rules of police behaviour.[39] There were in Ceylon informal community redistribution practices which were expected of authorities with power, including the low ranking policemen. The impartiality and rigidity of bureaucratic norms were not always consistent with justice, customary rights, or the needs of the more vulnerable members of a community who could traditionally have expected help from those in positions of power. For example, customary rights in Ceylon included practices such as a policeman taking 10% of recovered stolen goods, a practice that, in fact, the British administrators had allowed in the early years. Taking small bribes from merchants for protection would fall into the grey zone, whereas participation with criminals in robberies was clearly considered by the community to be wrong.[40]

The Giles Report suggested that salaries be increased for all ranks; however, funds for these salaries were not forthcoming. This was part of a longstanding pattern in which revenue officials, who were part of the executive branch of government, were continually at odds with the police administrators, reflecting the underlying problem of colonial economic priorities which favoured the development of infrastructure and insisted that the colony be self-funded. The Giles report was given favourable reviews in the local press and there were editorials arguing that it was regrettable that the government refused to devote more money to policing.[41] By the beginning of the twentieth century there were further calls in the press to increase the salaries of the police. And in 1905 angry letters in the newspapers suggested that the proposed small increase in police salaries from 15 Rs. to 18 Rs. per month was laughable and thus corruption would continue.[42] Letters continued to appear in the press in subsequent years arguing that this state of affairs would continue, as one editorial put it,

38 Campbell, G.W.R. Inspector General of Police, *ARC*, 1869, p. 249.
39 On situational definitions of corruption see Jeffrey, C. 2002. "Caste, class, and clientelism: a political economy of everyday corruption in rural North India." *Economic Geography*, Vol. 78, No. 1, p. 37.
40 See Makovick, N. and D. Henig. 2018. "Neither gift nor payment: the sociability of instrumentality," in Ledeneva, A. (ed.), *The Global Encyclopedia of Informality*, Vol 1. London: University College London Press; Graycar, A. and D. Jancsics. 2017. "Gift giving and corruption." *International Encyclopedia of Public Administration*, Vol. 4, No. 12, pp. 1013–23.
41 *Ceylon Observer*, May 10, 1889.
42 *Ceylon Observer*, February 22, 23, 1905.

"so long as a policeman earned less than an active rickshaw driver."[43] This situation revealed once again the relatively low priority the colonial administration gave to the "great gift" to Ceylon of rule of law.

In lieu of improved remuneration, attempts were made to foster pride in the organisation, by introducing blue serge uniforms like those of the London Metropolitan Police. However, there was a deep concern in certain quarters that such "mimicry," as they saw it, broke down racial distinctions. A blatantly racist editorial in the *Ceylon Times* opined:

> Nothing can be more absurd than the half monkey, half man appearance presented by these unfortunates cooped up in a dress which is disagreeable to themselves and renders them contemptible not only in the eyes of the community but in those of the infractors of the peace.[44]

While it is unclear what the Ceylonese actually thought of the uniforms, they prompted an increased number of assaults on the police by drunken British civilians and officers of the garrison at the Fort in Colombo.

The police in Colombo were also regularly drilled in public, doing military style marching, bayonet practice and rifle drills, more in order to "improve the men in mind and body and appearance and engrain a habit of instant and unhesitating obedience," than to prepare them for serious battle.[45] Campbell's approach was a classic example of governmentality in the Foucaudian sense of regulating the conduct of the police by attempting to instil pride and a sense of professional responsibility. Such a public spectacle of power was also thought to discipline the onlooking public.

There was a concern to make the police appear professional and impartial in their demeanour as well. They were warned against "putting on airs" and urged to "be particularly cautious not to interfere idly or unnecessarily in order to make a display of [their] authority." They were also reminded

> that there is no qualification so indispensable to a police officer as a perfect command of temper, never suffering himself to be irritated by any language or threats that may be used, but to do his duty in a quiet and determined manner.[46]

Campbell realised that these various measures taken to instil pride and loyalty were largely unsuccessful. Furthermore, he understood that the threat of dismissal was insufficient to control corruption. And so, as a tough utilitarian who thought of deterrence as the most rational and workable strategy, Campbell

43 *Ceylon Independent*, April 23, 1910.
44 *Ceylon Times*, February 9, 1844.
45 Inspector General of Police, *ARC*, 1879, p. 38B.
46 *Provisional Instructions*, p. 291.

made sure that some of the corrupt police were sentenced to prison.[47] His campaign included making an example of a constable who had received a mere one penny bribe by arranging that he be sentenced to receive fifteen lashes and three months' imprisonment.[48] The prospect of prison no doubt had a chilling effect on police crime, but the risk of getting caught not following regulations was fairly low given how few supervisory staff there were, and so the effect was temporary and limited.

The Ceylon government acknowledged that police violence was also a problem. By 1870, the earlier high expectations among the public for an efficient police force had given way to dissatisfaction as expressed in angry letters to the press complaining about police violence.[49] In one particularly egregious case reported in the papers, a policeman stripped an innocent man "stark naked, and dragged him to the station amid fearful blows dealt by his fellow constables."[50] Although police administrators officially condemned the violence, in effect they encouraged it by hiring Malays for their aggressiveness in crowd control wherever possible. One could further argue that the British model of power and authority entailed violence, as the British treated the Ceylonese brutally on a daily basis.[51] While this may have been especially true of planters in the rural areas, the British and other Europeans living in the cities were often cruel in their treatment of household servants.

Surveillance: the space-time grid

The administrators had reluctantly concluded that the inculcation of bureaucratic values was unlikely to be achieved, and so they devised numerous surveillance techniques which included various forms of red tape such as requiring police officers to keep records of their daily activities and constabulary leave forms in an effort to deal with the problem of absenteeism, as policemen would regularly leave without permission whenever family obligations drew them back to their villages.

The Colebrooke-Cameron reforms of 1833 sought to rationalise and make uniform the administration of the island by reconfiguring it spatially and temporally at a variety of scales. At the largest scale, the island was divided into provinces, each with a government agent and assistant government agent who

---

47 *Police Gazette*, Ceylon Government Printer, January, 1868.

48 *Ceylon Times*, 19th October, 1867.

49 *Ceylon Times*, January 11, June 17, July 12, September 6, 1870; *Ceylon Examiner*, February 9, 1870.

50 *Ceylon Times*, April 26, 1870.

51 On official and unofficial European violence see, Kolsky, E. 2010. *Colonial Justice in British India: White Violence and the Rule of Law*. Cambridge: Cambridge University Press; Sherman, T. C. 2010. *State Violence and Punishment in India*. London: Routledge; Saha, J. 2011. "Histories of everyday violence in south Asia." *History Compass*, Vol. 9, No. 11, pp. 844–53; Heath, "Bureaucracy, power," p. 365.

were to tour their districts on a regularly scheduled basis.[52] This geographical reorganisation and a network of new and improved roads linking all the major towns to Colombo was intended to facilitate communication and centralised governmental power throughout the entire, newly unified country and to break down the isolation and power of the Kandyan nobles who had over the centuries operated an anti-road strategy of defence against a succession of colonial powers on the coasts.[53]

The spatial-temporal reorganisation reached as far down as the micro level of the police beats in Colombo. Based on the belief that the more precisely quantified a system was, the more rational and effective it would be, forty districts of the town were to be patrolled by 100 men doing a nine-hour night shift and fifty patrolling a thirteen-hour day shift. Each man had every third night off, followed by day-duty, which meant that every third day a policeman walked the beat for twenty-two continuous hours. For their personal safety they patrolled in pairs, but only thirty batons and ten pairs of handcuffs were issued to the entire force.[54] The fact that just over half of the night patrols were armed with a baton and only 20% had handcuffs meant that policemen had to rely on their brawn to detain suspects. As the police were lightly armed and outnumbered by often hostile civilians who not infrequently used violence in resisting them, there was a high injury rate among policemen.[55] Given that the regulations specified that police on the beat must keep moving at all times unless they see a crime in progress, these shifts were a punishing workload and in the case of the twenty-two-hour shifts, beyond human endurance if done conscientiously.[56] Consequently, from the beginning there was concern that exhausted policemen would not, and in the case of the double shifts physically could not, patrol their beats as they were required. Sergeants were instructed to patrol their divisions every two hours; but as the sergeants were also not trusted to follow the regulations, inspectors were instructed to survey the sergeants in various places within their division every twenty-four hours, varying the times to retain an element of

---

52 Many of the traditional names and boundaries were purposely ignored in this territorial reshaping. Perera, *Society and Space*, pp. 41–45.

53 See Sivasundaram, S. 2007. "Tales of the land: British geography and Kandyan resistance in Sri Lanka, C. 1803–1850." *Modern Asian Studies*, Vol. 41, No. 5, p. 933; Perera, *Society and Space*, p. 43 on the spatial organisation of Ceylon by the British.

54 Pippet, *A History*, pp. 44–55. Of course, the metropolitan police in England were also only lightly armed at the time. In his review of colonial police forces, Arnold states that following the practice in England, the police in South Asia during the nineteenth century were normally unarmed except in certain areas or where there was periodic unrest. Arnold, D. 1977. "The armed police and colonial rule in south India, 1914–1947." *Modern Asian Studies*, Vol. 2, No. 1, pp. 101–25. In recently conquered territories in Africa in the early twentieth century the police were better armed Deflem, M. 1994. "Law enforcement in British colonial Africa: a comparative analysis of imperial policing in Nyasaland, the Gold Coast and Kenya." *Police Studies*, Vol. 17, No. 1, pp. 45–67.

55 Dep, *A History*, p. 166.

56 In England before 1870, it was the norm for police to work ten to twelve hour shifts. Taylor, *The New Police*, pp. 52–53.

surprise.[57] In practice this system was not very successful. During the day sergeants in charge of the beats could see and track their men from a distance, but they were often unable to locate their men on their unlit beats at night. Likewise, the inspectors in charge of supervising the sergeants had difficulty finding them or discovering if they were surveying their patrols. As a consequence, there were fewer and fewer inspections and the patrolmen were left to their own devices.

When it was suspected that house servants were regularly engaging in nighttime burglaries, the police patrols were urged to search the bundles carried by "suspicious-looking characters." However, so few bundles were turned in that administrators believed that constables were keeping the confiscated goods for themselves. It was also feared that patrolmen were using this initiative as an excuse to rob innocent passers-by. Consequently, in an effort to involve the general public in supervision, sergeants and inspectors were urged to "pay particular attention to all complaints made [by the public] against any individual of the police force."[58] However, because the largely Malay police were feared and hated by the Sinhalese and Tamil population, it was understood that complaints against the police might well be false.

Such was the concern that the patrol system had broken down that in 1844 the newly issued *Instructions and Orders for the Regulation of the Police Force in Ceylon* called for a dramatically increased level of surveillance.[59] It was believed that the police used their local knowledge, not so much in the detection of crime, as in seeking out hiding places where they could rest out of sight and "friendly" taverns where they could safely drink undisturbed. The answer from the point of view of police administrators was to shift to a system of surveillance similar to one employed in England at the time that included a temporal dimension. This was to be accomplished by specifying the speed at which constables were to walk on their beats. A constable was required to walk at 2 ½ miles per hour so that he can "see every part of his beat once at least every 15 or 20 minutes; and by so doing any person by remaining in one spot for that length of time will meet a constable." As in England, sergeants were expected to make sure the patrol men walked at the correct speed and could be found in specified places at specified times along the beat. Should a constable depart from his proper spatial-temporal coordinates, he had to "satisfy his officers that there was sufficient cause for such apparent irregularity." The punishing nature of this system can be seen if one does the maths. To follow regulations a policeman would have to walk 22.5 miles on a night shift, 32.5 miles on a day shift, and 55 miles on the days he did a changeover. Such

---

57 *Provisional Instructions*, p. 289.
58 *Ibid.*, p. 289.
59 See Legg, S. 2007. *Spaces of Colonialism: Delhi's Urban Governmentalities*. Oxford: Blackwell, p. 84, on spatial techniques of surveillance in India with the intention of creating "the impression of constant surveillance without the need for constant supervision."

a schedule was physically unsustainable even by someone committed to the institution, which the police were clearly not. As such, the British became complicit in the violations of the regulations by putting in place an unrealistic system. Unsurprisingly, both the constables and the sergeants resisted attempts to make town policing more legible and continued to systematically violate police regulations. This can be seen from regulations issued in 1857 in which administrators felt it necessary to warn officers of all ranks that they would be dismissed if they were intoxicated, extorted confessions from suspects, accepted gifts or bribes from a person they have arrested or were violent towards persons in custody.[60]

In the mid-1860s the newly appointed Inspector General of Police realised that the current system of regulating the beats failed to stop sergeants and constables colluding with criminals and so he supplemented it with another strategy. Henceforth, each constable was to change his beat every day until he had served on every beat and had taken a turn at guard duty as well. Given that Colombo expanded spatially during the 1860s with affluent Europeans, Burghers and Ceylonese moving out of the old Fort and Pettah residential areas into the new inner suburb of Cinnamon Gardens, the number of beats in the city increased greatly.[61] Consequently a constable would return to a particular beat only once a month. It was hoped that this would disrupt collusion with criminals and the setting up of systems of extortion. This worked, but at a cost, especially to detective work, as constables were unfamiliar with the residents on their beats.

In the early 1880s the physically demanding nature of the beat system was finally addressed. The Acting Inspector General of Police in 1882 argued that the shifts were "more than even the most robust constitution can stand for any length of time." He continued, "It is scarcely to be wondered at that the men are frequently found sitting down, and some even skulking in out of the way places where they can obtain rest."[62] He thought that attempting to deal with this problem was an important reason that senior police spent a disproportionate amount of time on the supervision of subordinate police, rather than on solving crime. But apparently nothing was done about this at the time, apart from the customary hand wringing.

Five years later, however, in response to mounting public pressure from editorials such as the one appearing in the *Ceylon Examiner*, which claimed that crime was actually encouraged by corruption among the police, the Legislative Council met again to discuss the general incompetence and lax supervision of

60 *Instructions and Orders for the Regulation of the Police Force.* 1857. Colombo: William Skeen, Government Printer, Ceylon.
61 Perera, *Society and Space*, pp. 50–51.
62 *ARC*, 1882, p. 25C. By the 1870s the eight hour shift was standard in England. Taylor, *The New Police*, p. 63.

the police force.[63] One member went so far as to suggest that the government was complicit in that the long shifts would encourage police to shirk their duty. He claimed:

> constables are not men of strong physique. These men have to do fourteen-hours work a day, ie twelve on their beat; an hour going there and an hour to get home; if they arrest anyone it involves an extra two or three hours duty to attend court and give evidence. Such a state of affairs is little better than a premium to constables for deliberately letting criminals go, rather than incur extra work, when already overtaxed by long hours of duty.[64]

As a result of these interventions, the twelve-hour shifts were reduced to eight hours and each beat was staffed by three constables who were to walk for four hours, take a short rest and then do another four-hour shift. This change somewhat reduced unscheduled rest stops, but instead of hiring more men, the administrators simply eliminated a number of day shifts in order to staff the night shifts, thereby reducing police effectiveness during the day. As we will see, this is but one of many examples where economic constraints defeated the plans of the administrators, no matter how carefully thought out. Unsurprisingly, the eliminated day shifts were all located in the poorer sections of the city.

In his report on the police in 1889, Giles was highly critical of the system of moving constables and sergeants from beat to beat as this made it "impossible for a man to become thoroughly acquainted with any beat or with the bad characters frequenting it." He acknowledged that, "there are objections to keeping a man perpetually in one place, but perpetually moving him is still more objectionable. Experience fixes two or three years as the period which a policeman may be advantageously kept on one beat," that is before he began to know the locals well enough to trust their discretion or willingness to collude.[65] He said that the decision to constantly shift beats was an example of how the higher echelons of the police had become so obsessed with surveying police corruption that they had lost sight of the primary goal of curbing non-police crime.[66] This is a rather dramatic example of how over time resistance by subordinates could transform the organisational goals of a bureaucracy and the very nature of policing.

Detectives

Throughout the nineteenth century, Malays were greatly overrepresented on the force because of the administration's focus on crowd control and

---

63 *Ceylon Examiner*, August 20, 1886.
64 *Ceylon Hansard (CH)*, 1886–87, p. 85.
65 Giles, *Report on the Administration*, p. 343.
66 *Ceylon Observer*, June 29, 1890.

deterrence.[67] The Malays, however, were found to be next to useless in solving crimes because the majority Sinhalese and Tamil residents of the town disliked them and were unwilling to divulge information.[68] And so in the mid-1860s a small number of Sinhalese and Tamils were hired to act as plainclothes detectives.[69] This was contentious, however, because in Colombo, as in Britain, there was a widespread fear that plainclothes detectives would infringe on privacy and abuse their authority.[70] The various strategies of surveying police were clearly unworkable for detectives, as they were expected to mix with the community in an unobtrusive manner and establish ties in the criminal underworld. This required that they be given autonomy and discretion, something the authorities were loath to grant, especially as they were assumed to have divided loyalties.[71] The recruitment of Sinhalese and Tamil detectives exacerbated tensions within the force as they saw themselves as an elite branch that was superior to the Malays on the beat and their officers. The resentment was increased by the decision to allow detectives to wear civilian clothes and not acknowledge officers in public so as to better blend into the underworld.[72]

In 1887, in response to a perceived rise in crime, the number of detectives was increased from a small handful to several dozen.[73] However, a year later it was found that this increased force had not only solved few crimes, but had engaged in so many breaches of the rules, that all but a few had been dismissed.[74] Such was the extent of the corruption among detectives that, in his report, Giles recommended that the detective force be disbanded and constables and sergeants be assigned detective duties on a rotating basis. He even went so far as to propose that the names of those serving as detectives be kept secret so that other members of the police would not tip off criminals about investigations.[75] Giles' reorganisation of the detective branch reproduced the shifting of personnel that he had criticised in the organisation of police on the beat. Predictably, his new system failed to solve many crimes, as the rotating

67 The police in England at the time were also thought of primarily as guardians of the public order rather than as crime solvers. Gatrell, V.A.C. 1980. "The decline of theft and violence in Victorian and Edwardian England," in Gatrell, V.A.C., B. Lenman and G. Parker (eds.), *Crime and the Law: The Social History of Crime in Western Europe since 1500*. London: Europa Publications, p. 271.

68 The same policy was adopted in India. See Arnold, "Bureaucratic recruitment," p. 9.

69 Campbell, G.W.R. Inspector General of Police, *ARC*, 1867, p. 250.

70 Critchley, T.A. 1967. *A History of Police in England and Wales, 900–1966*. London: Constable, pp. 160–61.

71 Dep, *A History*, p. 150.

72 Dep, *A History*, p. 149.

73 Inspector General of Police, *ARC*, 1887, p. 39C.

74 Giles, *Report on the Administration, 346*. Similar problems were found in England. In 1877 three out of four chief inspectors in the Detective Branch were found guilty of corruption. Critchley, *A History*, p. 161.

75 Giles, *Report on the Administration*, p. 346.

detectives had few sources of information in the underworld. Consequently, in the mid-1890s the detective system was shut down temporarily.

However, with the growing obsession with the idea of "habitual" criminals among officials and the general public in the 1890s, in Ceylon as well as in Britain, the Inspector General of Police reluctantly formed the Criminal Investigation Department in Colombo in 1896.[76] He wrote,

> It was with great reluctance that I formed this department, as the extreme danger of anything like a detective service in the East is well known ... I hope by careful supervision of the officers and especially that of the inspector in charge, that we may avoid the evils to which such a department is liable.[77]

The revived system of plainclothes detectives in charge of identifying and getting to know professional criminals, as well as cultivating a network of informants, was once again tried, but soon found to be as corrupt as ever.

The traffic scam

The hopelessness of opting for increased surveillance while failing to pay a living wage was revealed dramatically when the news of a major police scandal in Colombo erupted in 1906. The press expressed shock at the scale and organisation of the corruption that had been taking place in the capital under the very noses of senior police officials. It emerged that for some years virtually all carters, drovers and rickshaw coolies had been forced to bribe the traffic police in order to enter into or work anywhere in Colombo. Small bribes were paid daily to intermediaries who then passed the money on to the constables after work or while they ate lunch.[78] It was obvious that such a spatially extensive network of crime entailed a good deal of community collusion and a conspiracy of silence. As I will show, in the countryside the dispersion of the population impeded surveillance; however, in Colombo it appears that the very density of population had increased the illegibility of corruption, for the scandal revealed a vast, previously undetected underlife of the lower levels of the police who ran an illegal tolling scheme on commercial movement in the city.[79]

76 It is not clear whether there were more criminals or if new ways of identifying and recording criminals accounted for the increase.

77 K.F. Knollys, *ARC*, 1896, pp. B1–2.

78 Report of the Committee on Illegal Gratifications Received by the Police. *Ceylon Sessional Papers (SP)*, 1906–1907.

79 Because of the scandal, the Inspector-General was able to convince the government to hire policemen from England to supervise traffic duty in Colombo; however, even this worked out badly for his department. Twenty arrived in May 1910, but when they realised that they could not live decently on their salaries as they had been promised, most resigned and returned home or went on to Australia. An editorial in the *Ceylon Observer* was sympathetic to these British recruits noting that they were deceived by the government about their pay. *Ceylon Observer*, November 11, 1910.

The British policy of paying the town police "coolie wages" while putting them in positions of authority almost guaranteed that corruption as self-payment would become the norm. An editorial in the *Ceylon Examiner* stated what the government had known for decades; that men "look upon the post of constable as something to fall back upon when every hope of securing honest employment is gone."[80] In spite of this, the government refused to substantially increase the wages of the police. They continued to see the failure of the detectives, sergeants and constables to follow regulations as a bureaucratic, technical problem to be solved by increasing supervision. But there were too few supervisors and too many spaces where the police could remain unobserved.

---

80 "A crying want," *The Ceylon Examiner*, September 14, 1900.

4   Governing at a distance

Policing the countryside

Although in principle the British were committed to ensuring law and order throughout the island of Ceylon, their main concern was to protect the safety of their economic interests.[1] Consequently, what few resources they were willing to devote to policing were concentrated in Colombo and a few other towns. Order in the villages was largely left in the hands of local elites, who were to ensure that the new laws of the land were observed and that village crimes were reported to the British authorities.[2] By the middle of the nineteenth century the rural social structure was being disrupted in certain regions by the impact of plantations, and the presence of traders and other outsiders. Village headmen in these areas often found themselves unable to cope with the rise in crime associated with this influx, and so the government began to establish police posts staffed by non-local police along important transportation routes, as well as in the bazaars and settlements that were arising to service the plantations.

Not surprisingly, as British economic interests penetrated the rural areas, so the British became more interested in policing them. However, there was concern that neither the village headmen nor the small handfuls of regular police in the outposts could be trusted to observe the rule of law. Such lack of cooperation stemmed in part from headmen pursuing their own interests, in part from a lack of knowledge of the law, but also from a different view of what constitutes law and order.[3] As we have seen, the British had come, albeit reluctantly, to the conclusion that surveillance was the only effective way to deter crime and to prevent the police from engaging in unlawful activities. But if it was difficult to adequately survey Colombo, the task was much more daunting in the thousands of villages that dotted the countryside. This, of course, is the

---

1 Killingray, D. 1986. "The maintenance of law and order in British colonial Africa." *African Affairs*, Vol. 85, No. 340, p. 414 argues that in British colonial Africa the government was likewise primarily interested in protecting European lives and property.
2 Emsley, C. 2014. "Policing the empire, policing the metropole: some thoughts on models and types." *Crime, Histoire & Sociétés/Crime, History & Societies*, Vol. 18, No. 2, p. 18 terms this policy which was put in place throughout the empire "franchise policing."
3 On this latter point in colonial Africa see Killingray, "Policing the empire," p. 413.

problem of establishing technologies for governing at a distance; the government was, as Nikolas Rose and Peter Miller put it in a different context, "seeking to create locales, entities and persons able to operate a regulated autonomy."[4] In this chapter I first explore the problems the British encountered in attempting to police the rural areas through indirect rule which, as it turned out, served to disrupt traditional authority structures, more than to make effective use of them. I then examine resistance on the part of the headmen and regular police to the rule of law. And finally I look at the effectiveness of the spatial grid of villages and police posts.

The problem of surveillance at a distance

In 1806, as part of the creation of a police force for the island, unpaid village police (*vidanes*) were appointed to assist village headmen in keeping order. The British thought that by choosing these officials themselves and giving them alone the power to make arrests they would at a stroke be able to have agents of British power in each village, thereby undermining the power of the local landed, elite families who had long dominated the posts of village headmen. However, this strategy was vigorously resisted by the village elites who openly opposed, at times physically attacked and occasionally even killed candidates not supported by themselves.[5] Consequently, few *vidanes* were chosen without the blessing of the powerful families in the villages and British hopes of having a trusted set of eyes and ears in each village came to nought. After the Kandyan Kingdom was conquered in 1815, village headmen in that region themselves took on the job of police *vidane* as an additional duty, thereby reducing British control there as well.[6]

The police *vidanes* were granted the authority to apprehend vagrants and those suspected of committing crimes, to report all murders, accidental deaths and suicides, and the presence of criminal gangs near their village.[7] The British, through an inability to do otherwise, largely left it to the *vidanes* to decide what constituted a crime, who was to be apprehended and whether or not to report any of this to the colonial authorities. As an editorial in *The Overland Times of*

4 Rose and Miller were referring to the general idea of governing at a distance. Rose, N. and Miller, P. 2010. "Political power beyond the state: the problematics of government." *British Journal of Sociology*, Vol. 61, No. 1, pp. 271–303.

5 Kannangara, P.D. 1966. *The History of the Ceylon Civil Service, 1802–1833*. Dehiwala: Tisara Prakasakayo, pp. 85–87.

6 While the British thought of the police *vidanes* as a version of the Indian system of village watchmen, in fact there were important differences. Whereas village watchmen in India were typically of low-status within their village, in Ceylon the high status headmen who became police *vidanes* in the Kandyan districts looked down on and were uncooperative with the policemen from the few scattered police posts in the rural areas. *Ceylon Sessional Papers (SP)*, 1889, pp. 69–70. For India see Bayly, C.A. 1996. *Empire and Information: Intelligence Gathering and Social Communication in India, 1780–1870*. Cambridge: Cambridge University Press, p. 16.

7 *SP*, 1884, 11.

Ceylon put it, "headmen are not only the eyes by which the Government look out upon the island, but they are also the hands by which every enactment is carried out and every law enforced."[8] The British attempted to exert some control by threatening to severely punish *vidanes* if they "arrested any person, or searched the house of any person through malice, or with a view of extorting money."[9] But the *vidanes* knew that this was a relatively empty threat and that the whole system was dependent on their willingness to perform their duties on a voluntary basis.[10]

The surveillance of village officials took the form of the regular circuits of the AGA of each district and less frequently of the GA of the province. Although these circuits were linked to the official calendar, choices were made as to which villages were visited. Those deemed to be especially important or where trouble had been reported were likely to receive frequent visits while small, remote villages might not be visited even annually. This strategy of surveillance was countered by headmen who withheld information on trouble in their villages, thereby reducing official visits. Upon arrival in a village the AGA would hold court, meeting with the headman and hearing complaints from peasants, usually in the presence of the headman. The format of these interactions was managed by headmen in such a way as to filter as much as possible the information available to outside officials.

Another form of surveillance, as we shall see, took the form of rewarding anonymous tips to officials about village criminal activity or violations of rules by headmen. These tips, which were often the reason for the inclusion of a village on a circuit, were difficult to follow up on, as villagers were reluctant to speak openly in the presence of headmen.[11] Furthermore, given the highly

---

8 "The Ceylonese and the civil service continued," *The Overland Times of Ceylon*, November 2, 1889.

9 Geo. Luisignan, Lieutenant-Governor, in Pippet, G.K. 1938. *A History of the Ceylon Police, Volume 1, 1795–1870*. Colombo: The Times of Ceylon, p. 152.

10 For comparable attitudes in India see Arnold, D. 1985. "Crime and crime control in Madras, 1858–1947," in A.A. Yang (ed.), *Crime and Criminality in British India*. Tucson: University of Arizona Press, p. 78; Freitag, S.B. 1985. "Collective crime and authority in North India," in A.A. Yang (ed.), *Crime and Criminality in British India*. Tucson: University of Arizona Press, p. 154. In England there was widespread local resistance on the part of magistrates to the establishment of county-wide police, whom they feared would undermine their authority. Ager, A.W. 2014. *Crime and Poverty in 19th Century England: The Economy of Makeshifts*. London: Bloomsbury, p. 121; Bailey, V. 1981. "Introduction," in V. Bailey (ed.), *Policing and Punishment in Nineteenth Century Britain*. London: Croom Helm, p. 13; Weinberger, B. 1981. "The police and the public in mid-nineteenth Century Warwickshire," in Bailey, V. (ed.), *Policing and Punishment in Nineteenth Century Britain*. London: Croom Helm, p. 65; Field, J. 1981. "Police, power and community in a provincial English town: Portsmouth 1815–1875," in V. Bailey (ed.), *Policing and Punishment in Nineteenth Century Britain*. London: Croom Helm, pp. 42–64. In England after 1860, district constables were thought of as inefficient because they were torn between local notables and the central administration. Nijhar, P. 2009. *Law and Imperialism: Criminality and Constitution in Colonial India and Victorian England*. London: Pickering and Chatto, p. 109.

11 K.F. Knollys, *Administration Reports, Ceylon (ARC)*, 1898, p. B12.

factionalised nature of villages, the villagers were often motivated, not by some abstract sense of duty to the government or hope of a small reward, but by the satisfaction of giving trouble to neighbours, whether or not the reported crimes had actually happened. As a consequence, there was so little effective supervision of the headmen and police *vidanes* that they ran their villages largely as they saw fit, ignoring official regulations. This would have been expected by the community, as traditionally headmen used their power not impartially, but to help those in most need or to do favours for those who supported them and their families or faction.

The spatially uneven spread of capitalism across the island resulted in flows of people, goods, services and money into certain villages, precipitating the breakdown of traditional forms of social control and weakening the authority of headmen.[12] Beginning in the 1840s, in response to what they believed was a rural power vacuum, the British began placing a number of police posts consisting of a few regular policemen each along main roads that were important to European commerce and near the bazaars that had sprung up to cater to the growing number of plantations. The establishment of small posts, staffed usually by Malay police from outside the locality, was resented by the Sinhalese villagers, who feared they would be exploited by these outsiders, and by the headmen, who worried that their authority would be usurped. Unlike the police *vidanes*, the small groups of regular police in their posts were sometimes armed. Although the British were wary of arming the Ceylonese, they did so on a number of grounds. Firstly, each post had only a few men, who, being outsiders, could not count on support from villagers, so they needed to be armed to deal with criminals who sometimes carried weapons. Secondly, given the hostility of villagers to their presence, they might need protection from the villagers themselves. However, access to arms among the police was partially restricted. Carbines were kept stored in major posts around the country and issued to the policemen by their officers only when needed. Arming the regular police was seen by the British as less of a risk than arming the *vidanes*, for the latter might have the influence to lead an armed insurrection, while the former were too alienated from the villagers to able to do so. In spite of the fact that the regular police were armed, the resistance to the establishment of posts sometimes

---

12 Administrators understood this in terms of social changes in Europe. During much of the nineteenth century vagrants and other rootless people were considered as a criminal type. Fears about the mobile poor were one of the principal reasons for the professionalization of the police in Western Europe. Lucassen, L., Willems, W. and Cottar, A. 1998. *Gypsies and Other Itinerant Groups: A Socio-Historical Approach*. New York: Palgrave Macmillan. The English Vagrancy Act was transferred to India as the Criminal Tribes Act. Radhakrishna, M. 2008. "Laws of metamorphosis: from nomad to offender," in K. Kannabiran and R. Singh (eds.), *Challenging the Rule(s) of Law: Colonialism, Criminology and Human Rights in India*. New Delhi: Sage, p. 12. While this act was not applied to Ceylon, as there were no "criminal tribes" on the island, there was a vagrancy act that was used as a catch-all for undesirables.

turned violent.[13] In 1857, when a string of posts was placed along the main road connecting Colombo to the coffee growing region near Kandy, the antipathy of the peasants living in roadside villages was so great that armed reinforcements had to be dispatched to protect the police from angry villagers.

Despite peasant resistance, posts continued to be established and the overhaul of the police in the mid-1860s accelerated the process. The authorities realised from the outset that, although the police posts were deemed necessary, they were also problematic in that the police were known to exploit local villagers. In 1854 William Macartney, the Superintendent of Police, issued the *Rules for the Guidance of Police in Rural Districts* in an effort to convert each police post in the country into a smaller version of the Colombo police force.[14] The men at every post were to parade in full uniform before going out in pairs on patrols within the territorial limits assigned to each post. The sergeant in charge of each post was to make sure that his men followed all of the rules and was to keep a diary to be shown to an inspector who would visit on a regular basis. Not surprisingly, in the smaller and more remote stations, these disciplinary regulations were totally ignored.[15] When J.S. Colepepper, the Superintendent of Police of the Central Province, went on tour the following year he reported that:

> there was not one station in all that I visited at my last inspection that I had no occasion to punish the men or the sergeant with dismissal. I have much reason to believe that the men placed at these outstations, if not sharply looked after by the superintendent and every offence by them instantly and severely punished, would become in short time a den of thieves.[16]

He further informed the government that the only way to keep the police from stealing was by frequent, unannounced inspections of all posts. While government officials accepted Colepepper's assessment, they responded that the government could only afford quarterly inspections. Upon his retirement four years later, Colepepper protested the handling of the outstations in the Central Province:

---

13 In Madras villagers similarly feared outside police as oppressors rather than guardians of the law. Arnold, "Crime and crime control," p. 81; Dhillon, K.S. 1998. *Defenders of the Establishment: Ruler-Supportive Police Forces of South Asia*. Shimla: Indian Institute of Advanced Study, pp. 154–55. In England at the time there was similar non-cooperation and occasional violence directed at outside police when they were first introduced. Weinberger, "The police," p. 65.

14 Macartney had served in the RIC before coming to Ceylon and beginning in 1860 he made the police increasingly resemble it. Dep, A.C. 1969. *A History of the Ceylon Police, Volume 2, 1866–1913*. Colombo: Police Amenities Fund, p. 1.

15 Pippet, *A History*, p. 192.

16 *Ibid.*, p. 194.

The police are commonly low-country people or Malays, men generally of enterprise and cunning and they will do their duty when they are well looked after and may be honest, but only when they are afraid to steal. At the same time, they are no more to be trusted to their own responsibility than a jackal in a hen roost. This establishment is a great expense to Government and has almost, if not entirely suppressed the offence of coffee stealing; but if left for a short time would not only be useless but mischievous in the extreme. Witness the state in which I found it upon my return from leave a few years ago; the greatest thieves in the country were the police and the better paid among them robbing their subordinates. It is not for the purpose of settling the reports of sergeants that the visits of superintendents are so necessary and important, but for the purpose of hearing the complaints, if any, against them from the inhabitants and examining the witnesses on the spot.[17]

Over the next decade, inspections of the posts became more infrequent and corruption continued unabated.[18] In 1867 the newly appointed Inspector General Campbell decreed that all posts should keep a file containing a list of all villages in their district and their distances from the post, the location of all gambling houses and taverns, and the names of the "bad characters in the district." In most posts such files were kept, but not necessarily acted upon. He also increased surveillance by ordering an inspector to travel by pony fifteen days per month, remaining at selected stations for a few days to see how they operated and to talk to local people about the behaviour of the police.[19] At first this schedule acted as a check on police corruption in the outposts, but within a year the inspections became less and less frequent, again on budgetary grounds, and so the police in the rural posts reverted to their former behaviour. Some officials thought that the only way to stop police corruption was to move policemen every two weeks to "prevent their becoming too intimate with the people around the station."[20] As one of the main duties of the police was to become familiar with the locality and its inhabitants, such proposals suggest that worries about police corruption overshadowed concerns about other types of crime.

In the late 1860s there arose a dispute between the ambitious new Inspector General of Police Campbell and the GAs about how the countryside should be policed. There was a general agreement that police posts were necessary near plantations and along roads in the districts where British and Ceylonese capitalism had penetrated, as village *vidanes* were unable to deal with outsiders.

17 *Ibid.*, pp. 195–96.
18 Saha, J. 2013. *Law, Disorder and the Colonial State: Corruption in Burma c. 1900*. London: Palgrave Macmillan, documents similar corruption in Burma.
19 Dep, *A History*, p. 130; Pippet, *A History*, p. 237.
20 L.F. Lee, *ARC*, 1869, p. 34.

However, there was a dispute over whether the police should be given the authority to visit all the villages throughout the island and oversee the *vidanes*. The Inspector General outlined his vision of police surveillance in his first annual report:

> Eventually small stations, of a sergeant and four men each, should be placed all over the island, at a distance of from twelve to twenty miles apart in the most populous, and at much longer intervals where the people are few and scattered. Were every hamlet and village visited, if only once a fortnight, by a trained policeman, and on the occurrence within it of any serious crime, by a party from some station within six or eight miles, few crimes would, as many do at present, altogether escape observation, and a fair percentage would be convicted. Spite or envy or revenge, or even mere gossip, would bring to the notice of the visiting policeman most crimes that might occur. The unpaid village headman would seldom take bribes with so great a risk of detection. On the contrary, acting in view of his paid co-adjudicators, and consequently of the public to a certain degree, his undoubted influence would probably be exercised in the cause of justice.[21]

This plan was attacked by a number of senior revenue officers for ignoring the corruption of the regular police and the unwillingness of *vidanes* and villagers to cooperate with them. Perhaps most outspoken was the highly respected GA of the Northern Province, who argued that,

> to indiscriminately place detachments of police over the whole country, as recommended by the Inspector General of Police, would, I believe, have a most demoralizing effect on the headmen and people ... What is required in the rural districts of the country is, not an armed police like the present police force, but the improvement and proper encouragement of the present rural police of the village headmen, supported and kept under the supervision of a sufficient number of magistrates.[22]

Some administrators understood that the presence of an outside police force challenged the already weakened authority of headmen. As one district judge argued,

> speaking as an administrative as well as a judicial officer, I venture to assert, that so intense would be the annoyance and irritation produced in

---

21 *ARC*, 1867, p. 256. The attempt to coordinate the efforts of the village police with the regular police through visits by the latter to villages was abandoned as a failure in Madras. Arnold, "The police," p. 5.

22 GA Northern Province, *ARC*, 1870, p. 82.

the minds of country people, by the inquisitorial visits and constant presence of men of an alien race, in their villages, where often the whole community is of one family, that it will be considered a curse.[23]

The reference here to "an alien race" expressed the concern that a force composed largely of Malays would be viewed as predatory outsiders by Sinhalese and Tamil villagers. However, despite the ambivalence of the various officials, Malays continued to be favoured by the British in Ceylon and various parts of the empire precisely because they were tough, aggressive, and politically safe, as they were ethnically different from the majority populations. The favouring of the Malays helped perpetuate, in Heath's words, "the system of everyday violence upon which colonial rule depended." She points out that while the British in India recognised some forms of police violence as a problem, they tended to regard them as simply "indigenous practices" rather than a manifestation of structural problems largely of their own making.[24] In this sense the British attitude served to legitimate and perpetuate violence. Again we find that by naturalising such practices as racial failure, the British failed to acknowledge their own complicity in colonial violence.

While believing the *vidanes* to be problematic, some government officials supported them over the regular police, due to their belief that the *vidanes* were drawn from the old ruling gentry, while the regular police came from the "disreputable" classes.[25] In 1870 a GA asked,

what are the police constables as a body? What has been their conduct in many instances, when detached on duty to a distance? I know from experience, that many of the men taken into the force are bad characters, and not to be trusted from under the immediate supervision of the inspector. Does Mr. Campbell really expect that respectable headmen will associate themselves with such men and co-operate with them in the detection of crime? ... the respectable men would resign their appointments; the police headmanships would no longer be looked up to and sought for by men of responsibility and position; and they would soon fall into the hands of the,

23 District Judge, Nuwarakalawiya, *ARC*, 1870, p. 245.
24 Even as late as the 1890s, Malays formed over one third of the force, even though they only constituted 0.3% of the island's population. Malays were more interested than other ethnic groups in enlisting in the force, in part because they had limited other occupational options but also because they considered it an expression of their martial tradition. Heath, D. 2016. "Bureaucracy, power and violence in colonial India: the role of Indian subalterns," in P. Crooks and T.H. Parsons (eds.), *Empires and Bureaucracy in World History: From Late Antiquity to the Twentieth Century*. Cambridge: Cambridge University Press, p. 365.
25 The English were ever ready to use categories that they employed at home. Brown argues, for example, that crime on the part of peasants was compared to the lower yeomanry of the English countryside responding to hard times. Brown, I. 2007. "South East Asia: reform and the colonial prison," in Dikotter, F. and I. Brown (eds.), *Cultures of Confinement: A History of the Prison in Africa, Asia and Latin America*. London: Hurst, p. 245.

in many instances, police constable stamp. The sending of such men as I have seen detached for duty to Jaffna, Mannar and Kurunegala, into the rural districts, would be a positive curse to the poor villagers who would have the misfortune to be near any of their stations.[26]

This dispute over who should police the countryside boiled down to a difference of opinion about which option was worse. Campbell thought that his regular police were more likely to be reliable as they could be given training in policing, while the *vidanes* were forever out of reach of his police bureaucracy, and the revenue officers felt that their social class and ethnicity impeded the police from solving rural crimes or supervising *vidanes*.

Revenue officers also opposed Campbell's plans because each police post would cost every male inhabitant within its limits two shillings a year.[27] At the prospect of a post being located in their area, many villages sent petitions signed by every villager, stating that they could not afford to be included within police limits.[28] The GA of Sabaramaguva, in the Kandyan highlands, wrote in his annual report for 1874 that because villagers saw their assessment tax as unjust, it was difficult to collect.[29] And so Campbell's ambitious vision of a disciplinary gaze by police extending across the whole country was rejected by the government. Not only was it costly, and likely to antagonise the villagers, it would have had the effect of weakening the traditional elite in the Kandyan highlands at the very moment when official policy was shifting towards supporting them as a counter weight to the emerging maritime elites.[30] Consequently, as late as 1891, only one percent of the territory of the island was under the jurisdiction of the regular police.[31] Shortly before his retirement, Campbell complained about the government's priorities:

> I have often reported that in my opinion our government compared with other governments, devote a very small and insufficient portion of their revenue to their police and to their prisons. It is a question of policy—a question whether a generation or two hence the people will be better off by

26 *ARC*, 1870, p. 82. The regular police were not only considered inferior by headmen because many were Malay, but even when they were Sinhalese they would have been lower caste than the headmen who were predominantly of the highest Goyigama caste. See Roberts, M. 1982. *Caste Conflict and Elite Formation: The Rise of the Karava Elite in Sri Lanka, 1500–1931*. Cambridge: Cambridge University Press, pp. 144–48.

27 *Ceylon Hansard (CH)*, 1871, pp. 74, 72.

28 *Ceylon Examiner*, November 19, 1870.

29 *ARC*, 1874, p. 71.

30 De Silva, K.M. 1979. "Resistance movements in nineteenth century Sri Lanka," in M. Roberts (ed.), *Collective Identities, Nationalisms and Protest in Modern Sri Lanka*. Colombo: Marga Institute, p. 143.

31 P. Ramanathan, Solicitor General, *ARC*, 1892, p. A2.

being provided early with railways and roads, and water tanks, or with comparative safety to life and property."[32]

Out of sight: following their own rules

Given the inability of the authorities to effectively oversee either the *vidanes* or the regular police in the posts, they had to count on them following the rules out of loyalty to the government, which they knew was unrealistic. Not only did the *vidanes* see the British as usurping their power, but the British also made little effort to hide their disdain for the headmen, especially those from non-elite families. While many administrators acknowledged that the *vidanes'* low level of cooperation stemmed in part from the lack of a salary, they seldom addressed questions surrounding their own legitimacy. Instead, they simply fell back on the belief that the Ceylonese were inherently untrustworthy.

Some administrators were aware that the situation was complicated by structural problems of their own making. A prime example is that of the *vidanes* and regular police not being clear about the laws they were to enforce. In this regard the British were complicit in the inefficiency of rural policing. The Acting GA of the North Western Province wrote in 1867,

> the knowledge they possess of the laws which most concern them as police officers is imperfectly picked up from others. It has often occurred to me that a codified epitome of such ordinances, printed in convenient shape, would be most useful, and ought to be freely distributed to headmen of all ranks. The Sinhalese translations of the ordinances, issued from the Government Press, from the endeavours made to render it literal, are unintelligible and their circulation too limited to meet the demands of the minor headmen.[33]

Two years later the Acting District Judge in the Tamil-speaking area of Mullaitivu complained that although it had been a year since he had requested Tamil translations of the ordinance to give to police *vidanes*, he had not received them, and consequently found the *vidanes* ignorant of the laws they were meant to enforce.[34] And three years after that the GAs of the Southern and Eastern Provinces complained that they still had not received copies of the ordinances. The latter wrote,

---

32 *ARC*, 1887, p. 39C. The government of India also greatly underfunded the police throughout the nineteenth century. Their policy of indirect rule at the village level through watchmen was guided by a wish not to be overly intrusive in peasant life and to save money. Dhillon, *Defenders*, pp. 17, 125. Similar reasons were given at the time for keeping the police out of the villages in Madras. Arnold, "Crime and crime control," p. 78.

33 E.N. Atherton, *ARC*, 1867, p. 54.

34 R. Massie, *ARC*, 1869, p. 200.

many of these ordinances entail duties on police headmen, the neglect of which is punishable and yet there are no copies of these ordinances in the native languages to be had; and the only answer that can be given to the repeated application for some guide to their duties, which are made to me by police headmen, is, in effect, "Copies of the laws are not procurable in the language of the country, and you must know intuitively the obligations they impose on you; if your instinct is at fault, you are liable to the penalties which the laws provide for your neglect of duty."[35]

And if the police *vidanes* did not know the laws of the country, certainly the villagers were even more in the dark. The Acting District Judge in Matara wrote that:

it is much to be regretted that the people generally should be allowed to remain ignorant of the provisions of the various Ordinances, as they are at present. The publication in the Gazette and newspapers of the Ordinances as they are passed ... are sufficient to provide persons who can read English with opportunities of becoming conversant with the statute law of the country, but the average Sinhalese has no chance of knowing what the law is unless he constantly, voluntarily attends court or unless he has the misfortune to be involved in litigation.[36]

A decade later in 1884, one headmen told his AGA in the Central Province that the villagers want to see copies of the penal code written in their own language, "in order that they might know what new crimes have been invented, and so learn to avoid them."[37] This poignant request reveals a significant violation of rule of law by the government in that a primary principle is that the law be communicated to the people as well as to those whose job it is to enforce it. Furthermore, the headman's choice of words is telling; the belief that new crimes were being "invented" by the British reveals the lack of legitimacy such laws held in the eyes of the Ceylonese. Such a view is not surprising given that the British had effectively criminalised the customary use of many of the common lands, thereby at a stroke legally depriving villages of an important component of their traditional livelihood.[38]

---

35 P.L. Templer, *ARC*, 1871, p. 137; R. Morris, *ARC*, 1871, p. 182. Whereas a poorly translated version of the Police ordinance was available in Sinhalese, there was no Tamil translation, thereby leaving much of the North and East of the country in the dark.
36 J.A. Swettenham, *ARC*, 1873, p. 28.
37 H.L. Moysey, *ARC*, 1884, p. 64A. Killingray, D. 1986. "The maintenance of law and order in British colonial Africa." *African Affairs*, Vol. 85, No. 340, p. 413, points out that in British colonial Africa colonial rule also "created new 'crimes', many of which were offences against the imposed structure of colonial management."
38 Killingray, "The maintenance of law and order," p. 413, argues that enforcement was made more difficult in Africa by all of the new crimes that the colonial government created.

I would argue that the British were also complicit in the corruption and lack of cooperation that characterised rural policing by their unwillingness to pay the *vidanes* a wage and their decision to pay the regular police in their posts starvation wages of approximately half of the very meagre wages paid to the town police. For financial reasons, the British reluctantly agreed to exempt *vidanes* from certain taxes and claim 10% of all stolen property that they recovered upon an arrest of the culprit, despite considering this a violation of bureaucratic norms of fixed salaries.[39] They were further concerned that allowing *vidanes* a percentage would encourage them to make false claims of theft.

Certain administrators had long argued that the *vidanes* might prove more honest if they were paid a salary.[40] As early as 1852, the Committee of the Executive Council accepted in principle that they should be paid, but argued that the government could not afford to do so.[41] In the mid-1870s the financial position of the colony had sufficiently improved such that civil servants were granted increases in salary, but the *vidanes* remained unpaid. The Tamil member of the Legislative Council argued that,

> to tell the headmen, therefore, "Be content with your exemptions from taxes in some cases and percentages in others", even when the colony could well afford to pay them would not be just. We know what the consequences of non-payment have been. In many cases they have paid themselves in an illegitimate manner. Such a corrupt state of things should not be tolerated.[42]

Some urged that in place of a salary, *vidanes* should be allowed to collect increased duties and receive greater commissions on their services.[43] This, after all, was a traditional way in which headmen had been rewarded.[44] But even such a meagre income was regularly reduced, as it was common for headmen to be forced to kick back as much as half of their legitimate fees to superior headmen and government clerks. If they refused, they might have to wait months or even years to get paid.[45] Others argued that the percentage system was ineffective because in poor districts there was little possibility of larger fees being generated and "in districts where the fees are small, it is almost impossible to get candidates to come forward and fill a vacancy."[46]

---

39 Pippet, *A History*, pp. 9, 151–52.
40 Saha, *Law, Disorder and the Colonial State*, p. 31, argues that in Burma commissioned investigators reasoned that the pervasive corruption among the lower ranks was in large part due to insufficient wages.
41 *Ibid.*, p. 122.
42 Coomaraswamy, *CH*, 1876–77, p. 200.
43 R.W. Ievers, AGA, Kegalla, *ARC*, 1879, p. 36.
44 Arnold, "Bureaucratic recruitment," p. 40, argues that among the Madras Constabulary, corruption could largely be accounted for by the combination of colonial parsimony and indigenous custom.
45 *Colombo Observer*, May 16, 1863.
46 C.M. Lushington, Acting AGA, Puttalam, *ARC*, 1885, p. 32A.

By the late 1870s, it was commonly understood that the desirability of the post of *vidane* varied geographically,

> among the Kandyans, traces of the old traditions still linger, and make the possession of office an object of ambition to respectable men for its own sake, but in the Maritime districts these things have passed away and the native has learned the folly to work except for hire. ... How can a low country Sinhalese be expected to seek an unpaid appointment, the due discharge of the duties of which can benefit him nothing, and will most likely make him many enemies among his fellow villagers.[47]

It was clear however, that while there was widespread feeling that headmen were corrupt because they were not properly rewarded, either financially or with meaningful positions with real power attached, there was still no will to pay them.

Nevertheless, the police *vidanes*, many of whom were landowners, tended to live well in comparison to the regular police. In the late 1850s the Superintendent of Police complained that he had trouble retaining rural policemen. On touring the Central Province, he reported that "all the station houses require repair, some falling down and unfit for use; no cattle sheds and in many places no cookhouses or privies."[48] Six years later conditions in the posts were, if anything, worse.

> At Welimada the police station was in such a dangerous state that the men were afraid to enter it and were living in a small, dirty *maduwa* (shed for storing grain) outside. At Haputale the police were living in one room in a bazaar man's house which was unfit for the commonest cooly to live in. It is impossible to suppose that they can have any authority or respect from the people until they are put in a proper position there. At Gingathena Gap I found everything in a most disgraceful state. The shed the police are put to live in is not fit to tie a dog in and the two constables stationed there look more like *veddahs* than policemen ... At every station that I visited I had one universal complaint from the constables, that they are starving, their pay being one half sufficient to pay for their living, and this I know to be true.[49]

While the physical conditions of the posts were improved over the decades, remuneration continued to remain so inadequate that it was an inducement to supplement it illegally.

---

47 P.A. Templer, AGA, Puttalam, *ARC*, 1877, p. 30.
48 Pippet, *A History*, p. 198.
49 *Ibid.*, pp. 221–22.

Officials assumed, and not always without reason, that a *vidane* would know of practically every crime that took place in his village, and would furthermore have a reasonable idea of where to find the responsible party. Consequently, if a crime wasn't reported, or a suspect wasn't apprehended within his village, the assumption was that the *vidane* was lazy or was shielding someone. It was a common understanding among officials that the *vidanes* placed loyalty to family and village faction over their duties to the state. The Acting District Judge in Trincomalee in 1867 wrote that *vidanes* "are utterly unable, for want of proper training and native prejudice and their own family affairs" to enforce the law.[50] Rogers points out that few headmen were willing to prosecute their relatives and, when faced with the choice between protecting a relative and doing their duty, chose the former. According to community standards, the headman owed loyalty first to his relatives, second to the faction in the village to which he was allied and third, if at all, to his bureaucratic position. This hierarchy was accepted as normal and morally correct by the Ceylonese.[51] As such this is but one of many situations I have come across where rule of law clashed with local moral codes.

Such was the concern that *vidanes* used their official position to further their own partisan agendas that in 1893, when the government was considering whether to issue handcuffs to *vidanes*, the Inspector General argued that "handcuffs are very dangerous weapons to entrust to police headmen. Charges are often brought against persons for the mere sake of getting handcuffs put on them and that satisfaction having been made are afterwards abandoned."[52] What is being recognised here, although not fully acknowledged, is that, having a lack of respect for British institutions, the Ceylonese regularly turned the power of the state to their own ends. The fact that these ends were often petty and personal is understandable given the limited scope of power they could access.

The British were equally concerned that *vidanes* were profiting economically from illegal activities.[53] It was widely assumed that they protected criminals in return for bribes and at times participated in the crimes they were supposed to prevent.[54] As the Inspector General wrote in 1890, "the village headmen rarely

50 *ARC*, 1867, p. 185.
51 Rogers found that in England at the time there were similar concerns that constables showed more loyalty to their friends than to their bureaucratic responsibilities. Rogers, J.D. 1987. *Crime, Justice and Society in Colonial Sri Lanka*. London: Curzon Press, p. 122.
52 Inspector General of Police to the Colonial Secretary, January 19, 1893, in Dep, *A History*, p. 302.
53 G.W.P Campbell, Inspector General of Police, *ARC*, 1869, p. 230; M. De Saram, Magistrate, Colombo, *ARC*, 1867, p. 256; M. Fisher, Police Magistrate, Galadera, *ARC*, 1867, p. 200.
54 *Ceylon Times*, February 25, 1870; "Crime and bribery in the Southern Province," *Ceylon Examiner*, May 1, 1894. It was common knowledge in Bengal that police and watchmen often turned a blind eye to gang activities or sometimes were in league with them. Mclane, J.R. 1985. "Bengali bandits, police and landlords after the permanent settlement," in Yang, A.A. (ed.), *Crime and Criminality in British India*. Tucson: University of Arizona, p. 40.

point out the criminals, though in many cases they know them well, but rather screen them and participate in their gains."[55] In particular, the police *vidanes* were thought to turn a blind eye to gambling and the illegal sale of arrack and toddy in exchange for bribes.[56] This was of great concern to the British, not only because they held gambling and drink to be among the root causes of other crimes such as theft and assault, but because taxes on the sale of liquor were a very important source of income for the government. Both of these crimes were difficult to suppress, in that the former was a traditional village occupation and the latter was considered by the people to be harmless.[57] Again we see a lack of legitimacy at issue. Rogers argues that the law against gambling was probably seen by the Ceylonese as simply a regulation put in place to allow *vidanes* and regular police to take bribes.[58] If this is the case, and it seems likely, then it speaks volumes about the extent to which the Ceylonese normalised or even accepted what the British saw as corruption at all levels. Another source of illicit income was the covering up of serious crime such as murder in exchange for a bribe. As M. Selby, the Police Magistrate of Mallakam wrote in 1867, "murder and manslaughter are very common in Ceylon … When men could get off the punishment of murder by, at worst a heavy fine in the shape of a bribe to a venal headman, they were not very fearful of committing the crime."[59]

Another common source of income was accepting bribes from villagers who illegally cut timber or cultivated *chenas* on crown land.[60] The prohibition against cultivation on crown land was considered to be a crime unfairly invented by the British. There was outrage on the part of peasants to the criminalisation of customary usage of common lands in the early 1840s and they ques-

55 G.W.R Campbell, *ARC*, 1880, 28B. for India, see Robinson, F.B. 1985. "Bandits and rebellion in nineteenth century western India," in Yang, A.A. (ed.), *Crime and Criminality in British India*. Tucson: University of Arizona, p. 55.

56 A.R. Dawson AGA Kegalla, *ARC*, 1873, p. 119; C.M. Lushington, Acting GA, Puttalam, *ARC*, 1885, p. A 32; F.R. Dias, Acting Crown Council Midland Circuit, *ARC*, 1892, p. A16. Saha, J. 2013. "Colonization, criminalization and complicity: policing gambling in Burma c. 1880–1920." *South East Asia Research*, Vol. 21, No. 4, Special Issue: Colonial Histories in South East Asia—Papers in Honour of Ian Brown, pp. 655–672, similarly found that gambling accounted for a regular source of bribes to the police in Burma.

57 In England at the time there was also resistance to laws being enforced that contravened local customary practice. Weinberger, "The police," p. 65.

58 Rogers, J.D. 1991. "Cultural and social resistance: gambling in colonial Sri Lanka," in Haynes, D. and G. Prakash (eds.), *Contesting Power: Resistance and Everyday Social Relations in South Asia*. Berkeley: University of California, p. 202.

59 *ARC* 1867, p. 202. Charges that murders were covered up by headmen who had been bribed continued over the decades. See "Crime and bribery in the Southern Provinces." *The Ceylon Examiner*, May 1, 1894.

60 C.M. Lushington Acting AGA, Puttalam *ARC*, 1885, p. 32A.

tioned the legitimacy of laws forbidding such activities.[61] Such laws legally, but inhumanely, dispossessed peasants of their rightful resources in order to sell land to European and Ceylonese capitalists.[62] The criminalisation of the use of common lands was a clear example of British property law contradicting Ceylonese customary rights to food and fuel. Linebaugh, writing about commons more generally, argues that conceptually commons are outside state government. They provide their own security with commoners rather than a police force regulating their use. It is, as he says, "custom rather than law [that] safeguards and defines the commons."[63]

The crime most associated with officials, from the most senior in charge of a large rural area to the police *vidane*, was participation in the theft of cattle.[64] This crime was so widespread in certain parts of the country, and the *vidanes* so unable or unwilling to stop it, that the government instituted a "Sale Certificate System" to make it more difficult to sell stolen cattle. Under this system, no one was allowed to buy cattle unless a certificate of ownership was issued by their village headman. Almost immediately, one district judge reported that "three headmen who were authorized to issue these certificates were themselves convicted of cattle stealing. The grants of false certificates by some of the headmen for a pecuniary consideration is I fear more extensively practiced than is suspected."[65] As an AGA in the Central Province in his annual report stated, "the cattle stealers and receivers of stolen cattle are, it is found, generally under the protection of the headmen, without whose connivance the crime could not be practiced with impunity. Every effort is being made to awaken the headmen to their duty, or at least to make them afraid of connivance with cattle stealers."[66]

The British officials thought that *vidanes* were apt to commit violations either out of self-interest or when they believed that village affairs were best resolved

---

61 See De Silva, "Resistance movements," p. 139. Meyer, however, points out that there was less open peasant resistance to the British expropriation of land than there was in India. Meyer, E. 1998. "Forests, chena cultivation, plantations and the colonial state in Ceylon 1840–1940," in Grove, R.H., Damodaran, V. and Sangwan, S. (eds.), *Nature and the Orient: The Environmental History of South and Southeast Asia*. Delhi: Oxford University Press, p. 815. For a general discussion of the privatisation of common lands and the criminalization of commoners whose customary rights to the products of the commons were expropriated see: Linebaugh, P. 2014. *Stop, Thief! The Commons, Enclosures, and Resistance*. Oakland: PM Press.

62 Roberts argues that this selling off of Crown land had profound effects upon the social structure. For not only did it cut back on the resources available to poor villagers but also undermined traditional patronage systems and opened up economic opportunities for elites from the maritime lowlands to invest in the Kandyan highlands. Although the majority of plantations in the highlands were owned by Europeans, there was a higher percentage of "native" ownership of plantations than was found in British India or Dutch Indonesia. Roberts, *Caste Conflict and Elite Formation*, pp. 99–101.

63 Linebaugh, *Stop, Thief*, p. 19.

64 Rogers, *Crime, Justice and Society*, pp. 83–121 provides a detailed analysis of cattle theft.

65 D.E. De Saram, Kurunegala, *ARC*, 1869, p. 198.

66 *ARC*, 1873, p. 16.

internally. On the other hand, the regular police in the posts were thought to prey upon villagers or encourage criminal behaviour among them. The AGA in Matale summed up the majority view when he wrote, "in remote spots where no effectual supervision can be exercised, small detachments soon teach the Sinhalese (and very little instruction is needed) the arts of tyranny and extortion, and of getting up false cases."[67] In 1878 the Tamil member of the Legislative Council in a debate on police in the rural districts, stated that he had "received complaints on that subject from all classes" and that the police "instead of repressing crime they rather tended to encourage it."[68] It was widely believed that the police from the posts not only extorted the peasants, but were sometimes in collusion with criminals in doing so. In a notorious case in 1879, a well-known thief named Kadiravelu was apprehended after a robbery in Nawalapitiya in the Central Province. When questioned he claimed that Constable Sahit was on his beat during the robbery and that they shared the proceeds. Upon searching Sahit's house the stolen goods were found and the constable admitted his involvement. He defended himself saying, "it was the custom for the police for a long time back to share stolen goods." He then implicated two sergeants and six other constables, and searches subsequently revealed stolen property in their houses as well.[69]

The fact that British officials tolerated such a level of corruption among the rural police suggests that maintaining a professional police force in the areas of the country where there were few Europeans remained a low priority.[70] The financial investment necessary for adequate remuneration of the police and police *vidanes* and to pay for the degree of surveillance that would have been necessary for adequate crime prevention was far more than the British were willing to commit to.

Fixed boundaries and mobile crime

The bureaucratic technique of dividing spatial units into discrete subunits, each with their own chains of command, was an example of British efficiency that ended up exacerbating the problems of rural policing. While the spatialisation of administration is central to imperialism in that it literally "grounds" new laws and regulations, in this particular instance it was a failure. The British converted traditional village boundaries into legal demarcations and gave *vidanes* responsibility for law and order within those boundaries.[71] Likewise, the

---

67 G.S. Williams, *ARC*, 1870, p. 56.
68 *CH*, 1878, p. 61.
69 Dep, *A History*, p. 171.
70 Killingray refers to such a policy as "selective administration." Killingray, "The maintenance of law and order," p. 413.
71 On attempts to render a population spatially legible, see Scott, J.C. 1998. *Seeing like a State: How Certain Schemes to Improve the Human Condition Have Failed*. New Haven: Yale University Press.

jurisdictions of the regular police were constrained by the territorial limits surrounding their posts. The territories of the regular police posts' limits covered
only 205 square miles, leaving the remaining 25,128 square miles of the island
in the hands of the *vidanes*.[72] Given that people frequently absconded after
committing a crime, such a spatial strategy could only have been effective if
there had existed a high level of cooperation between the *vidanes* and the regular police, which could not have been further from the case. Being protective of
their territorial power, village headmen tended to be uncooperative with headmen from other villages while meeting the regular police from the posts with
silence and misdirection when they dared to leave their limits to make arrests.
The GA of the Western Province complained in 1889 that:

> one of the principal defects in the administration of justice in Ceylon
> is ... the entire absence of any machinery for hunting down absconding
> criminals. At present, if a man who has committed an offence can succeed
> in leaving his village before he is arrested ... he has every chance in his
> favour of not being arrested ... His own headman of course, will not look
> for him beyond the limits of his village, and is rather pleased than other-
> wise at getting rid of a troublesome character. The headman of the village
> where the criminal has gone to does not know the man, or the fact that he
> has committed an offence, and takes very little notice of the arrival of
> a stranger. Unless the injured person is very much in earnest about securing
> the criminal, and is sufficiently well off to bear the expense of making
> inquiries, taking warrants to other districts, and paying headmen and villa-
> gers to assist him in arresting the accused, the chances of his being cap-
> tured are very remote indeed ... He remains in his new home for a period
> of varying from six months to two or three years, according to the gravity
> of the offence: and when the complainant's zeal to prosecute him has
> cooled down, the friends of the accused have little difficulty in compound-
> ing with the injured person to allow the criminal to return to his village
> and to say no more about it. It is obvious that such immunity from punish-
> ment is an incentive to crime.[73]

For decades the British were unwilling to officially change this territorial
system, but instead urged *vidanes* and the regular police to range illegally outside their territorial limits and at times they reluctantly complied.[74] The revenue
officers in charge were often casual about legal boundaries and at times blamed
the regular police for not venturing out, for example, "to prosecute gamblers
and drunkards with whom the *Arachchies* will not interfere."[75] The Solicitor

---

72 *ARC*, 1892, p. A2.
73 F.R. Saunders, *ARC*, 1889, p. 11B.
74 Rogers, *Crime, Justice and Society*, p. 53.
75 *ARC*, 1873, p. 119.

General provides examples of how requests by magistrates and the regular police for cooperation were appropriated by headmen in order to profit themselves or settle scores in the village; he wrote that it was common for police *vidanes*, when asked by the regular police to provide the names of villagers involved in illegal gambling, to hold an auction to decide which villagers' names would be included on the list: "A's brother hearing that A's name has been inserted gives five rupees to have it struck out; B gives ten rupees to have the name of his enemy inserted, etc."[76]

But lack of cooperation wasn't the only risk the regular police took when they left their limits. At times they were attacked by villagers, often with the blessing of the *vidanes*. In an example from 1878, the Inspector General complained bitterly to the Colonial Secretary that two of his Malay constables who had been sent to a village twenty miles from Galle with a warrant to arrest a man were attacked by his supporters with knives and clubs. "In a minute [the officer's] skull was laid open and one eye burst and utterly destroyed. His comrade Omar Abdoola, who came to his assistance also had his head laid open to the bone and his arm broken." These two men were hospitalised for a long period, but surprisingly, given that it was common practice for judges to overlook these extra-legal forays by the police, the judge acquitted the accused on the grounds that by law the police should not have left their limits.[77] That this was not an isolated occurrence is evidenced by the fact that during the early 1880s the annual injury rate in the force ran at 20%.[78]

The regular police began to be sent outside their limits less and less often, except for serious cases such as murder, and in such cases the police were often ineffective, for as Campbell put it "many murders are not reported to us until the available evidence has been so neglected or corrupted as to be valueless and the criminals go unpunished."[79] So exasperated did the acting Inspector General of Police become that he wrote in his annual report that "it is unfair to in any way hold the police responsible for crimes committed outside police limits."[80] A new Inspector General of Police wrote:

> This system could never be successful: it is one thing for a constable to perform his duty in a country which he knows, and among people whom he understands, who are used to him, and who recognise his authority; but it is quite a different matter when he is suddenly sent forth on a raid into a strange country, among strangers who are unaccustomed to him and are not used to regard him as acting on behalf of and under the protection of the law.[81]

76 *ARC*, 1892, p. A3.
77 Inspector General of Police to the Colonial Secretary, August 15, 1878, in Dep, *A History*, p. 129.
78 Dep. *A History*, p. 166.
79 *ARC*, 1886, p. 32C.
80 F.R. Ellis, *ARC*, 1890, p. B3.
81 L.F. Knollys, *ARC*, 1892, p. B2.

It is interesting that the Inspector General reveals that he understood that villagers refused to recognise the legitimacy of the police from the posts and consequently at times attacked them.

On the other hand, in his annual report of 1888, the Solicitor General stated, "In actual practice, police officers and police headmen have rendered most valuable service in inquiring into and prosecuting cases; this is, however, done without legal sanction, and it seems to me very desirable that the police should be invested by law with special powers to take evidence."[82] So, finally, in 1891, the Criminal Code was revised allowing *vidanes* to legally pursue suspects beyond their village limits, but many *vidanes* were either unaware of the change in the law or simply preferred to remain in their villages.[83]

Rather than understanding the failure of rural policing as stemming from the lack of police legitimacy, Giles in his report of 1889 described it a technical, bureaucratic problem caused by the fact that the *vidanes* were under the authority of the AGAs and the regular police were under the Inspector General of Police. His solution was to place them both under the authority of the AGAs in the belief that a shared channel of authority would integrate the two branches of rural policing. In order to secure more cooperation from the *vidanes*, a layer of paid village sergeants was inserted as a buffer between the *vidanes* and the regular police. It was hoped that the *vidanes* would more readily accept these sergeants as they were from their own region and of the same ethnicity, but this plan failed to account for the intensity of the localism of the village perspective.[84] Consequently, by the late 1890s, this new system was abandoned. Resistance by the *vidanes* and villagers had effectively led to the return of a system that had been considered ineffective since its inception in 1806. The village, once again, had successfully repelled the intrusion of the centralised police bureaucracy.

But the government persisted in trying to fix the system in other ways and so in 1898 Governor Ridgeway proposed paying *vidanes* for their expenses and setting up a reward system both of honours and of money;[85] but the new reward system was difficult to administer. When rewards were offered to headmen for keeping their villages crime free, they covered up crime; and once it became known that headmen could be dismissed for misbehaviour, their enemies reported false cases.[86] Despite all the large and small changes over the decades, the *vidanes* remained unpaid and as many British officials had warned, they were inclined to reward themselves.

Frustrated with the lack of progress, in 1904 the Governor received sanction from the Secretary of State for the Colonies to place the regular police in the

---

82 C.P. Layard, *ARC*, 1888, p. C2.

83 M. Carbery, *ARC*, 1897, p. E19.

84 Dep, *A History*, p. 301.

85 *CH*, 1897–98, p. xii.

86 F.R. Ellis, *ARC*, 1900, p. B12, cited in Rogers, *Crime, Justice and Society*, 52.

countryside once again under the control of the Inspector General who could devote more time to solving police problems than the AGAs, a decision that was applauded in the press.[87] And in 1906 the Inspector General instituted yet another reorganisation. Thirty new police posts were created in the most high crime places with the police who staffed them receiving better training and slightly higher salaries. A programme of increased rewards for *vidanes* appeared to make them more cooperative, except in cases where the accused was a relation of the headman, in which case little assistance was forthcoming.[88] However, in areas remote from posts where the reward system was little used, the *vidanes* continued to be uncooperative and to accept bribes.[89]

In spite of all the many adjustments to the system of policing, surveillance at a distance continued to be a problem for the British administration. The reporting and prosecuting of village crime was largely carried out on the villagers' own terms, an outcome that clearly reflected the success of their resistance tactics. Despite a very few small gains, the British were unable to create a rationalised police force committed to modern bureaucratic ideals and the police continued to play the system, distorting police administration to such a degree that it was forced to focus primarily on combatting police corruption rather than on crime more generally.[90]

Conclusion

In the century following the founding of the Ceylon police in 1806, none of the efforts made by British administrators to create an effective police force could be considered successful. There was relatively little that the British could do in the face of covert resistance to the prevailing order that took the form of subterfuge.[91] As neither the regular police nor the *vidanes* could openly

---

87 Governor Sir Henry Blake to Rt. Hon. Alfred Lyttelton, Secretary of State for the Colonies. September 28, 1904. Lyttelton to Blake, December 30, 1904, *SP* 1904–05; *Ceylon Observer*, September 21, 1905. A police commission in India likewise found very widespread oppression and extortion by the police. Dhillon, *Defenders*, p. 72.

88 T.A. Carey, AGA Hambantota, *ARC*, 1911–12, p. D20; R.B. Hellings, GA Southern Province, *ARC*, 1914, p. C4.

89 W.L. Kindersley, GA North-Central Province, *ARC*, 1911–12, p. G3; B. Hill, GA Eastern Province, *ARC*, 1914, p. E4.

90 In 1903, the 1,803 policemen on the island received 4,711 punishments for disciplinary infractions. In 1914, eight years after the latest reorganisation, the 2,614 man force received 5,071 punishments (*ARC*, 1903, p. B1; 1914, p. B3). While there is no way of knowing how often the police got away with illegal behaviour, it is safe to assume that these known infractions only represented the tip of the iceberg.

91 I should note here that although I am using an expansive notion of resistance, I fully acknowledge that resistance on the part of the most oppressed is not equivalent to the resistance of those in positions of greater power, such as the police *vidanes*, or as we shall see, those with other positions of authority in the various bureaucracies. The resistance of the most subaltern groups may be more necessary as a survival strategy than those more privileged in structures of power, but this fact does not make their resistance more authentic.

challenge the British bureaucracy of which they were a part, they appropriated what power they had to benefit their own needs and goals. Their resistance was of the everyday, unspectacular and uncoordinated sort. Scott describes this type of resistance as a "quiet, piecemeal process which when multiplied many thousand-fold may, in the end, make an utter shambles of policies dreamed up by their would-be superiors in the capital."[92]

The British tended to interpret piecemeal resistance on the part of the police as evidence of racial failings. As a consequence, they effectively depoliticised this resistance by dealing with it through bureaucratic reform. They failed to adequately acknowledge the larger fact of their oppressive rule and the legitimacy they were never going to be able to achieve. As Neocleous argues, "the genius of liberalism was to make the police appear as an independent, non-partisan agency simply enforcing the law and protecting all citizens."[93] It creates the illusion that it is rules and not politics that govern.[94] It would appear that, from the point of view of the British and members of the Ceylonese elite, liberal bureaucratic ideology obscured the extent to which the power of capital and property are embedded in law and the administration of criminal justice. However, it is likely that the Ceylonese general population, including the lower echelons of the police, were not so deceived. Most likely they saw the police as the enforcement arm of a government illegitimately imposed on them from the outside. The ideas of justice that were enforced by the British too often failed to align with those of the Ceylonese. The difference between standpoints inevitably led to much righteous frustration on the one side and passive resistance on the other. The result was, as Comaroff put it, that "far from being a crushingly over-determined, monolithic historical force, colonialism was often an undetermined, chaotic business."[95] The lack of an effective surveillance system meant that the British effectively ceded control of 99% of the countryside to *vidane* headmen, even though they officially granted them only the power to apprehend, not to judge and punish.[96] This is because, as we shall see in the next chapter, the police *vidanes* inserted themselves into the court system in ways such that they were able to influence the outcome of criminal cases which they had chosen, or were forced, to report.

92 Scott, J.C. 1986. "Introduction," in Scott, J.C, J. Benedict and T. Kerkvliet (eds.), *Everyday Forms of Peasant Resistance in South-East Asia*, Library of Peasant Studies no. 9. London: Frank Cass and Co, p. 8.

93 Neocleous, M. 2000. *The Fabrication of Social Order: A Critical Theory of Police Power*. London: Pluto Press, p. xiv.

94 Personal communication from David Nally.

95 Comaroff, J. 2006. "Colonialism, culture and law: a foreword." *Law and Social Inquiry*, Vol. 2, p. 311.

96 This policy of unwilling dependence upon traditional village structures of authority paralleled the British experience in India. For a discussion of British dependence upon village officials in India see Freitag, S.B. 1991. "Crime in the social order of colonial North India." *Modern Asian Studies*, Vol. 25, No. 2, pp. 227–61; Arnold, "Crime and crime control"; Bayly, *Empire and Information*, p. 164, pp. 333–34.

Part III

The courts and the arts
of dissembling

5   Taking liberties

The court bureaucracy and its discontents

T.B. Macaulay, an early architect of colonial law in India, espoused uniformity and certainty as central tenets of rule of law; he famously declared "Uniformity when you can have it; diversity when you must have it; but in all cases certainty."[1] The British saw this liberal legal system as their greatest gift to their colonies. There can be no doubt, however, that in practice this law was applied neither certainly nor uniformly across race or space in the colonies.[2] From the beginning of British rule, the criminal law was a mixture of Roman-Dutch law, English common law, and later in the 1880s, the Indian Penal Code. Civil law supplemented the above with customary and local laws such as *Thesavalami* inheritance and marriage laws. Such legal pluralism complicated the management and enforcement of the whole legal system.[3]

Rule of law, as constitutionally constrained government and individual legal equality impartially administered, was intended to guarantee the protection of life and property while standing as an object lesson to colonised peoples of the superiority of

1 Macaulay, T.B. 1867. *Speeches and Poems with the Report and Notes on the Indian Penal Code.* New York: Hurd and Houghton, p. 200.
2 On the contradictions between the theory and practice of law in India see Kolsky, E. 2010. *Colonial Justice in British India: White Violence and the Rule of Law.* Cambridge: Cambridge University Press; Wiener, M.J. 2009. *An Empire on Trial: Race, Murder and Justice under British Rule, 1870–1935.* Cambridge: Cambridge University Press; Killingray, D. 1986. "The maintenance of law and order in British colonial Africa." *African Affairs,* Vol. 85, No. 340, pp. 411–37. Also see Saha, J. 2013. *Law, Disorder and the Colonial State: Corruption in Burma c. 1900.* London: Palgrave Macmillan, on how the colonial state was actually experienced on a day to day basis.
3 However, as Barkey argues, "this lack of unity often allowed imperial systems to function more pragmatically but also made them vulnerable to competition and conflict." She further argues that "legal pluralism was a tool for the management of diversity." Barkey, K. 2013. "Aspects of legal pluralism in the Ottoman empire," in Benton, L. and R.J. Ross (eds.), *Legal Pluralism and Empires, 1500–1850.* New York: New York University Press, pp. 83, 85. As Benton has argued, legal pluralism which allowed for customary laws that applied differently to the various religious groups legitimised the concept of difference within an overarching liberal system based on universal principles. Difference for Benton is never absolute as some post-colonial theorists posit, but a process of mutual, albeit unequal, constitution. Benton, L. 1999. "Colonial law and cultural difference: jurisdictional politics and the formation of the colonial state." *Comparative Studies in Society and History,* Vol. 41, No. 3, p. 572.

British governance.[4] But far from serving as merely an ideological prop for imperialists to justify their domination of a foreign people, the law in a fundamental sense was constitutive of a new political order incorporating subjects within the domain of bureaucratic administration.[5] Having said this, ideological props are important, for as Mann points out, ideology not only works towards incorporating the masses into a new political order, but also towards unifying and giving moral authority to ruling elites.[6] When a chasm develops between ideological claims and "on the ground" results of their implementation, the bureaucrats charged with enforcing policies are often demoralised, as we will see in the case of Leonard Woolf.

Guha, Mattei, Nader and others have argued that the rule of law has a dark side as it was systematically employed by the British to dispossess indigenous people of their land and to create the legal conditions whereby British capitalists could exploit the land, labour and the natural resources of occupied territories.[7] Guha goes further in arguing that colonial law was little more than rule of force and that rule of law never really existed in colonial India. He states his view in the starkest of terms: "None of its noble achievements—liberalism, liberty, rule of law and so on—can survive the inexorable urge of capital to expand and reproduce itself by means of the politics of extra-territorial colonial dominance." Guha's assertion hangs on the belief that rule of law is dependent upon hegemony. This, he argues, was absent in India, as Indians were subjects rather than citizens who had not chosen their own laws. He further suggests that rule of law was not observed in India due to a double standard based in racism.[8] This latter point resonates in the wider literature on colonialism. Chatterjee has

---

4 Travers, R. 2007. *Ideology and Empire in Eighteenth Century India.* Cambridge: Cambridge University Press; Den Otter, S. 2012. "Law, authority and colonial rule," in Peers, D.M. and N. Gooptu (eds.), *India and the British Empire.* Oxford: Oxford University Press, p. 168; Wiener, *An Empire*, pp. 2–5. On Africa see Hynd, S. 2008. "Killing the condemned: the practice and process of capital punishment in British Africa, 1900–1950s." *Journal of African History*, Vol. 49, pp. 403–18; Hynd, S. 2012. "Murder and mercy: capital punishment in colonial Kenya, ca. 1909–1956." *The International Journal of African Historical Studies*, Vol. 45, No. 1, pp. 81–101.

5 Fitzpatrick, *Law as Resistance*, p. 26. However as Wiener, Benton and others argue, the rule of law can also be used to oppose imperial power. Wiener, *An Empire*; Benton, L. 2002. *Law and Colonial Cultures: Legal Regimes in World History; 1400–1900.* Cambridge: Cambridge University Press; Evans, J. 2005. "Colonialism and the rule of law: the case of South Australia," in Godfrey, B.S. and G. Dunstall (eds.), *Crime and Empire, 1840–1940: Criminal Justice in Local and Global Contexts.* Cullompton: Willan, pp. 58–59.

6 Mann, M. 1986. *The Sources of Social Power, Volume One.* Cambridge: Cambridge University Press, p. 24; Osborne, T. 1994. "Bureaucracy as a vocation: governmentality and administration in nineteenth century Britain." *Journal of Historical Sociology*, Vol. 7, No. 3, p. 290.

7 Guha, R. 1997. *Dominance without Hegemony: History and Power in Colonial India.* Cambridge, MA: Harvard University Press, p. 57; Mattei, U. and Nader, L. 2008. *Plunder: When the Rule of Law Is Illegal.* Oxford: Blackwell; Mehta, U. 1999. *Liberalism, and Empire: A Study in Nineteenth-Century British Liberal Thought.* Chicago: University of Chicago Press; Pitts, J. 2005. *A Turn to Empire: The Rise of Imperial Liberalism in in Britain and France.* Princeton: Princeton University Press.

8 Guha, *Dominance*, pp. 67, 72, 85, 256, 275–77.

argued that a "rule of colonial difference" prevailed in colonial India whereby the universalising discourse of liberalism was fractured by hierarchical theories of race.[9] Others have demonstrated that, contrary to the ideal, the implementation of the rule of law was tensile; its meaning dependent upon local interpretations which varied from colony to colony, and it was routinely bent or broken in favour of white settlers or European capitalist interests.[10] As Saha points out, "the law was formally fixed but informally fluid."[11] While rule of law is undoubtedly a "mythic image" systematically distorted along racial lines, the administration of law is, as Valverde argues, "a key site for the reproduction and contestation of various forms of power relations."[12] I will show that there is no question that it was used strategically by the Ceylonese, not so much to openly contest, but to undermine their oppression.[13] The Ceylonese used the courts extensively, and one could argue that the court system played a much greater role in Ceylon society at all social levels both symbolically and practically than it did in India, Britain, or anywhere else at that time.

Parkin's and Scott's understandings of power are more nuanced in comparison with Guha's relatively Manichean perspective. They believe that coercion is usually unnecessary as there often exists a weak form of hegemony in which subalterns are compliant not because they believe the system is just, but because they believe it is inevitable.[14] Weak hegemony is fragile and tentative, leaving

---

9 Chatterjee, P. 1993. *The Nation and Its Fragments: Colonial and Postcolonial Histories*. Princeton: Princeton University Press.

10 Hunter, I. and S. Dorsett. 2010. "Introduction," in Dorsett, S. and I. Hunter (eds.), *Law and Politics in British Colonial Thought*. New York: Palgrave Macmillan, pp. 1–10; McLaren, J. 2010. "The uses of the rule of law in British colonial societies in the nineteenth century," in Dorsett, S. and I. Hunter (eds.), *Law and Politics in British Colonial Thought*. New York: Palgrave Macmillan, pp. 74–75, 84; Raman, K.K. 1994. "Utilitarianism and the criminal law in colonial India: a study of the practical limits of Utilitarian jurisprudence." Modern Asian Studies, Vol. 28, No. 4, pp. 739–91; Kolsky, E. 2005. "Codification and the rule of colonial difference: criminal procedure in British India." *Law and History Review*, Vol. 23, No. 3 (Fall), pp. 631–83; Kolsky, E. 2010. *Colonial Justice in British India: White Violence and the Rule of Law*. Cambridge: Cambridge University Press; Fischer-Tine, H. 2009. *Low and Licentious Europeans: Race, Class and White Subalternity in Colonial India*. New Delhi: Orient Longman, p. 498; Ibhawoh, B. 2013. *Imperial Justice*. Oxford: Oxford University Press, pp. 10, 17.

11 Saha, J. 2013. "Colonization, criminalization and complicity: policing gambling in Burma c. 1880–1920." *South East Asia Research*, Vol. 21, No. 4, Special Issue: Colonial Histories in South East Asia-Papers in Honour of Ian Brown, p. 672.

12 Valverde, M. 2003. *Law's Dream of a Common Knowledge*. Princeton: Princeton University Press, p. 1.

13 Benton, L. 1999. "Colonial law and cultural difference: jurisdictional politics and the formation of the colonial state." *Comparative Studies in Society and History*, Vol. 41, No. 3, p. 564. Mattei and Nader (*Plunder*) while acknowledging that rule of law can empower the oppressed, nevertheless argue that rule of law has systematically being used to legalise and thereby attempt to legitimate exploitation of the weak and the plundering of resources. In their view the dark side of rule of law has greatly outweighed the good it has done.

14 Parkin, F. 1971. *Class Inequality and Political Order*. New York: Praeger; Scott, *Weapons of the Weak*, pp. 314–51.

space for a complex interplay of acceptance, resistance and coercion. As Merry states, "the dominant ideology itself establishes the terms for acts of resistance." She cites Scott's work in Malaysia where he shows that "resistance takes place in terms established by the dominant ideology itself." Again citing Scott, Merry suggests that "every dominant ideology must contain elements which appeal to subordinate groups or it would not be accepted. It is these elements which provide grounds for resistance within the dominant ideology itself."[15] A good example is offered by Sharafi, who shows how members of a Parsi elite in colonial India were able to systematically appropriate colonial law by embedding themselves within the judicial system and sponsoring legislation favourable to themselves.[16] Similarly, I will show that people in nineteenth-century Ceylon, through their high level of participation in the colonial court system as plaintiffs, witnesses and minor court officials, were able to systematically subvert the rule of law and also use it creatively to resist British authority. In the process, they shaped the evolving structure of the colonial judicial system itself.

Benton's and Comaroff and Comaroff's complex views of colonial law are helpful in understanding attitudes to law in nineteenth-century Ceylon. They argue are that while there is often no hegemony in the narrowly defined sense of freely-given consent, "the constructs and conventions that have come to be shared and naturalised throughout a political community" reveal that a loose hegemony in the context of colonial law prevailed.[17] Despite being codified, colonial law was a social practice that was continually produced and reproduced in struggles among various groups, some admittedly much more powerful than others.[18] I found that an alternating embrace of the court system and resistance to it on the part of the general public shaped the ongoing practice of the rule of law. These struggles added another dimension to the pluralism which was codified in law. In practice the judicial system did not emanate solely from the state; rather it was cross-cut by a number of other unofficial normative orders—secular, religious and ethnic—each with their own rules and enforcement mechanisms. In the case of Ceylon, local people enthusiastically participated in the state legal system, while artfully applying the values of other normative orders.

15 Merry, S.E. 1990. *Getting Justice and Getting Even: Legal Consciousness among Working-Class Americans*. Chicago: University of Chicago Press, p. 8; Wiener, *An Empire*; Benton, *Law and Colonial Cultures*; Evans, "Colonialism," pp. 58–59.
16 Sharafi, M. 2014. *Law and Identity in Colonial South Asia: Parsi Legal Culture, 1772–1947*. New York: Cambridge University Press.
17 Comaroff, J. and Comaroff, J. 1991. *Of Revelation and Revolution: Christianity, Colonialism and Consciousness*. Chicago: University of Chicago Press, p. 24; Benton, *Law and Colonial Cultures*, pp. 254–60.
18 Merry argues that resistance was greatly facilitated by the fact that colonial courts were few relative to large populations and judges had limited knowledge of local people and languages and were dependent upon local assistants. Merry, S.E. 2000. *Colonizing Hawaii: The Cultural Power of Law*. Princeton: Princeton University Press; Merry, S.E. 2010. "Colonial law and its uncertainties." *Law and History Review*, Vol. 28, No. 4 November, p. 1067.

Rule of law as a tool of oppression, especially when it dispossesses people of land causing hunger and potentially starvation, is what Sen calls "legality with a vengeance."[19] While it is beyond question that the Ceylonese were regulated in their property relations and other interpersonal relations by English law, and so were incorporated bodily and otherwise into the colonial state, this incorporation was far from complete. There was always, to use Foucault's term, a "swarm of points of resistance" to official power.[20] On the one hand, the law legitimated oppressive practices such as enforcing exploitative contracts with plantation owners and the seizing of common *chena* lands from the peasantry, while, on the other hand, the court system was also used by the Ceylonese in ways that undermined the ideals of rule of law.[21] As Comaroff points out, the law is "a vehicle simultaneously of governmentality and of its subversion, of subjection and emancipation, of dispossession and re-appropriation."[22]

There were also myriad points of resistance to rule of law among some judges and revenue officers. J.F. Stephen, the Law Member of the Governor-General's Council in India, observed in 1870, "there is an obvious difference between the judicial and executive temper. A judge must go by strict rules ... an executive officer, on the other hand, must constantly look beyond the rules."[23] Pressure was regularly applied by members of the executive to loosen the rule of law in the name of pragmatism. This was facilitated, as we shall see, by the fact that the lower levels of the judiciary were stocked with members of the executive branch, who had little or no legal training and consequently thought more like administrators than judges. The mostly British, upper level officials, like the Ceylonese lower level officials and members of the general population who came into contact with the court system, were largely unconcerned with the cumulative effects of such practices. Most were simply doing their job in the most expedient way in order to just "get through" their day. I will argue that such practices, especially when normalised, constituted a form of resistance. Of course, not all resistance was equal; its impact depended on what position in the power structure one resisted from and what personal and collective gains were possible.

19 Sen, A. 1981. *Poverty and Famine: An Essay on Entitlement and Deprivation.* Oxford: Clarendon Press, p. 166. Also see Nally, *Human Encumbrances*, p. 13; Elkins, J. 1996. "Legality with a vengeance: famines and humanitarian relief in 'complex emergencies.'" *Millennium: Journal of International Studies*, Vol. 25, No. 3, pp. 547–75.

20 Foucault, M. 1990. *The History of Sexuality. Volume 1: An Introduction.* Translated by R. Hurley. New York: Vintage, pp. 96–97.

21 Duncan, J.S. 2016. *In the Shadows of the Tropics: Climate Race and Biopower in Nineteenth Century Ceylon.* London: Routledge, pp. 92–94, 147–50.

22 Comaroff, J. 2006. "Colonialism, culture and law: a foreword." *Law and Social Inquiry*, Vol. 2, p. 307.

23 Dhillon, K.S. 1998. *Defenders of the Establishment: Ruler-Supportive Police Forces of South Asia.* Shimla: Indian Institute of Advanced Study, p. 128.

The pre-British system of justice

To more fully understand the origins of the hybrid nineteenth-century system of criminal justice, I will briefly outline the history of criminal justice prior to the British conquest. In the early sixteenth century the Portuguese captured much of the island's littoral, from Jaffna in the north to the Kingdom of Kotte in the south. From there they launched periodic raids against the Kandyan Kingdom in the central highlands, but were never able to subdue the Kandyans. While they were very keen to proselytise for the Roman Catholic faith, they left the administrative and judicial structures of the littoral largely intact, leaving Sinhalese officials in charge of the various provinces in the south. Later in the century they replaced only the most senior posts with Portuguese officials, leaving the administrative structure and laws of Jaffna largely untouched.[24]

The Dutch captured the southern littoral from the Portuguese in 1656 and two years later took Jaffna. Unlike the Portuguese, the Dutch had no interest in proselytising. They divided their possessions into three *commanderies*, Colombo, Galle, and Jaffna and established major courts of justice in each place, with a circuit court or *Land Raad* to sit in different districts presided over by local officials. While the Dutch made Roman-Dutch law the law of the land, they also recognised local customary law except when it conflicted significantly with their own. Thus, Tamil customary law (*Thesawalamai*) was recognised in the Jaffna peninsula, as were the Tamil customary laws of the coastal districts of Batticaloa and Puttalam.[25] Muslim customary law prevailed where there were concentrations of Muslims and traditional Kandyan law prevailed in the highlands. There was a lot of local variation in customary law in Sinhalese areas in the south and southwest, but as this region became the centre of Dutch power on the island, increasingly Roman-Dutch law came to prevail, especially among the Christian Sinhalese. The influence of Roman-Dutch law was greatly increased by its use for all criminal and property law. The Dutch had three types of courts: High Courts of Justice, District Courts, and Civil or Town Courts. The most important of the High Courts was in Colombo and it had criminal jurisdiction over all the inhabitants, both European and Ceylonese on the island. There were also High Courts in Galle and Jaffna.[26]

The Kandyan Kingdom, which had remained free of European rule until 1815 when the British defeated the Kandyans, kept no written laws or records of judicial proceedings; nevertheless, it had a longstanding judicial system with

24 Collins, C. 1951. *Public Administration in Ceylon.* London: Royal Institute of International Affairs, pp. 5–9.
25 Tambiah, H.W. 1954. *The Laws and Customs of the Tamils of Ceylon.* Colombo: Tamil Cultural Society of Ceylon.
26 On the Dutch legal system in Ceylon see, De Silva, M.U. 2006. "Litigiousness in Sri Lankans: an examination of judicial change and its consequence during the late Dutch and early British administration in the Maritime Provinces of Sri Lanka." *Journal of the Royal Asiatic Society of Sri Lanka*, New Series, Vol. 52, pp. 127–42.

well-accepted legal norms.[27] While ultimate authority rested with the King, most justice was administered at the village level. There existed a hierarchy of courts, ranging from the King down through various provincial officials to the village level. At each level trials could be held with the possibility of appeal to the succeeding level. The *gansabhavas*, or village councils, consisted of the most influential men in the village who attempted to resolve minor civil and criminal disputes. They operated less as a punitive institution than a place of conflict resolution. The law was considered to have a supernatural foundation and when there was conflicting evidence or no evidence whatsoever, judgement could be reached by oaths or ordeals.

In theory, if not always in practice, each party appeared before the presiding official to give their testimony.[28] Although officially there was no fee for the administration of justice, it was customary to bring the judge a small symbolic present. By the late eighteenth century such symbolic presents had often increased to the point that they functioned as bribes. Although the King forbad bribing, the practice was common, especially in the outlying districts far from Kandy. Officials received no salary for the judicial functions they performed, but they were able to keep any fines they levied. At times powerful or wealthy individuals took their cases directly to a high official who, upon being bribed, would judge in favour of the man without hearing the evidence of the other party.[29]

The sentences for serious crimes such as murder and treason were delivered either directly or indirectly by the King. The penalty for murder depended not only on the nature of the murder, but also on the status of the accused. Punishments ranged from imprisonment, to flogging, mutilation, degradation, banishment and death. For most other crimes a fine or light corporal punishment was the norm. On the whole, the punishments were relatively mild in comparison to European corporal punishments at the time.[30] The fact that most trials were held at the village level, where all parties and their activities were known to each other, had a dampening effect on the level of false accusations.

The organisation of the courts during the British period

The British captured the Dutch possessions in Ceylon in 1796, placing them under the control of the East India Company. Three years later the island was taken away from the East India Company and placed under the Crown. A Royal Charter of Justice was proclaimed, and the Dutch High Courts of

27 Dewaraja, L.S. 1972. *The Kandyan Kingdom of Ceylon, 1707–1780.* Colombo: Lake House, p. 203.
28 Pieris, R. 1956. *Sinhalese Social Organization.* Colombo: Ceylon University Press Board, pp. 149–60.
29 Dewaraja, *The Kandyan Kingdom*, pp. 201–04.
30 Pieris, *Sinhalese Social Organization*, pp. 145–55.

Justice were replaced by a Supreme Court of Judicature and High Court of Appeal with a Chief Justice and one Puisne Judge appointed by the Colonial Office rather than the Governor. In addition, five provincial courts were created at Colombo, Matara, Puttalam, Jaffna and Trincomalee. It was also decreed that all court documents be written in English. In 1802 a second layer of courts was created called the Courts of the Justices of the Peace. The justices of the peace were sometimes British civil servants, but more often were Burghers.[31] In 1811 the Blackstonian trial-by-jury system was introduced for criminal sessions in the Supreme Court. Juries could be composed of either Europeans or Ceylonese and this was the first time this privilege had been extended to a local population in Asia.[32] This seemingly progressive use of juries and assessors to aid judges was largely an attempt by the British to deal with their own lack of knowledge of local customs.[33] The system was also progressive in that the death sentence was imposed in Ceylon only for treason and murder, while in Britain it continued to be imposed for a number of lesser crimes until 1830.[34]

After the capture of the Kandyan Kingdom in 1815, its judicial system was initially left largely in place and was partially staffed by Kandyan officials. The only important modifications were the abolition of torture and the death penalty except by order of the Governor. However, after the uprising in 1818, the government rescinded the Kandyan privilege of trying criminals except for minor crimes. This demotion was compounded by the principle of equality and uniformity in law, which the Kandyan chiefs deemed an assault upon their prestige and discretionary power.[35] The court system in the Kandyan provinces remained separate from that of the Maritime Provinces until 1833.[36] The hybridity of the legal system in the early nineteenth century was thus most pronounced in the

31 Jayawardena, K. 2000. *Nobodies to Somebodies: The Rise of the Colonial Bourgeoisie in Sri Lanka*. Colombo: Social Scientist's Association and Sanjiva Books, p. 233; Roberts, M., I. Raheem and P. Colin-Thome. 1989. *People Inbetween: Burghers and the Middle Class in the Transformations within Sri Lanka, 1790s-1960s*, Volume 1. Ratmalana: Sarvodaya Books, p. 56.

32 Nanaraja, T. 1972. *The Legal System of Ceylon in Its Historical Setting*. Leiden: E.J. Brill, p. 16.

33 Mendis, G.C. 1948. *Ceylon under the British*. Colombo: The Colombo Apothecaries, p. 39; Rogers, J.D. 1987. *Crime, Justice and Society in Colonial Sri Lanka*. London: Curzon Press, p. 67. Assessors were important in the early years as they not only advised judges on cultural practices but on "native" psychology as well, such as on what constituted provocation. On the role of assessors in Africa see Ibhawoh, B. 2009. "Historical globalization and colonial legal culture: African assessors, customary law, and criminal justice in British Africa." *Journal of Global History*, Vol. 4, pp. 429–51.

34 Rogers, *Crime, Justice and Society*, p. 7.

35 Mendis, *Ceylon under the British*, p. 22.

36 Mills, L.A. 1964. *Ceylon under British Rule, 1795–1932*. London: Frank Cass and Co., pp. 35–37, 58; Nanaraja, *The Legal System*, pp. 57–66. On separate court systems in Egypt see Brown, N. 2018. "Politics over doctrine: the evolution of sharia-based state institutions in Egypt and Saudi Arabia," *James A. Baker III Institute for Public Policy of Rice University*, pp. 1–15.

central Kandyan highlands where a combination of English law and Kandyan law as interpreted and codified by British officials prevailed.[37] Sandra Joireman's argument regarding the amalgamation of customary and British law in colonial India and Kenya applies equally well to the Kandyan highlands.[38]

> The traditional school of thought, advanced by the imperial powers, suggested that customary and religious law continued in traditional form, alongside but subordinate to the law of the metropole. An alternative and more historically accurate view is that customary law was constructed in the context of colonisation by local elites empowered by the colonisers, developed in a fashion to both suit the needs of the colonising power and to promote the interests of the elite ... Thus it would be incorrect to view customary law as pre-existing colonisation in the form that it became institutionalised by the colonising powers. Instead, customary law can be viewed as both opportunistic and constructed by political circumstances.[39]

In 1832 the Colebrooke-Cameron Commission Report recommended that the administration of the island be reorganised along utilitarian lines.[40] The

---

37 As part of the British policy of relying upon the customary law of their possessions where possible, John D'Oyly, the British resident in Kandy after the conquest produced a summary of the Kandyan constitution and law. See, D'Oyly, J. 1929. (1832). *A Sketch of the Constitution of the Kandyan Kingdom.* Colombo: Government Printer. An excellent biography of D'Oyly can be found in Gooneratne, B. and Y. Gooneratne. 1999. *This Inscrutable Englishman: Sir John D'Oyly (1774–1824).* London: Cassell. A later compilation of Kandyan law entitled the Niti Nighanduwa was published in 1880 (*Niti Niganduva or the Vocabulary of the Law as it Existed in the Days of the Kandyan Kingdom.* 1880. Trans. T.B. Panabokke and C.J.R. Le Mesurier. Colombo: Ceylon Government Press); Marasinghe, M.L. 1979. "Kandyan law and British colonial law: a conflict of tradition and modernity—an early stage of colonial development in Sri Lanka." *Verfassung und Recht in Übersee/Law and Politics in Africa, Asia and Latin America,* Vol. 12, No. 2, pp. 115–27. Sinhalese law in the Maritime provinces had been greatly eroded during the Dutch period and the British found there was little in the way of customary law for them to implement. Tambiah, H.W. 1968. *Sinhala Laws and Customs.* Colombo: Lake House, p. 29.

38 In Egypt, Islamic law and other religious influences on the state judiciary were contained and limited to customary law having to do with marriage, the family and personal status among Muslims and Milli courts for other religions (Brown, N. 2007. *Rule of Law in the Arab World: Courts in Egypt and the Gulf.* Cambridge: Cambridge University Press). This separation of customary and European law was common across the British Empire. In her discussion of legal pluralism in the Ottoman empire, Barkey makes a case that the separation of customary from European law provided flexibility and the ability to manage diversity and allow for in-group policing. Barkey, K. 2013. "Aspects of legal pluralism in the Ottoman empire," in L. Benton and R.J. Ross (eds.), *Legal Pluralism and Empires, 1500–1850.* New York: New York University Press, pp. 83–108. Benton, "Colonial law and cultural difference," p. 563, reminds us, however, that this hybrid system was hierarchically ordered with "native courts" controlled by and subservient to state law.

39 Joireman, S. 2006. "The evolution of common law: legal development in Kenya and India." *Journal of Commonwealth and Comparative Politics,* Vol. 44, No. 2, p. 194.

40 Samaraweera, V. 1974. "The Ceylon Charter of Justice of 1833: A Benthamite blueprint for judicial reform." *The Journal of Imperial and Commonwealth History,* Vol. 2, No. 3, pp. 263–77.

resulting Royal Charter of Justice in 1833 put in place a uniform judicial system for the island and separated the revenue branch of government from the judicial, thereby denying the governor the right to establish courts. The report also created a uniform court system for both Europeans and Ceylonese. In practice, however, as we shall see, Europeans and Ceylonese rarely received equal justice.[41]

The Supreme Court was composed of a Chief Justice, two associate justices, and thirteen jurors in criminal cases.[42] The court was divided into the District of Colombo and three circuits, each divided into districts with a court in each. It also had an appellate jurisdiction.[43] The Supreme Court handled the gravest cases and was the only court to try cases where the punishment carried the death penalty. Advocates and proctors were allowed to argue cases before the district courts and decisions were handed down by judges without the benefit of juries.

As there was only one district court per district during the 1830s, these courts were soon unable to handle the large number of cases that flooded in. Governor Campbell wrote to Whitehall in 1842. "The condition of the system of justice is the subject of universal complaint and with the greatest reason. The delays and the practical denial of justice both in civil and criminal matters are unparalleled in any country."[44] The following year, in order to deal with the backlog, a lower level system was created to handle minor offences. These inferior courts had very limited powers of punishment, and representation was forbidden, as were appeals. Plaintiffs and defendants were left to argue their own cases before the judge, and as the official language of all courts was English, all proceedings operated through interpreters.[45] By 1845 the creation of these minor courts sufficiently relieved pressure on the district courts so that their numbers could be reduced. People now flooded these courts with minor cases. Subsequently, the police courts and district courts were granted greater power to punish, thereby relieving some of the pressure on the Supreme Court.[46]

---

41 Mendis, *Ceylon Under the British*, p. 39.
42 Puisne judges were associate judges.
43 De Silva, C.R. 1953. *Ceylon Under the British Occupation, 1795–1833*, Volume 1. Colombo: Colombo Apothecaries, p. 339.
44 CO 54/196. Campbell to Stanley, 56 of 18 April 1842.
45 The court system was attacked by the Queen's Advocate for banning representation in minor courts as this was seen to deny justice to litigants. Nanaraja, *The Legal System*, pp. 96–101. The name Police Court was subsequently changed to Magistrate's Court. De Silva, K.M. 1973. "The courts," in De Silva, K.M. (ed.), *University of Ceylon. History of Ceylon*, Volume 3. Colombo: Colombo Apothecaries, p. 326; Digby, W. 1879. *Forty Years of Official and Unofficial Life in an Oriental Crown Colony; Being the Life of Sir Richard Morgan, Kt., Queen's Advocate and Acting Chief Justice of Ceylon*, Volume 2. Madras: Higginbotham and Co, pp. 305–27.
46 O. Morgan, Crown Council for the Midland Circuit, *Administration Reports, Ceylon (ARC)*, 1886, C20; Mendis, *Ceylon Under the British*, p. 56.

Figure 5.1 Court House, Kandy, 1852.
(Source: The British Library Board, F. Fiebig Collection)

The Charter of Justice of 1833 preserved the right of people to use the traditional village courts (*gansabhavas*). But over the next two decades the government systematically undermined the power of village headmen, thereby greatly reducing the enforcement power of these courts. The desire to undermine rival, traditional forms of authority was based on the growing belief that civilisation could only be advanced by the replacement of traditional, collectivistic judicial practice with utilitarian, individualistic, modern practice.

However, by the mid-nineteenth century, in a softening of their views on local power, the British began to be more concerned about the social impact of the breakdown of traditional values in the face of economic and cultural change. Whereas, in general, the British welcomed the modernisation of institutions as inevitable and progressive, they nevertheless began to recognise that some changes had negative consequences. Of particular concern was the continuing collapse of irrigation systems throughout the island due to a breakdown in the ability of village headmen to enforce customary rules regarding their maintenance. Governor Sir Henry Ward in 1856 then sought to carve out a niche for tradition and collectivism within villages by reviving the *gansabha-vas* and tasking them with organising cooperative irrigation and settling disputes over land and water at the village level.[47] The subsequent success of the *gansabhavas* in encouraging village cooperation in agriculture led to calls to further expand their jurisdiction to enforce village rules and resolve minor village disputes.[48]

It was within the context of several decades of heavy use of the British court system by peasants and the recent revival of the *gansahavas* that an illegal trial took place in a village in Negombo district in January 1864. Two boys quarrelled with a villager on whose land they were cutting fodder without permission. After angry words were exchanged, the landowner cut his hand attempting to grab one of the boy's sickles. Instead of turning the boys over to a magistrate, the *vidane*, who was a friend of the landowner, brought them to the landlord's house to be tried for assault. In imitation of a Supreme Court trial, the *vidane* appointed himself judge and assembled a jury of thirteen villagers. After hearing the boys' account of the incident, he immediately pronounced them guilty and asked his jury to suggest a punishment. When the jury couldn't decide on a punishment, the landowner's wife stepped forward and said they should be slapped in the face by a low caste person, a traditional Sinhalese form of degrading punishment. This was done in front of the assembled village and the boys released. The case is a striking example of legal pluralism, a hybrid mixture of the English and Sinhalese court systems. The *vidane*, with his imperfect knowledge of English court system, was attempting to legitimate his impromptu trial while circumventing the court in Negombo. The unofficial trial can clearly be seen as a product of the cultural contact zone of the maritime region.[49] The British reaction to what they termed a "mock trial" was condescending amusement at the *vidane's* appropriation of court procedure. The father of the two boys was less amused and instituted an action in the district court in Negombo against the headmen and fourteen villagers for damages.[50]

Seven years after this unofficial village trial, a bill was passed allowing trials to take place in villages. In introducing the bill in the Legislative Council in 1871 Governor Robinson said,

> Our rule has destroyed every vestige of the system of village government, and has given the people in its place about forty minor courts … presided over by European magistrates and conducted according to European forms of civil and criminal procedure … What is wanted is some inexpensive,

47 Samaraweera, V. 1978. "The 'village community' and reform in colonial Sri Lanka." *The Ceylon Journal of Historical and Social Studies, New Series*, Vol. VIII, No. 1, January–June, p. 71.

48 Digby, *Forty Years*, Vol. 2, pp. 101–03. On separate "native" courts in Africa see Saho, B. 2018. *Contours of Change: Muslim Courts, Women, and Islamic Society in Colonial Bathurst, The Gambia, 1905–1965*. East Lansing: Michigan State University Press.

49 Pratt, M.L. 1992. *Imperial Eyes: Travel Writing and Transculturation*. New York: Routledge.

50 "A novel case," *The Ceylon Examiner*, January 20, 1864.

prompt and popular means of settling village disputes on the spot. This would tend to arrest in the very germ the growth of those contentions which at present develop into such a prolific crop of both real and false petty charges.[51]

Robinson said that he hoped that the *gansabhavas* would take pressure off the police courts while rendering "detection and punishment of every false charge a matter of certainty."[52]

The ensuing Ordinance of 1871 established village committees to frame rules for village affairs and the establishment of village tribunals to try petty offences. Each tribunal represented a number of villages, and consisted of a president appointed by the governor and five locally elected councillors. All proceedings were to be in the local languages and no representation was permitted. Henceforth, the tribunals had criminal powers, but these were limited to minor assaults, petty thefts, and malicious injury, and they had the authority to sentence people to fines of up to twenty rupees and two weeks' rigorous imprisonment. These sentences were enforced by police courts. More serious cases were to be directly referred to police court. The tribunal's jurisdiction was purely local in that no outsiders, Burghers or Europeans had standing in them.[53] The tribunal's decision could be appealed to the GA.[54] This piece of legislation was described by J. Parsons, the GA of the Central Province, as an "experiment of conservative reform," in the sense that the British retained control, as the president of each tribunal was appointed by the governor and appeals were to the GA.[55]

The revival of the *gansabhavas* can be seen as a classic case of governmentality as the "conduct of conduct," for they were intended to instil in villagers a sense of individual property rights, obedience to authority, and self-control. For example, the regulations the *gansabavas* were to enforce specified that property boundaries be clearly marked and that villagers should not leave rubbish next to or allow their children to play on a public road, villagers should not slaughter their animals without their headman's permission, carts should not be raced or driven without a light at night, and no gambling, cock fighting or abusive language was allowed. All violations were subject to a fine.[56] In this sense, although the *gansabhavas* were an attempt to recreate traditional village government, they

---

51 Governor Robinson, Address to the Legislative Council, 3 October 1871, *Ceylon Hansard (CH)*, pp. 218–19. Furthermore, in the interests of financial expediency, the British realised that indirect rule could be achieved by reviving, reshaping and repurposing local institutions.

52 *Addresses Delivered in the Legislative Council of Ceylon by Governors of the Colony together with Replies of the Council.* 1880. Colombo: Government Printer, p. 219.

53 *Ceylon Sessional Papers (SP)*, 1889, p. 398.

54 Nanaraja, T. *The Legal System*, pp. 101, 116.

55 *ARC*, 1872, p. 47. By the end of the century, every head of a village tribunal was drawn from the aristocracy as were all sixteen Sinhalese police magistrates. This was a conscious policy by the British to divert power away from the new educated elite. De Silva, "The development," p. 221.

56 *ARC*, 1871, p. 24.

were in fact a comprehensive intrusion of rationalised governmentality and contemporary British town planning notions into Ceylon's villages.

H. Neville, the AGA at Trincomalee, wrote of the tribunals,

> the villagers still require some years' careful education in self-government before they will spontaneously exert themselves for their own benefit. It appears to me desirable to foster such spontaneous action very vigilantly and equally to hold back or conceal guidance from the *kachcheri*, if the full benefit hoped for is to be obtained.[57]

It was symptomatic of the officials' ambivalent attitudes that on both pragmatic and paternalistic grounds they sought to devolve power to local officials, while retaining a good deal of control for higher level administrators.

A mere decade after the *gansabhavas* were reintroduced, the government agents were surveyed about whether they considered them worth retaining. Their conclusion was that they were most successful in the Central Province and in more remote areas where traditional patterns of authority still held sway, whereas they were "eminently unsuitable to the neighbourhood of large towns and to some parts of the maritime districts, where the communal feeling has been lost and been replaced by a counterfeit semblance of municipal organization."[58] In spite of the generally favourable view of the tribunals, it was felt that the presidents must be supervised by Europeans to keep them honest. The consensus was that the tribunals did in fact help to suppress false evidence as it was difficult to put forward false cases at the village level where the facts of the matter could be more easily discovered. They also removed some minor cases that otherwise should have come to the police courts. John Dickman, the Acting GA of the Western Province, wrote,

> the Ordinance has substituted law for custom; it has given the villagers a means of redressing their small wrongs and grievances by properly constituted village courts, formed on the ancient native model, in place of the irresponsible and uncontrolled authority of the village headman.[59]

However, the fact that the government continued to appoint the presidents of the *gansabhavas* rather than allowing villagers to elect them, resulted in much resistance and abuse of the system which lasted throughout the remainder of the century. It also transpired that there was widespread bribery of *gansabhava* officials and regular complaints from disgruntled litigants.[60] In spite of this, the

---

57 *ARC*, 1883, p. A104.
58 Village Tribunals. Papers Relating to the Working of the Tribunals, *SP*, 1880, p. 179.
59 *Ibid.*, p. 184.
60 Goonesekere, R.K.W. 1958. "Eclipse of the village court." *The Ceylon Journal of Historical and Social Studies*, Vol. 1, pp. 145, 153.

tribunals were considered by the government to be worthwhile in terms of efficiency because they reduced the number of cases in the police courts.

Throughout the first half of the nineteenth century, there existed what Benton has termed "strong" legal pluralism.[61] Officially English law was dominant, with the exception of various customary religious laws concerning marriage and family. But in practice English law was cross-cut by Ceylonese understandings of justice and attitudes about the production of truth, which often stood in opposition to the liberal rule of law. The actual functioning of the courts therefore was the result of an ongoing struggle between rule of law and the subversive manipulation of the courts by both the public and government officials. The Ceylonese neither accepted the courts, in any strong hegemonic sense, nor rejected them. Rather, as we shall see, they used them extensively, but strategically. Cumulatively, their individual acts of appropriation both reaffirmed the courts' centrality to the colonial order, while undermining their smooth functioning.[62]

The judges

While the judges of the Supreme Court tended to be quite strict in following the letter of the law, untrained lower court judges were not; sometimes this was out of ignorance of the law itself and sometimes out of exasperation with legal procedures which they viewed as shielding defendants whom they considered guilty despite lacking proof. While the government officially condemned these violations of rule of law, in practice it was complicit due to its policy of appointing untrained civil servants rather than lawyers to the bench. This was done on financial grounds and also in an attempt to ensure that judicial decisions aligned with government policy.[63]

At times there was a yawning gulf between the codification and the practice of the law. Cameron, in the Colebrooke-Cameron Commission Report of 1832, condemned Ceylon's judges as "gentlemen not only unconnected with the profession of the law, but whose education has been in no degree adapted to the special purpose of qualifying them for the administration of justice."[64] The resultant 1833 Charter ordered that barristers rather than civil servants fill all district judgeships, not only in order that judges be familiar with the law, but to

61 Benton, *Law and Colonial Cultures*, p. 256.

62 For India, see Den Otter, "Law," p. 170. Merry, *Getting Justice*, p. 180 found that in late twentieth-century America, litigants similarly fought to shape legal systems to suit their needs.

63 Of course, lack of formal independence did not always mean that judges ruled as the government wished. Chandrachud, A. 2015. *An Independent, Colonial Judiciary: A History of the Bombay High Court during the British Raj, 1862–1947.* Oxford: Oxford University Press, found a similar pattern of judicial behavior in Bombay.

64 Report of Charles H. Cameron Esq. upon the Judicial Establishments and Procedure in Ceylon. 31 January 1832 in Mendis, G.C. ed. 1956. *The Colebrook-Cameron Papers: Documents in British Colonial Policy in Ceylon, 1796–1833*, Volume 1. Oxford: Oxford University Press, p. 123.

create a separation between the judicial and the administrative branches of government. In addition, it specified that each district court have three local assessors to provide judges with knowledge of the cultural traditions of the people.[65] The Ceylon Government, however, resisted pressure from Whitehall to appoint judges with legal training, preferring to save money and retain control over the courts by appointing their own civil servants. Given the professed ideological importance to the British of the separation of the executive and the judicial powers and more generally of bringing rule of law to their occupied territories, it struck officials in the Colonial Office as unfortunate that so many judges had no legal training. As Sirr pointed out in 1850, the vast majority of police magistrates and most district court judges had "never opened a law book until they received their appointments."[66] This woeful situation was reluctantly accepted by Whitehall on the grounds that there existed a paucity of British-trained lawyers in Ceylon, and that the only locally trained lawyers were Ceylonese. It was not until 1856 that the Colonial Office ordered that at least the district judgeship in Colombo be filled by a member of the Colombo Bar. And it was not until 1872 that the same was ordered for the district court at Kandy. While paying lip service to the separation of the administrative and judicial branches, the government continued to fill all of the other district courts with revenue officers, most with no legal training.[67]

Judges of the lower courts were often woefully ignorant of the laws they were meant to enforce. In theory, a new recruit from Britain to the Ceylon Civil Service was upon arrival to pass a local language test and an examination in law. He was then to be supervised by a police magistrate for six months before presiding in a police court. But due to a shortage of funds, the civil servants were too thin on the ground for these rules to be followed with regularity. More commonly, a young civil servant was immediately assigned a temporary police magistracy as part of his training. In 1869 the newly arrived Inspector General of Police, G.W.R. Campbell, launched a swingeing attack on this policy:

> The ordinary practice at present is, that the young members of the civil service, as soon as they have acquired the imperfect knowledge of their first native language, which suffices them to pass the prescribed examination, are sent out as Magistrates and Justices of the Peace, often to perfectly isolated districts. From the very first day of the young magistrate's arrival, the court is full of cases which he has to summarily dispose of with the help of his clerk and such native lawyers as his station may possess. He has nobody over him—none to teach him ... At this time he is hard at work

---

65 De Silva, *Ceylon under the British*, p. 339.
66 Sirr, H.C. 1850. *Ceylon and the Cingalese*, Volume 2. London: William Shoberl, p. 240.
67 Mills, *Ceylon Under British Rule*, p. 97. There were heated debates into the early twentieth century about the wisdom of having members of the revenue branch serve as judges. *Ceylon Observer*, July 14, July 17, 1905.

learning another language and studying the law and revenue books required for his second examination. A large district may have, and often does have a succession of such untaught young magistrates.[68]

However, Campbell's criticism is mild in comparison to that of an editorial in the *Ceylon Examiner*:

A young man, or rather a boy, just emancipated from the control of perhaps a severe schoolmaster, makes his appearance in the country with a writer's commission in his pocket ... He finds himself the presiding judge of an outstation court, without any experience in judicial matters, and without any knowledge of the customs, habits and prejudices of a large Asiatic population, over whose destinies he finds himself called upon to preside. He is required to perform the duties of a police magistrate, commissioner of requests and justice of the peace; this is not such a power as should be placed in the hands of inexperienced and often very self-sufficient boys, as in such hands it is liable to be exercised in a manner hurtful to the feelings and prejudices of the people ... I think it is the paramount duty of government to see that these places are filled by experienced officers.[69]

In 1871 the governor promised to place older, more experienced men in charge of minor courts, but for budgetary reasons this policy was not followed through.[70] In 1883 a new penal code spelled out the law in greater detail in order to reduce the tendency of judges with little legal training to misinterpret it.[71] But in 1889, twenty years after the newspaper editorial, Giles wrote an official report on the police and the courts in which he made a nearly identical complaint,

young men with little training or experience, and a very imperfect knowledge of the natives, their language and customs, are placed early in positions of much importance with little to guide them but strange Ordinances, rebukes for blunders made in cases which happen to go before Superior Courts, and their own crude ideas of criminal law, and how it should be administered.[72]

---

68 G.W.R. Campbell, Inspector General of Police, *ARC*, 1869, p. 227.
69 *The Ceylon Examiner*. Quoted in *ARC*, 1869, p. 228.
70 The Queen's Advocate, *CH*, 1871, p. 25. Editorial. "Covenanted writers versus uncovenanted experienced colonists as police magistrates," *Ceylon Observer*, September 25, 1880; "Our Junior Magistrates." *Weekly Ceylon Observer*, October 8, 1886.
71 Basnayake, S. 1973. "The Anglo-Indian codes in Ceylon." *The International and Comparative Law Quarterly*, Vol. 22, No. 2, p. 288.
72 Giles, A.H. 1889. *Report on the Administration of Police, Including the Actions of the Courts and the Punishment of Criminals in Ceylon*. Colombo: J.A. Skeen, Government Printer, p. 370.

But it was not simply young, newly arrived civil servants who violated the law. In 1893 the Chief Justice of the Supreme Court criticised the "illegality" of decisions by experienced district judges who had no formal legal training.[73] As an editorial in the *Ceylon Examiner* put it, "It is monstrous that we have hitherto patently submitted to life and liberty and the rights of property being placed in amateur hands."[74] Still, as late as 1903 newly appointed magistrates could preside for up to nine months before having to pass an examination demonstrating that they were familiar with Ceylon law.[75]

The experience in 1905 of the newly arrived Leonard Woolf, a civil service cadet in Jaffna, was typical:

> They have made me Additional Police Magistrate now: I spend the evenings in trying to learn something about the law; in the day the work is something of a horror and a relief. At first it is a mere whirl: sitting in sheer ignorance up there in front of the court, writing down the evidence, listening to proctors and witnesses, thinking of questions to ask, trying to make up your mind—all at the same time. I felt that at any moment I might raise your old cry: "I resign."[76]

In the scene in Figure 5.2 a woman gives evidence to a judge about mistreatment by her husband who stands near the door, while a clerk takes notes. A policeman stands in the doorways and headmen are seated in the courthouse. Villagers lounge around outside witnessing the spectacle.

The problem of ignorant judges was compounded by the civil service attitude that the judiciary was of less importance than the revenue branch and so could be used as a dumping ground for British recruits who were not talented enough for the revenue and executive branches.[77] In 1894 the *Ceylon Examiner* put it in the strongest terms, "We have to complain of the total want of appreciation on the part of the government of the importance of the administration of justice in the Island."[78] The judiciary was also seen as a suitable place for those, such as Burghers and British-born in Ceylon, who were not trusted with the more important administrative posts.[79] As the century progressed, increasing numbers of educated Sinhalese and Tamils were allowed to join the judiciary as this was seen as a safe way of generating support for the legal system among the elites.[80] As Governor Longden wrote to the Colonial Office,

73 "Not fit to be a magistrate," *Ceylon Times*, March 1, 1893.

74 "Judgeships and lawyers," *Ceylon Examiner*, February 13, 1893.

75 This period could be extended for a further six months by the governor. White, H. 1903. *The Ceylon Manual for the Use of Officials*. Colombo: GJA Skeen, Government Printer, Ceylon, p. 68.

76 Woolf, *Growing*, p. 61.

77 In India the same hierarchy prevailed. Wiener, *An Empire*, p. 151; Dhillon, *Defenders*, p. 130.

78 *The Ceylon Examiner*, "The district court of Colombo." May 8, 1894.

79 Mills, *Ceylon under British Rule*, p. 80; De Silva, "The Development," p. 214.

80 For a similar situation in India see, Singha, R. 1998. *A Despotism of Law: Crime and Justice in Early Colonial India*. New Delhi: Oxford University Press, p. 303.

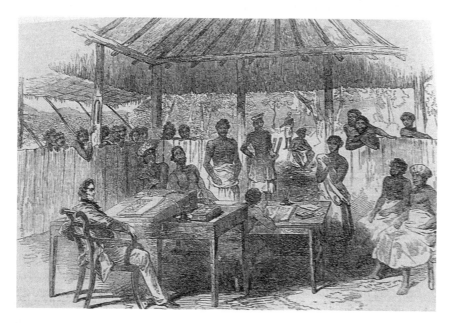

Figure 5.2 Itinerating Police Court, Central Province, 1850.
(Source: *London Illustrated News*, August 17, 1850).

Natives make tolerable magistrates and rise sometimes to District Judge-ships, but the jealousies of race are so strong, natives view each other with such distrust, and are so incapable of governing Europeans that a great change must take place in the social condition of the country before natives can be entrusted with the administration of political and revenue appoint-ments such as different agencies.[81]

Increasingly over the years, Ceylonese appointees were trained lawyers, but there were also local candidates appointed through patronage, who had been employed in other occupations before being appointed magistrates and district judges. Those who proved to be unusually incompetent were often simply moved to another station rather than discharged.[82] Ceylonese judges, whether they were Burgher, Sinhalese or Tamil tended to be middle class, anglicised,

81 Longden to Kimberly, 8 February 1881 in CO54/331. Local judges were rarer in Ceylon courts than in India in the late nineteenth century Clarence, L.B. 1896–1897. "Application of European law to natives of Ceylon." *Journal of the Society of Comparative Legislation*, Vol. 1, p. 229; Fein, H. 1977. *Imperial Crime and Punishment: The Massacre at Jallianwala Bagh and British Judgement, 1919–1920*. Honolulu: University Press of Hawaii, p. 55.

82 Giles, *Report on the Administration*, pp. 397–98.

and ran their courts in a similar fashion to those run by British civil servants.[83] Their views of villagers were often not significantly different from those of European civil servants. Crucially, however, they differed by casting their views of the peasantry in class rather than racial terms. Their prejudices and explanations were therefore somewhat more nuanced, and they lacked the sense of unfamiliarity which so strongly characterised British emotional reactions to Ceylonese culture.

The Chief Justice, himself a trained lawyer, in 1878 characterised many magistrates as ignorant not only of the law, but even of the actual workings of the courts.[84] Ironically, the new Code of Criminal Procedure of 1883, which sought to reduce ambiguity and help untrained magistrates follow the law, was found by many justices of the peace and magistrates in the police courts to be too complex. Many therefore disregarded those parts of the law that they struggled to understand.[85] The government's response was to revise the ordinance and simplify it so that the untrained judges could understand it. In the mid-1880s Governor Arthur Gordon, never a man to mince words, wrote, "the District Judges are those who are too stupid or incapable to hold Revenue offices."[86] There were so many of the latter sort, he claimed, that the trained judges of the Supreme Court showed their contempt for the district judges by "reversing their decisions by the hundreds as not observing the letter of the law."[87]

Judges and their staff sometimes appeared incompetent through no fault of their own, because the government neglected to inform them of changes in the law. In the early 1890s the Acting Crown Council for the Midland Circuit wrote in his annual report that

> a new volume of our existing ordinances is a great desideratum. Considering the number of ordinances amended, repealed and re-amended, it is now a matter of the greatest difficulty to discover what is or is not in force, and this has sometimes led to the perpetration of the most ludicrous mistakes by both judges and practitioners.[88]

83 Rogers, *Crime, Justice and Society*, p. 60.
84 Quoted in Longden to Hicks-Beach, 24 July 1878, CO54/514 (243), cited in Rogers, *Crime, Justice and Society*, p. 65.
85 *The Weekly Ceylon Observer and Summary of Intelligence*, January 9, 1886. The Penal Code of India Act XLV of 1860 is based on the Code of 1837 drafted by a commission of four and chaired by Lord Macaulay. Jennings, I. and H.W. Tambiah. 1952. *The Dominion of Ceylon: The Development of its laws and Constitution*. London: Stevens and Sons, p. 282. Nadaraja, *The Legal System*, p. 232; Basnayake, The Anglo-Indian Codes.

    For a discussion of unofficial European opposition to the Code of Criminal Procedure in India, see Kolsky, "Codification and the Rule of Colonial Difference," pp. 631–83.
86 Stanmore Papers, Add. Mss. 49218, Gordon to Selbourne, 24 April 1884, fol. 218. Cited in Peebles,"Governor Arthur Gordon," p. 96.
87 Peebles, "Governor Arthur Gordon," p. 96.
88 F.R. Dias, Acting Crown Council, Midland Circuit, *ARC*, 1892, p. A18.

Secondly, the government further undermined the rule of law by encouraging pragmatic decision-making on the part of judges over strict adherence to the law.[89] As
Samaraweera put it, "far from being eschewed, the absence of professionalism in
the judges was looked upon as a great virtue by the colonial government."[90] What
were valued were judges who handed out pragmatic sentences, irrespective of legal
"technicalities," for these were seen to increase the efficient running of the
colony.[91] As many judges wished to be promoted to the more prestigious revenue
branch, they strove to be recognised as efficient rather than as adhering strictly to
the letter of the law. An editorial in the *Ceylon Examiner* was highly critical of
civil servants serving as judges, arguing that this practice violated the separation of
the judicial from the administrative branches of government. It cited as evidence
a circular sent by the government to judges "requiring them not to reprimand the
police or headmen publicly or reflect on their evidence." The editorial went on to
observe "how seldom it is that a civilian judge decides against the government in
a civil case."[92] Although occasionally judges were criticised by administrators for
being too lenient, more often they were criticised for being "technical," the term
used to describe judges who allowed themselves to be constrained by the law. In
other cases the government and the press were critical of judges who evaded the
law by throwing out complicated cases in order to avoid having a flawed decision
overturned by the Supreme Court.[93] In 1869 Inspector General of Police Campbell
went so far as to claim in his annual report that "the faulty working of the courts
not only fails adequately to put down crime, but actually fosters many phases of
it."[94] While this claim was hotly denied by the judiciary, and mocked in the press,
the charge resonated widely.[95] Despite his hyperbole, Campbell was expressing
what many administrators and non-official European residents believed, namely
that what Ceylon needed was a disciplinarian, but paternalistic, system of justice
that wasn't excessively tied to the letter of the law.[96] In his government report,
Giles also argued that, "the cumbrous methods of English criminal administration,

89 The same attitude was prevalent in India. Wiener, *An Empire*, pp. 150–52. The view that it was pref-
   erable for judges to be "practical men" rather than students of the law was hardly restricted to
   Ceylon. For, as Nijhar points out, "on the ground, despite the form of law, in justices' courts of
   Victorian England, the vast majority of the population was relegated to a legal territory in which
   untrained magistrates often dispensed justice largely as they felt appropriate, given local exigencies
   of resources such as incarceration and policing manpower." Nijhar, *Law and Imperialism*, p. 30.
90 Samaraweera, "British justice," p. 112.
91 McLaren, "The uses of the rule of law," p. 82, points out that in the Caribbean at this time, although
   judges were formally independent they were expected to be compliant with government goals.
92 "Professional v. unprofessional," *Ceylon Examiner*, January 20, 1893.
93 *Colombo Overland Observer*, April 30, 1863.
94 G.W.R. Campbell, Inspector General of Police, *ARC*, 1869, p. 230.
95 L. Liesching, District Judge, Nuwarakalawiya, *ARC*, 1870, p. 245; *Ceylon Times* October 26, 1870.
96 Nearly four decades earlier, Cameron in his Commission report on the judiciary suggested as
   much when he wrote, "The moral and intellectual condition of the natives is such, that the European magistrate who is to distribute justice among them, can only do so effectively by the exercise
   of something like paternal authority." Mendis, *The Colebrook-Cameron Papers*, p. 172.

have been too much imitated in Ceylon ... When crime has made undue headway, many legal technicalities devised for the protection of the innocent, have to give way to a simplification of procedure."[97]

There was in fact no consensus about the extent to which rule of law should pertain in the courts. For example, in an editorial in the *Ceylon Times* in 1893 the author argued that when "legal quibbles" are "utilised to allow hardened criminals to escape sentences they have justly earned, we think justice is sacrificed to technicality."[98] It is important to note that an important implication in the debate over "technical judgements" is that the constraints dictated by the law do not necessarily result in justice. In this respect the editor of the conservative *Ceylon Times* appears to share the widespread Ceylonese belief that English law often did not produce justice. By contrast, an editorial in the *Ceylon Examiner* in 1900 attacked judges who deviated from rule of law.

> No improvement is possible in the proceedings of our lower courts till District Judges and Police Magistrates have realised that their duty is to administer the law as it has been laid down and desist from absurd endeavours to give effect through means and methods of their own invention to what they conceive to be the aims and object of the law.[99]

Contrary to Guha's blanket assertion that rule of law was a vehicle of oppression, it is clear that rule of law, by constraining judges, in fact provided the Ceylonese with the space to turn judicial proceedings to their own advantage. Of course, this was most likely to happen when people were aware of their legal rights. For example, Roberts points out that although Ordinance 5 of 1841 gave plantation labourers numerous rights, they were largely unaware of the existence of this ordinance.[100] And so the planters had a real interest in making sure their labourers remained unaware of their rights. Although lower court judges sometimes violated legal constraints, the police were the primary resistors. An editorial in the *Overland Ceylon Observer* in 1894 mocked the Inspector General of Police for referring to persons charged with a crime as

> the criminal [who] by anticipation, ought to be convicted, and that the Police Court, the Attorney-General's Department, appeal to the Supreme Court, or trial by jury, are simply so many agencies by which the attainment of that desirable and due end is obstructed.[101]

---

97 Giles, *Report on the Administration*, p. 372.
98 "Technicality vs. justice," *Ceylon Times* June 26, 1893.
99 "Magistrates censured," *Ceylon Examiner*, March 5, 1900.
100 Roberts, M. 1965. "The master servant laws of 1841 and the 1860s and immigrant labour in Ceylon." *Ceylon Journal of Historical and Social Studies*, Vol. 8, p. 25.
101 *Overland Ceylon Observer*, August 15, 1894.

The plantation districts were notorious for summary justice, as it was not uncommon for planters with no legal training to be unofficial justices of the peace, creating an unjust situation by which plantation owners and managers sat in judgement of their own Tamil labourers accused of breaking their contracts. These justices of the peace could not try cases, but could issue warrants and turn suspects over to police magistrates. Thus the practices of officials in the highlands were an example of corruption under cover of the law itself.[102]

So transparent was the departure from the rule of law in the highlands that it eventually became an embarrassment to the government. As early as the mid-1840s, J.S. Colepepper, Superintendent of Police for Kandy, wrote in his report to a Commission of Inquiry that plantation workers who threatened legal action over unpaid wages were often jailed by planter justices of the peace. He cited the case of a labourer who was arrested on trumped up charges and jailed for eight days for demanding back wages. When he returned to the plantation he was flogged for good measure. Although he arrived back at the police station bleeding from the nose and mouth, no charges were filed against the planter by the justice of the peace.[103] Relatively little, however, was done about such abuses until pressure was brought to bear on the government by an editorial campaign in the press against abuses of the law by planters acting as justices of the peace. Some, it was claimed, issued blank arrest warrants to other planters to be filled in as they saw fit, and it was common for apprehended labourers never to be brought to trial at all, but simply handed over to any planter who needed labourers. An inquiry revealed that planter justices of the peace also were not keeping records as required, in order to hide their violations of law. One editorial stated that "the sooner they resign an office which many of them appear to have abused, the better for the Colony."[104] In large part, as a result of this campaign, Ordinance 1 of 1864 was passed requiring that JP warrants be signed by the superintendent of police before going into effect. But this ordinance was widely ignored.[105] So little control was exerted over these violations of the law that in 1884 a Committee of the Legislative Council reported systematic abuses in the use of warrants to recapture deserters from the plantations.[106] The upshot was that justices of the peace were no longer allowed to issue warrants.

102 Duncan, *In the Shadows*, p. 94; Ludowyk, E.F.C. 1966. *The Modern History of Ceylon*. London: Weidenfeld and Nicolson, p. 70; Bandarage, A. 1983. *Colonialism in Sri Lanka: The Political Economy of the Kandyan Highlands, 1833–1886*. Berlin: Mouton Publishers, p. 233.

103 CO 54.235. April 21 1847. Tennent to Grey Enclosure.

104 *Colombo Observer*, July 10, 1863; "Unofficial justices," *The Ceylon Examiner*, May 6, 1863; *The Ceylon Examiner*, June 18, 1863; "Justices of the peace," *The Ceylon Examiner*, February 10, 1864.

105 *Ceylon Times*, October 18, 1870. A.A. King, District Judge, Badulla, *ARC*, 1869, p. 195. See Roberts, "The master servant laws," pp. 28–29.

106 *SP*, 1884, p. 33.

The context for the ongoing struggles between the planters, the Tamil labourers and the Sinhalese peasants was the legal expropriation of common *chena* lands by the government in 1840. In the beginning there was some sporadic open resistance to this expropriation. The planter George Ackland testified before a British Parliamentary Commission looking into the 1848 rebellion that the Kandyans burned down a house on his new plantation and threatened to murder his workers. He claimed that a headman told him that the planters would never be allowed to settle on these lands.[107] Although the peasants were in fact unable to halt the British occupation of the highlands, it was common in the early years for villagers to tear down fences and destroy coffee trees under cover of darkness.[108] While this type of vandalism continued over the decades, more commonly villagers simply stole coffee from the plantations. According to a police officer,

> in some villages the young men are brought up from their childhood with the idea that according to the ancient Kandyan custom the highland on which the [coffee] estate has been opened belongs by right of inheritance to the village below and in helping themselves they are only collecting rent on the ground.[109]

Throughout the nineteenth century, planters believed that the theft of their crops by villagers was a collective undertaking, protected by a conspiracy of silence. The Chairman of the Planter's Association wrote to his fellow planters that "the native thief enjoys a passive if not active encouragement from his own people, more especially where Europeans are the victims!"[110] The government response was to punish theft with jail sentences and flogging. But as peasants grew coffee next to the plantations and all raw coffee beans look alike, it was nearly impossible to establish that coffee seized from a villager was stolen from a plantation. While the *vidanes* tended to turn a blind eye to their fellow villagers' thefts, the regular police in posts strategically placed near the plantations were able to check the practice somewhat.

In their quest to stop the theft of raw coffee beans, planters customarily searched villagers' houses without warrants. Although they could get justices of the peace to charge labourers and villagers on suspicion of theft, these charges were frequently thrown out by magistrates for lack of evidence. Planters were outraged that the mere suspicions of a European were insufficient to convict a "native." As one planter put it,

107 *British Parliamentary Papers* 1850, Vol. 12, p. 19.
108 Vanden Driesen, L.H. 1957. "Land sales policy and some aspects of the problem of tenure, 1836–86 Part 2." *University of Ceylon Review*, Vol. 15, p. 42.
109 Dep, *A History*, p. 101.
110 *Proceedings of the Planters Association*, 1883–84, p. 153.

the cause of failure [of most cases] was due to lack of evidence, as in these cases so much evidence is required to convict the thieves, unless they are caught in the act; estate parchment [raw coffee beans] being all so much of a sameness, that you cannot swear to your own property.[111]

Another planter wrote,

there are few planters who think it of any use taking cases to court, for it is next to impossible to get a conviction. Most of the magistrates are mere boys, without legal knowledge or experience of any kind, and if they do convict, the punishment is so inadequate that it encourages rather than deters coffee stealers.[112]

We can see from such statements that the rule of law did in fact sometimes provide labourers and villagers with some protection against planters. For example, Sivasundaram cites a case from 1841 where a Sinhalese landowner took a planter who wanted to build a road across his land to court and won.[113]

The planters became so frustrated with the British norm of presumed innocence until proven guilty that they were able to convince the Ceylon government to overlook this commonly presumed right in the case of coffee stealing.[114] Thus the government passed the Coffee Stealing Ordinance of 1874, which included a controversial clause making it a crime for anyone to possess coffee that he could not prove was his. Punishment upon conviction was three months prison and up to twenty lashes. The legislation was unsuccessfully challenged in the Legislative Council by the Sinhalese member M. Alwis, who claimed that the clause "subverted the fundamental principles of the common law and justice" and that "assuming a man to be guilty until he could prove his innocence, threw the onus of proof upon the individual charged."[115] This law placed a tremendous burden especially on illiterate peasant cultivators. It was a clear case of the rule of racial difference trumping liberal principles, as only Ceylonese were considered guilty until proven innocent. However, as this ordinance proved ineffective, many planters wanted to take the law into their own hands. The chairman of the Planters' Association argued that while the government might frame laws based on a set of British ideals, the planters (as one official claimed) had to

---

111 W. Bisset, Correspondance on the subject of coffee stealing. *SP*, 1873, p. 172.

112 A.F. Harper, *Ibid.*, p. 174.

113 Sivasundaram, "Tales of the land," p. 954.

114 The maxim "innocent until proven guilty" was a traditional norm in many legal systems such as English common law rather than being a constitutional right. It was in the 20th century that it was proclaimed by the UN to be a human right. Pennington, K. 2013. "Innocent until proven guilty: the origins of a legal maxim." *The Jurist*, Vol. 63, pp. 106–24.

115 *CH*, 1876, p. 186.

endure the consequences of laws, conceived in the spirit of the times, which being interpreted, means the spirit of an advanced and highly cultured civilisation, applied to a heterogeneous people in a far lower social grade, whose habits and ideas are wholly diverse from those of the framers of that Code.[116]

And so the government continued to turn a blind eye to planters administering rough justice to their workers and villagers.[117] In 1885, with the collapse of the coffee plantation industry due to disease, the law was extended to all plantation crops as the Praedial Products Ordinance Number 9. This broadening of the law created such an uproar among the peasantry across the island that the Supreme Court overturned the clause as a violation of rule of law.

Underlying the violations of law by planters and justices of the peace was a distain for the Ceylonese, such contempt being the dark side of paternalism. Leonard Woolf argued that this contempt was normalised in Ceylon.

There are many things in the manners and methods of a Sinhalese or Tamil who comes to the *kachcheri* to get a cart licence or to buy a piece of crown land or to protect himself against a dishonest and malignant headman or to ruin a hated neighbour, which are exasperating and distasteful to a European, and many civil servants never really got over this initial annoyance and distrust. However much they liked their work and, up to a point, the people of Ceylon, as they walked into their office in the morning there was below the surface of their minds, when they passed through the crowd on the veranda, a feeling of irritation and contempt.[118]

This attitude resulted in harsh sentences by some magistrates who relished adding flogging to incarceration. In 1890 the Solicitor General reminded the judiciary that "no police magistrate is above the law,"[119] and Leonard Woolf, reflecting on his time as a novice magistrate at the beginning of the twentieth century, reached the following damning conclusion. He wrote that his

actual experience from the inside of the administration of law and what is called justice produced in me an ineradicable and melancholy disillusionment with those whose duty it is to do justice and protect law and order. Too often one watches the line between the criminal and the policeman or the judge growing thinner and thinner ... the faces of the eminent Judges

---

116 *Proceedings of the Planters' Association* 1883–84, p. 152.
117 For a discussion of the tactics employed by the Ceylonese to steal coffee and the strategies employed by the planters and authorities to prevent it, see Duncan, *In the Shadows of the Tropics*, pp. 154–67.
118 Woolf, L. 1961. *Growing: An Autobiography of the Years 1904–1911*. New York: Harcourt Brace Jovanovich, pp. 52–53; A.A. King, District Judge, Badulla, *ARC*, 1869, p. 195.
119 C.P. Layard, Solicitor General, *ARC*, 1890, p. A3.

of the High Court of Justice suggest that nastiness and brutishness are found upon the Bench as well as in the Dock.[120]

The mix of judicial scrupulousness on the part of some and ignorance, pragmatism, and ruthlessness on the part of others, failed to produce the liberal consistency prescribed by Macaulay. The Ceylonese who brought cases before the courts were very aware of differences among judges and they often took into consideration the reputation of individual judges when bringing cases to specific courts. C.M. Lushington, the Acting AGA at Puttalam claimed that newly arrived magistrates were often tested by local people to see what kinds of judgements they would reach. He wrote,

> it may happen that, when he first comes to the district, a lot of false cases come on for trial ending in acquittals, or nominal punishments. When this happens, the notion is spread that the Magistrate is technical or lenient and crime becomes rampant.[121]

The Solicitor General noted in 1892 that the paucity of convictions in Ceylon was due in part to magistrates not being satisfied with the evidence presented in cases and that some magistrates developed such a reputation for disbelieving evidence that

> suitors in general have been found to avoid him, and either take the law in their own hands, or travel great distances in search of any itinerating Magistrate whom the Government may have appointed in the district for the purpose of repressing some especially prevalent crime.[122]

The most flagrant and widespread violation of rule of law by judges was the differential treatment of Ceylonese and Europeans. Although there was pretence of equality before the law, it was impossible for judges to conceal their racial bias. One could argue that such a stance was made necessary by the quotidian violence on the part of British employers. Casual assault upon labourers by the British was such a common occurrence that it was normalised and unless it was

---

120 Woolf, *Growing*, p. 79. On contemporary critics of British imperialism see, Porter, B. 1968. *Critics of Empire: British Radicals and the Imperial Challenge.* London: Bloomsbury; Claeys, G. 2010. *Imperial Sceptics: British Critics of Empire, 1850–1910.* Cambridge: Cambridge University Press. As we can see, it was in the times and spaces of governance such as the turn of the century court, that some administrators such as Woolf developed a critique of colonial power from their own experience. See David Nally and Gerry Kearns on the need to pay more attention to the actual spaces of administration such as the judiciary "as distinctive sites of anti-colonial praxis." Kearns, G. and D. Nally. 2019. "An accumulated wrong: Roger Casement and the anticolonial moments within imperial governance." *Journal of Historical Geography*, Vol. 64, p. 1.

121 C.M. Lushington, Acting AGA Puttalam, *ARC*, 1885, p. A32.

122 P. Ramanathan, Solicitor General, *ARC*, 1892, p. A9.

life-threatening, it was rarely prosecuted.[123] Underpinning this complicity of the courts[124] was not only the racist view that assault inflicted on a "native" was not a serious matter, but also the reluctance of judges to prosecute Europeans to the full extent of the law.[125] Planters regularly assaulted and sometimes flogged their Tamil labourers for mistakes that they made or for insubordination, which could include looking a planter in the eye or speaking without having been spoken to. Planters might be reprimanded by the local magistrate privately if word got out that they had severely injured a worker, but it would be most unusual for them to be charged in court, even for inflicting grievous bodily harm and, when they were, they normally escaped with a fine.

If on a rare occasion a planter was jailed for an assault, the planters saw it as a travesty. For example, a letter to the editor of the *Overland Ceylon Observer* expressed outrage that a planter had been sentenced to a few days in jail for assaulting one of his labourers because, the writer claimed, the labourers would no longer respect planters.[126] Another example was a planter who attacked a policeman for not showing proper respect when greeting him. He was charged, but the case was dismissed as the magistrate believed that proper respect had not in fact been shown.[127] In contrast to the lenient treatment of British planters, it was reported that a magistrate had sentenced a *kangany* to seven months imprisonment for using insulting language to a planter.[128] Insolence was a form of symbolic resistance which was not tolerated by the British as it was seen to undermine their authority and superior status.[129]

Even seeking justice for an injury by a European was high risk for a Ceylonese. For example, a fiscal peon took a planter to court, claiming that he shot at him and then set his dogs on him as he tried to deliver a warrant on the plantation. The magistrate disbelieved the charge and sentenced the peon to two years hard labour for perjury. In another case, a Burgher charged a high

---

123 An interesting article in this regard is Bailkin, J. 2006. "The boot and the spleen: when was murder possible in British India?" *Comparative Studies in Society and History*, Vol. 48, No. 2, pp. 462–93. She argues that British violence towards indigenes was a matter of great debate and that the dividing line between legitimate and illegitimate violence shifted constantly over the colonial period. She quotes (p. 462) Amita Bazar Patrika from 1880, "judicial officers should also be aware that for Europeans to commit murders is an impossibility." As Bailkin shows, murder was often downgraded to accidental death. The Patrika quote was obviously an exaggeration to make an important point. In Ceylon, planters and other members of the British community did often get away with murder, but a few were convicted.

124 Collingham, E.M. 2001. *Imperial Bodies: The Physical Experience of the Raj, c. 1800–1947*. Cambridge: Polity, p. 142; Kolsky, *Colonial Justice*.

125 A similar situation existed in India. Wiener, *An Empire*, p. 130; Kolsky, *Colonial Justice*.

126 *Overland Ceylon Observer*, March 24, 1873. Wiener, *An Empire*, p. 132; Collingham, *Imperial Bodies*, pp. 141–49.

127 *Ceylon Observer*, April 15, 1915.

128 *Ceylon Observer*, September 18, 1905.

129 For a discussion of the politics of insolence, see, Shutt, S.K. 2007. "'The natives are getting out of hand': legislating manners, insolence and contemptuous behaviour in Southern Rhodesia, c. 1910–1963." *Journal of Southern African Studies*, Vol. 33, No. 3, pp. 653–72.

ranking member of the civil service with assault. The judge found the European guilty of the lesser charge of "wrongfully restraining the complainant" and fined him one rupee. The Burgher was then charged with perjury and bail was set at 150 rupees. As he was unable to pay he was sent to jail.[130] As we shall see in the following chapter, perjury in cases between Ceylonese was very seldom prosecuted. However, the knowledge that perjury charges could be laid against Ceylonese who prosecuted Europeans for violence had a dampening effect on this practice. Given the high levels of European violence against the Ceylonese, the laying of perjury charges differentially was crucial in allowing this violence to continue unchecked.

In the case of murder committed by a European there was a convenient clause that if medical evidence indicated that the victim's death was caused in part by an existing medical condition the assailant could be charged with a lesser crime in district court rather than homicide in the Supreme Court.[131] This defence was frequently used and normally the charge was reduced and the sentence limited to a fine or the case was dismissed. For example, Paul MacBae of Springmount Estate, Ratota, was accused of murdering his Tamil mistress with a rice pounder, but "the post-mortem revealed that every organ of her body was diseased. She had been given to much drinking ... the case could not be maintained."[132] Kolsky's observation about India could be applied to Ceylon:

> Despite a rhetorical stance of equality, legal practice and conventions placed most Europeans in India above the law and, in effect, tolerated and condoned widespread physical assault and abuse. This violated the theory of equal protection that undergirded the rule of law and made law complicit in acts of racial violence rather than a guard against them.[133]

The failure of magistrates to adequately punish planters at times drove labourers and villagers to take the law into their own hands by assaulting planters. Although they were punished to the full extent of the law, they managed to avenge their wrongful treatment.[134] The rule of law in practice was clearly of

---

130 "A fiscal peon punished for perjury against a planter," *Ceylon Times*, October 8, 1889; "The alleged assault by Mr. King CCS, disproved," *Ceylon Times*, July 22, 1889.

131 Rogers, *Crime, Justice and Society*, p. 143. The usual medical condition sited was an enlarged spleen caused by repeated bouts of malarial fever. In this regard, medical opinion was complicit in helping decriminalise European violence. Bailkin, "The boot and the spleen," p. 476. In cases where such a defense seemed implausible, a murder charge was usually reduced to manslaughter, on the grounds of provocation. Provocation was not limited to threat of physical force, but also included "failing to work in a speedy manner and using insulting language." Bailkin, "The boot and the spleen," p. 476.

132 Dep, *A History*, p. 191.

133 Kolsky, *Colonial Justice*, p. 4.

134 "Assault on a planter by natives," *Ceylon Times*, May 13, 1891; "Assault on Mr. Vizard," *Ceylon Times*, December 17, 1891; "Murderous assault on a planter," *Ceylon Times*, July 11, 1893; "Serious assault on a planter on Vogan Estate," *Ceylon Times*, June 30, 1897.

less importance than the rule of racial difference; we see racism triumphing over liberalism and administrative goals trumping judicial principles.

Advocates, proctors, petition drawers and outdoor proctors

Prior to the Colebrooke-Cameron Reforms of 1833, only British and Burgher advocates were allowed to practice before Colombo courts. After the reforms, other Ceylonese were allowed to practice as there were too few British lawyers willing to come to Ceylon. Although the profession continued to be dominated by Burghers, some members of elite Sinhalese and Tamil families did practice in the courts, especially after mid-century.[135] Two types of legal advisers were certified by the Supreme Court: advocates and proctors.[136] The former had more education and charged higher fees.[137] Affluent parties hired advocates and proctors to prepare and argue their cases, while most litigants could only afford petition drawers and "outdoor" proctors, who set themselves up outside the court house, but were not recognised by the courts.[138]

The newly created police courts in 1843 forbade advocates and proctors to appear in court on the grounds that such representation would slow down the administration of justice. But the ban simply led to a substantial growth in unofficial "outdoor" proctors who, although they could not appear in court, helped to prepare and coach litigants and witnesses on what to say in the dock. These outdoor proctors were frequently disbarred proctors, proctor's clerks and notaries whose licences had been withdrawn, and translators with only a smattering of legal knowledge.[139] By the mid-1850s the judiciary decided that having such people informally preparing cases was worse than allowing advocates and proctors in the police courts. The Queen's Advocate, Richard Morgan, argued forcefully in the Legislative Council that the lawful rights of the accused were violated by not allowing trained representation in police courts.

---

135 Jayawardena, *Nobodies*, pp. 231–33; Peebles, P. 1995. *Social Change in Nineteenth Century Ceylon*. New Delhi: Navrang, p. 184 By 1901 there were 355 Ceylonese and 10 European Advocates on the Island. By 1911 there were 553 and 9. Fernando, P.T.M. 1969–70. "The legal profession of Ceylon in the early twentieth century: official attitudes to Ceylonese aspirations." *The Ceylon Historical Journal*, Vol. 19, p. 2.

136 These were trained by apprenticeship with examinations for admission to the Bar conducted by the Supreme Court. After 1874 they were trained in the Ceylon Law College. Udagama, D. 2012. "The Sri Lankan legal complex and the liberal project: only thus far and no more," in Halliday, T. C., L. Karpik and M.M. Feeley (eds.), *Fates of Political Liberalism in the British Post-Colony: The Politics of the Legal Complex*. Cambridge: Cambridge University Press, pp. 219–44.

137 Rogers, *Crime, Justice and Society*, p. 44.

138 *Weekly Ceylon Observer*, March 27, 1886.

139 Giles, *Report on the Administration*, p. 399. It was not uncommon for proctors to be struck off not only for unethical behaviour but also for "an ignorance of the elementary procedures of law and pleading." *Weekly Ceylon Observer*, March 27, 1886; *The Ceylon Examiner*, September 7, 1886.

Defendants, he said, had no protection against gross violations of law by the judges. He argued that had legal advice in court been available,

> it would have prevented a man being tried and convicted in his absence, he not having been summoned; it would have prevented a man being charged with one offence, summoned for another and convicted of a third ... it would have prevented the court finding and declaring that there was no evidence against the accused and nevertheless convicting him upon the written report of a headman ... it would have prevented the court from convicting a man without hearing his evidence, on the ground that, as the magistrate believed the complainant's case, it was unnecessary to hear the defendant's witnesses; it would have prevented convictions and punishments upon repealed Ordinances; it would have prevented the court convicting a man, aye, even flogging him, notwithstanding that he appealed, without a tittle of evidence in the case.[140]

Those members of Council who opposed representation countered that trained representatives not only delayed trials, but were able to confuse untrained magistrates and then appeal their flawed rulings to higher courts! Morgan, himself a trained lawyer, demolished these arguments in his speeches to the Legislative Council. He embarrassed the opposition by arguing that it was shameful to deny defendants trained legal counsel on the grounds that they might catch out untrained judges and expose their miscarriages of justice. He asked the members of the Council whether they would think it just if they themselves were put on trial with the possibility of being sentenced to jail without benefit of council or knowledge of the language in which the trial was being conducted. He said that often the accused was "not able to comprehend the proceedings at his trial, indeed may remain comparatively ignorant of what takes place until the sentence is pronounced." He concluded with the following plea:

> If you do not follow these substantial forms and take means calculated to secure correct conclusions, if parties do not obtain a fair hearing for themselves and their witnesses, but if instead of this, their cases are galloped over, their evidence not well sifted and the points not well considered, you are not then dispensing justice; it is a mockery to use the word in connection with such a mode of proceeding.[141]

These powerful arguments were crucial to the decision to allow representation in the police courts (Ordinance 7 of 1854). However, there remained tensions between members of the revenue branch who advocated swift, authoritarian

---

140 Digby, *Forty Years*, Vol. 2, p. 324.
141 *Ibid.*, pp. 307, 320.

delivery of summary justice and senior members of the judiciary who believed in the restraints provided by the rule of law.

In the case of district courts, where representation was permitted in court, outdoor proctors continued to find work as touts for petition drawers and proctors.[142] In fact, petition writers drew up as much as 90% of all criminal complaints.[143] Petition writers and outdoor proctors prospered, as it was customary for court proctors to turn over 20 to 50% of their fees in return for the custom they received from these groups.[144] While British officials had a low opinion of the outdoor proctors, comparing them to "birds of prey,"[145] their view of the petition writers was not much higher. F.R. Saunders, the AGA of Sabaragamuwa complained that

> these pests waylay the villager as he is coming to Court or *Kachcheri*, and tender their services. No matter what the real complaint is, or whether he has a complaint or not, all they require is the name of his enemy, and they readily invent for him a charge, advise his actions, and even find him witnesses. This payment is grudgingly given, because it produces iniquitous results, and compels the adversary to make similar or larger payments to escape injury.[146]

In 1894 the government attempted to undermine the petition writers by passing the "Touts Ordinance," making it an offence to accost litigants or loiter around the law courts. The petition drawers and outdoor proctors dealt with this by moving out of sight of the courts, strategically placing themselves along routes that litigants normally took.[147] In response to this tactic by the petition drawers in 1898 the government passed an ordinance requiring all complaints to be presented orally, but this simply transferred the work from petition writers to outdoor proctors; it did not stop false cases from being prepared and rehearsed before trial and it gave increased power to minor court clerks who could control who was allowed to present a case to the magistrate. The unintended consequence of the government's highly questionable tactic of denying representation by trained lawyers was a whole assembly of amateur practitioners who remained outside the court but who filled the gap of knowledge and suffused court proceedings with personnel over whom the courts could exert little formal control.

---

142 *ARC*, 1869, p. 199.
143 Peter de Saram to Colonial Secretary, 23 April 1894, SLNA59/19, cited in Rogers, *Crime, Justice and Society*, p. 58.
144 Pagden to Advocate General, 6 November, 1893, SLNA59/19. Cited in Rogers, *Crime, Justice and Society*, p. 57.
145 *SP*, 23 1884, p. 15.
146 *ARC*, 1869, p. 24.
147 *The Ceylon Examiner*, September 1, 1894.

Fiscal peons and process servers

The heavy use of the courts by the Ceylonese created a great deal of pressure on minor court functionaries. In 1869 there were only six fiscals and their deputies and peons to serve all civil and most criminal processes.[148] The Queen's Advocate complained to the Legislative Council that each time a complaint was filed

> a summons (in some cases a warrant) was issued against the defendant; and the Clerk of Court is further required to issue as many subpoenas as the parties call for. All this costs the party nothing, and just because it costs him nothing, he does not care how many summonses or subpoenas he calls for—the officers of the Court are to prepare the same, to prepare the translations, and the Fiscal's Officers are to go from one end of the district to the other to execute the process.[149]

The pressure on the courts, caused by the large numbers of Ceylonese taking full advantage of their rights to use the courts, was compounded by the British government's unwillingness to pay a proper living wage to minor court officials. F.R. Saunders argued:

> It daily happens that a peon has to bring in criminals 20 to 60 miles and feed and keep them without being allowed to charge a penny. The natural consequence is, that a peon avoids arresting anyone. If he does so far act on the warrant, he will always release his prisoner for a consideration, or on a promise to him to surrender at the right time; if the arrested person fails to keep his promise, "the accused is not to be found"; if he surrenders, "the accused is herewith sent to court". Should the peon be at last obliged to detain a man who can't or won't pay, he ill-treats him—locks him up somewhere whilst he goes to serve other process; and he compels all persons to support himself and his prisoner under threats, which his victims know he is well able to carry out. But how else is he to do his work under the present system? These evils and many others even worse, I respectfully submit, arise from the manner in which "cheap justice" is administered in criminal cases.[150]

Six years later, the GA of the North-West Province argued that the gains from bribery were in fact considered a necessary supplement to the low salary.

---

148 *ARC, 1869*, p. 214.
149 *CH*, 1871, p. 25.
150 G.A. Sabaragamuva, *ARC*, 1869, p. 24; a quarter of a century later the situation had not changed. Fiscal's peons were still getting Rs. 15 per month, the wage of a common labourer, and continued to be corrupt. *Ceylon Observer*, August 20, 1894.

The men know perfectly well that they enjoy absolute impunity from pun-ishment, the worst that can befall them being the loss of their posts, which but for the opportunities for bribery they present, it would be difficult if not impossible to fill. It is impossible to prevent grievous hardship to suitors from the difficulties experienced in the due service of process on dishonest debtors, when it is worth their while to bribe the process servers with a few cents to make a false report.[151]

One aggrieved litigant wrote a letter to the editor of the *Colombo Observer* call-ing process servers "a set of liars; they demand and receive money, etc. from suitors and make false reports."[152] Although many peasants suffered from injustice due to corrupt fiscal peons, other peasants, knowing the peons to be corrupt, used this to their advantage in disputes with their enemies. The District Judge at Ratnapura described how it worked:

A man can institute a false case, fee the fiscal's peon to report the sum-mons served, and then get judgement by default, and afterwards take out his writ, when he thinks proper; and the enemy may know nothing about it until his property is seized by the Fiscal. This goes on regularly I believe.[153]

Occasionally process servers were violently attacked. The *Ceylon Times* reported that when a party sent by a magistrate, which included a fiscal's peon, two police sergeants and two constables, went to arrest eight villagers who had been involved in a violent incident near Colombo, they were attacked by a group of villagers and fled, leaving one of the constables locked in a village house. Shortly after a police inspector accompanied by a detective and twelve constables returned to the village to rescue the constable and make the arrests. But they in turn were attacked by a large group of villagers and the police inspector was assaulted with a rice pounder and his horse slashed with a bill hook (sickle-shaped tool). The police returned later in still greater force and finally made the arrests.[154] This kind of violent reaction to attempted arrests by fiscal peons was relatively rare and most likely to occur either in the case of an armed gang or when the police intervened in a violent quarrel between different castes or religious groups.

Another strategy of resistance to process servers was to turn the power of the spatial organisation of the court system against itself. It was not uncommon for people living near the boundaries of a district court to cross into the adjacent

---

151 R. Morris, *ARC*, 1875, p. 80.
152 *Colombo Observer*, April 30, 1863.
153 R. Reid, *ARC*, 1871, p. 344. In 1905 the process servers were still producing bogus reports for a bribe. *Colombo Observer*, September 26, 1905.
154 *Ceylon Times*, September 21, 1875.

district to evade a warrant. For example, a man living in the Negombo district, upon learning he was to be served a warrant, crossed the river into the Chilaw district. As the Negombo process server had no jurisdiction in Chilaw, the warrant had to be taken back to Negombo and then sent to Chilaw, thirty miles away. But the Chilaw process server, on searching for the man, discovered that he had crossed back into Negombo and so he was stymied as well.[155]

Clerks

There were also various clerks and peons who controlled access to the court, guarded prisoners, and recorded judgements. It was widely known that in many courts the clerks, who also were poorly paid, would accept bribes and if a plaintiff wished not to have his case interminably delayed, he would have to bribe the clerk. Such was the concern about bribery in the lower courts that a commission was formed in 1883. The Commission concluded that clerks were placing "delays and obstructions—at almost every stage of a suit or prosecution for the purpose of obtaining a bribe."[156] Campbell, the Inspector General of Police, singled out what he believed was the undue power of the clerks. He felt that the magistrate's courts were so overwhelmed with cases that clerks often assumed an authority they did not have. He said that it was not uncommon for a clerk, without reference to a magistrate or preliminary enquiry, to send summons, as he put it,

> to a cloud of witnesses ... It does not matter if these people know anything of the case or not, or whether one might not do as well as a dozen, nor does it matter how far they may have to come at their own expense, or how ruinous to them or beneficial to their accuser to leave their homes at the appointed date may be—they are bound to come.[157]

It is clear that low-level clerks managed to increase their own power and income through the inattentiveness of judges. Letters were written to the newspapers complaining that litigants who won judgements had to bribe the court clerks in order to collect what was owed them. So normalised had bribery become among clerks that, when one Burgher clerk was dismissed for taking bribes, an indignant letter was written to the editor of the *Colombo Observer* claiming this was unjust, as bribes were "a privilege sanctioned by custom."[158] This raises the question of the relationship between custom and law. A custom is an established practice considered acceptable within a community even when

---

155 "Negombo. From our correspondent," *The Ceylon Examiner*, June 18, 1863.
156 Report of the Minor Courts' Commission, *SP*, 1883–84, pp. 201–02.
157 *ARC*, 1869, p. 225.
158 Letter to the editor from "A Suitor"; Letter to the editor from "WN," *Colombo Observer*, April 30, 1863.

it is contrary to law. Whether or not a community sees the exchange of money to be a privilege, a gift, a commission for a service, or a bribe depends upon local moral codes, on who loses and who benefits from such an exchange, but most importantly upon whether the law of the land is considered legitimate.

Translators

Although in theory all judges were to learn the local languages, few were fluent. Phillip Anstruther, the Colonial Secretary, reported in 1850 that "there is a complete curtain drawn in Ceylon between the government and the governed; no person concerned with the government understands the language [of the people]." Although this situation improved after 1870, all areas of governance were heavily dependent on interpreters.[159] All pleadings and proceedings were presented in English, even though the vast majority of litigants did not speak the language. All documents in the vernacular had to be translated. As a result, official translators became powerful figures within the court system.[160] T. Berwick, District Judge of Colombo defended the use of English in the courts as follows:

> The educational and civilizing impetus of the use of the English language in our courts has been of enormous value. Doubtless it has drawbacks; the interpretation, and often double interpretation, adding very greatly to the work and time of the courts, not to speak of the deficiencies it is possible it occasions in the actual doing of justice.[161]

The language policy of the courts is a prime example of systemic inequality in the judicial system, for it made it expensive and complicated for non-English speakers who had to pay fees for translation. Despite the fact that the very idea that the use of English language was a "civilizing impetus" was alienating, English did in fact become a widely accepted mark of distinction and superior education. In the case of the court translators, fluency in English afforded them with power over litigants with no English and judges with no local language. Not only was the language policy contrary to the fiction that everyone is equal before the law, it proved to be a high price to pay for all concerned, given the endless delays and costs of translation.

There was also concern that the widespread bribery of translators distorted evidence presented to judges. The Inspector General of Police believed that

---

159 Select Committee on Ceylon Second Report. 1850. London: HMSO, p. 344; Coperehewa, Colonialism p. 35; Samaraweera, V. 1985. "The legal system, language and elitism: the colonial experience in Sri Lanka," in Marasinghe, L.M. and W.E. Conklin (eds.), *Law, Language and Development*. Colombo: Lake House Investments, p. 98.

160 Clarence, "Application of European law," p. 230.

161 *ARC*, 1871, pp. 281–82.

judges who were solely reliant on interpreters were too easily taken in by false evidence fabricated by them.[162] The Queen's Advocate, Sir Richard Morgan, clearly shared this view when he described someone who, "though an interpreter of a court, [was] perfectly honest."[163] Translators and court clerks who were familiar with the workings of the court and the prejudices of individual judges found their services in great demand and therefore lucrative. They could especially exercise influence on cases where the translator was also an unofficial proctor of one of the litigants. All of this, being perfectly legal, was done quite openly with minor court officials arriving at court followed by a crowd of clients whom they had been coaching.[164]

Having introduced the structure of the courts and the various official players, I will show how their corruption and incompetence opened spaces in the judicial system for the general public to use it to their own ends. I will describe how the public made extensive use of the courts and by doing helped to shape the institution.

162 *ARC*, 1869, p. 225.
163 Digby, *Forty Years*, Vol. 1, p. 303, quoted in Rogers, *Crime, Justice and Society*, p. 65.
164 Pippet, *A History*, pp. 82–83.

6 Speaking lies to power

How the Ceylonese used the courts

Hostile takeover: the Ceylonese use of the courts

The peasants who regularly went to court, making use of its structure to suit their own local purposes, reshaped it, but not fundamentally.[1] They co-opted some of law's power, but in doing so they reinforced its centrality in Ceylonese society. So heavily did they use the courts that the British considered the Ceylonese to be excessively litigious.[2] They knew this from the crowds that flocked to the courts daily and from the delays caused by a judicial system overwhelmed by litigants. They were convinced, and not without reason, that false cases abounded and that perjury was rampant. All of this, they thought, accounted for the extraordinarily high percentage of cases being withdrawn before trial, thrown out as frivolous or for insufficient evidence, and non-convictions in cases that proceeded to trial.[3]

To give a sense of the scale of the problem, the Inspector General of Police wrote that in 1869, out of 168,426 cases 112,367 were dismissed without trial and 26,487 remained untried at the end of the year.[4] In that same year, H. de Saram, District Judge in Kegalla, wrote that there was so much frivolous litigation in his court that out of 3,300 persons charged with having committed

---

1 On this point more generally see Hirsch, S. and M. Lazarus-Black. 1994. "Introduction: performance and paradox: exploring law's rule in hegemony and resistance," in Lazarus-Black, M. and S. Hirsch (eds.), *Contested States: Law, Hegemony and Resistance*. New York: Routledge, p. 1.

2 Samaraweera, V. 1979. "Litigation and legal reform in colonial Sri Lanka." *South Asia: Journal of South Asian Studies*, Vol. 2, No. 1–2, pp. 78–90; De Silva, M.U. 2006. "Litigiousness in Sri Lankans: an examination of judicial change and its consequence during the late Dutch and early British administration in the Maritime Provinces of Sri Lanka." Journal of the Royal Asiatic Society of Sri Lanka, New Series, Vol. 52, pp. 127–42.

3 Merry, S.E. 1990. *Getting Justice and Getting Even: Legal Consciousness among Working-Class Americans*. Chicago: University of Chicago Press. In her study of working class use of courts in the US in the twentieth century, Merry found a similar pattern of behaviour, with court officials complaining about frivolous cases and litigants using courts as extensions of petty quarrels.

4 *Administration Reports, Ceylon (ARC)*, 1869, p. 226. While the rate of actions commenced that proceeded to trial was much higher in Victorian England, between the sixteenth and eighteenth centuries it was only 10%. Brooks, C.W. 1998. *Lawyers, Litigation and English Society since 1450*. London: Hambleton Press, pp. 70–71.

various offences, more than 2,500 were discharged without trial and that only 189 people were convicted.[5] Judges were divided over what percentage of the cases were legitimate. The police magistrate at Galagedara thought that two-thirds of his dismissed cases were bona fide, but had been compromised before-hand. And the police magistrate at Point Pedro thought that 75% of his dis-missed cases were brought either to annoy or extort neighbours.[6] The *Ceylon Examiner* published an editorial arguing that magistrates were so unable to "dis-tinguish a true case from those that are entirely false, that the clause in a local ordinance, having reference to the institution of false and vexatious cases, has necessarily become almost a dead letter."[7] Thirty years later, the Solicitor Gen-eral admitted that the extent of crime was unknown, not only because so much was unreported, but because "the reported cases of crime really include both true and false cases, and who can say how much of it is true and how much false?"[8] The Inspector General of Police described false cases as the "crying evil of Ceylon" not only because innocent people were falsely accused, but because there were so many such cases filed that they prevented the courts from having the time to thoroughly investigate legitimate ones. He cited as evidence a magistrate in Matara, who in 1868 had to review 4,909 cases involving 17,764 accused people under review. He estimated that the magistrate would have to average more than nineteen cases with seventy-one accused persons per day.[9] Clearly, such a caseload was impossible to manage and so cases were postponed, and in those that were tried, justice tended to be much more sum-mary than was desirable. In the 1860s the conviction rate averaged less than 10% and slowly rose to around 25% in the early twentieth century. This increase was due to a higher percentage of people being charged with regulatory offences that were easier to prosecute, and also somewhat better evidence collection.[10]

In the 1880s the Queen's Advocate despaired that "the machinery of criminal law is almost invariably used for improper purposes and is employed as an engine of oppression rather than redress;" this he claimed is "fatal to justice" and "to every principle of social civilization."[11] In 1892 the Solicitor General, in his review, surmised that a third of the population was involved annually in a court case either as a principal or witness.[12] Rogers estimated that, since very few women were accused of crimes and a large percentage of the population were children, as much as 20% of the adult male population were charged each year. And if one adds witnesses subpoenaed, who again were

---

5 *ARC*, 1869, p. 186.
6 *ARC*, 1867, p. 200; *ARC*, 1867, p. 199.
7 *ARC*, 1869, p. 225.
8 *SP*, 1897, p. 333.
9 *ARC*, 1869, p. 225.
10 Rogers, J.D. 1987. *Crime, Justice and Society in Colonial Sri Lanka*. London: Curzon Press, p. 62.
11 *ARC*, 1880, p. 2B.
12 *ARC*, 1892, p. A1.

overwhelming male, then the figure amazingly rises to 50% of the adult male population involved in a court case in any given year. By the late nineteenth century, when comparative statistical data became available to government, it was clear that the Ceylonese brought criminal charges against each other many times per capita more than did Indians who also had a reputation for being litigious.[13] In 1849 Thomas Skinner, pioneer roadbuilder, spoke before the Special Working Committee of the House of Commons about the state of the courts in Ceylon. He stated,

> probably in no people in the world does there exist so great a love of litigation as in the Sinhalese. Perjury is made so complete a business that cases are regularly rehearsed in all their various scenes by professional perjurers as a dramatic piece is at a theatre.

Skinner blamed British maladministration and policies that disrupted traditional community traditions for this state of affairs.[14]

In his insightful analysis of crime in nineteenth-century Ceylon, Rogers argues that the Ceylonese never accepted the moral authority of the British criminal justice system. Rather they saw it as a structure through which they could gain access to the power of the state to pursue their own goals. Sharafi, similarly, found that the Parsis' heavy use of colonial courts in India was purely strategic and did not imply that they had "internalised colonial legal values."[15] In Ceylon, the structure of the court system, coupled with the lack of effective policing and detective work, meant that the system could be systematically co-opted. Much of the Ceylonese use of the courts, which officials viewed as abuse, was, from the Ceylonese perspective, a perfectly legitimate way to further their interests or pressure an adversary into settling a dispute.[16] In other words, they believed the ends justified the means. But becoming involved in a court case usually entailed significant costs: witnesses and headmen often needed to be bribed, as did clerks and translators, and petition writers and outdoor proctors required fees.

---

13 Rogers, *Crime, Justice and Society*, p. 61. In the 1980s Galanter coined the term hyperlexis to describe a society with very high litigation rates. Washington, D.C. for example had a lodgement rate of 20,321 per 100,000 inhabitants (Galanter, M. 1983. "Mega-law and mega-lawyering in the contemporary United States," in Dingwell, R. and P. Lewis (eds.), *The Sociology of the Professions: Lawyers, Doctors and Others.* London: MacMillan, pp. 6–11). Civil actions in England and Wales from 1860 to 1911 hovered in the 4,000 to 5,000 per 100,000 people (Brooks, C.W. 2004. "The longitudinal study of civil litigation in England 1200–1996," in Anleu, S.R. and W. Prest (eds.), *Litigation: Past and Present.* Sydney: University of New South Wales Press, p. 38.)

14 Skinner, T. 1974. (1891). *Fifty Years in Ceylon: An Autobiography.* Dehiwala: Tisara Prakasakayo, p. 221.

15 Sharafi, M. 2014. *Law and Identity in Colonial South Asia: Parsi Legal Culture, 1772–1947.* New York: Cambridge University Press, p. 315.

16 Rogers, *Crime, Justice and Society*, pp. 9, 71, 40.

Over the decades, officials tried to understand why the Ceylonese flooded the courts with cases despite the costs. The most common assumption was a racial/cultural one: that the population as a whole was excitable and vengeful. Each sub-population was thought to possess different negative characteristics that inexperienced British judges should be made aware of. For example, the Kandyans were thought to be obsessed with hereditary land. A district judge summed up the official view of the Kandyans thus:

> The attachment of a proprietor to his patrimony is almost a passion; it is unintelligible to a foreigner. Rob him of his money, and he will bear the loss with comparative equanimity; but deprive him of a foot of his hereditary land, and he will spend the last farthing he has to recover it, steep himself in debt, and bequeath the lawsuit to his heirs as their inheritance. At times if his adversary be rich and influential, and the Court so distant, or so deep in arrears that he sees no prospect of speedy redress, he will brood over his injury till life is a burden, and he will either hang himself, that his death may lie at his enemy's door, or he will slay him and take the consequences.[17]

The Moors (Muslims), whom the British associated with sharp business practices, were thought to be more likely to use the courts for extortion. For example,

> it appears to have become a regular trade with the Moormen to enter utterly false civil actions against the Sinhalese people. These actions are sometimes entered simultaneously in two or three different courts by accomplices, who get them fixed for trial on the same day. The consequence is, that the defendant must either hurry all over the country, and retain council and file answer, or else suffer judgement by default. He has no chance of recovering his costs, even if he succeeds in shewing that the claim is untrue, for the plaintiff in that event has recourse to the sanctuary afforded to him by the neighbouring coast of India. There are no less than three such cases before me now.[18]

Even more troubling to officials was perjury, as it suggested a fundamental lack of respect for the system. The most common official view was that the courts were used, not so much as a final arbiter of legitimate cases, but as a means to punish enemies by bringing false charges against them. Although it was rarely

---

17 L. Liesching, District Judge, Nuwarakalawiya, *ARC*, 1870, p. 247. As Sivasundaram (Sivasundaram, S. 2013. *Islanded: Britain, Sri Lanka and the Bounds of an Indian Ocean Colony*. Chicago: University of Chicago Press) points out, the British romanticised the Kandyans as mountain folk and sharply differentiated them from the maritime Sinhalese.

18 District Judge, Nuwara Eliya, *ARC*, 1868, p. 71.

clear to judges how the defendants became enemies in the first place, it is obvious that the Ceylonese general public used the courts to achieve their goals, according to their own sense of justice, by punishing their enemies for crimes and perceived crimes often unrelated to the case at hand and possibly committed years before. F. Jayetileke, the District Judge at Chilaw, wrote, "Nearly every villager has his 'enemy', and to be able to triumph over him in a court of law is the *summum bonum* of his earthly happiness."[19]

Thirty years later, J.O. Murty, the AGA at Matara claimed that

> love of litigation is hereditary with the people, and is the very breath in their nostrils. Wealth is only desirable in so far as it enables them to indulge in litigation, and the height of their ambition is to succeed in a case, especially if their enemy is punished.[20]

It was widely believed that the Ceylonese even made false accusations that they knew they were unlikely to be accepted by the court in order to punish enemies. It was common for cases to be instituted and then dropped before a trial, in order to threaten a neighbour into informally negotiating a settlement or even to simply harass him.[21] But many cases were carried much further than this. A judge at Matara wrote in frustration,

> A great deal of the Justice's time is wasted in investigating utterly false charges, and it is a great defect of the law, I think, that persons falsely accused before the Justice of the Peace and who are made to come backwards and forwards to the court for no other reason than because the complainant entertains some spite against them, cannot recover a farthing for the loss of time and expense they are put to, without the tedious process of formal civil proceedings.[22]

David de Saram, the District Judge at Kurunegala, explained how the courts could be used to humiliate an enemy.

> It very often happens that if A charges B with an assault or any other offence, he includes in that charge the members of B's family or B's friends also. Most of these cases result in acquittals, but a heartless man's cause is gained when he has succeeded in procuring the attendance of his adversary's wife and daughters in a public court.[23]

---

19 *ARC*, 1871, p. 329.
20 *ARC*, 1902, p. E37.
21 David de Saram, District Judge, Kurunegala, *ARC*, 1867, p. 197.
22 R.W.D. Moir, Acting District Judge, Matara, *ARC*, 1867, p. 186.
23 *ARC*, 1867, p. 197.

Judges were also struck by how minor the charges often were. A.A. King wrote that

> The large proportion of charges in Police Court are for petty assault; many of them are dismissed as frivolous—usually such as are brought by members of the same family against each other (it being by no means an unusual thing for a man to appeal to the Police Court for redress in the instance of his brother's dealing him a slap on the cheek); and instances have come to my notice of men travelling a whole day's journey to lay such a grievance before a magistrate.[24]

Some saw these cases as release valves for a population considered "excitable." As one put it, "the court house [in Badulla in the Central Province] is the arena chosen by popular consent in which the greater part of the superfluous excitement and passion of the native is exercised and worked off." Another wrote that

> the people [around Mannar in the North] ... come to court and institute a case; but as soon as the excitement of the quarrel is over, they revert to their normal state of lethargy; and forget the case, which is accordingly struck off. That most of the cases, both criminal and civil, have their origin in this kind of temporary disagreement, is shewn by the large proportion which those struck off bear to those which come to trial.[25]

There was also concern that the Ceylonese might take pleasure in attending court for its own sake. In trying to explain this to themselves, officials usually resorted to racist stereotypes. R. Massie, the Acting District Judge at Mullaittivu, thought that local people flocked to his newly established court with trivial cases simply because they considered it a "novelty."[26] M. Gillman, the District Judge at Kurunegala, thought "the natives find both profit and amusement in bringing false cases, and that they have a natural talent for lying."[27] Some, like the AGA of Kurunegala, wrote that

> with the indolent Kandyan population, entirely devoid of all those sources of amusement found in more civilized communities, a suit in court seems to be looked on as the answer to a want met elsewhere by the theatre, opera, music halls, etc., and has the advantage of cheapness.[28]

24 District Judge, Badulla, *ARC*, 1869, p. 195.
25 A.A. King, District Judge, Badulla, *ARC*, 1869, 192; P.A. Templar, AGA Mannar, *ARC*, 1870, p. 104.
26 *ARC*, 1869, p. 199.
27 *ARC*, 1867, p. 198.
28 R. Morris, *ARC*, 1869, p. 120.

The Queen's Advocate claimed in the Legislative Council that "the natives are fast getting demoralized by the love of litigation … which takes them away from their fields and gardens and leads them to pass their time in court, gratifying the worst passions of their nature."[29] One AGA even went so far as to consider the use of the courts as a form of gambling.

> The low-country Sinhalese and boutique-keepers from Panadure, Galle, and Matara, are great frequenters of Court. Gambling of some kind—either with law or with dice—is their hobby, and they seem equally pleased whether they stand in the position of complainant or that of defendant.[30]

This composite picture closely corresponds to the stereotypes of the "indolent" poor in Britain, who were seen as too lazy to work hard and interested only in amusement. As such it represents the projection of commonly held British class attitudes onto race in the colonies.

Although the officials attempted time and again to convince the Ceylonese to use the courts in the officially approved manner, they were unable to stop them from appropriating them in other ways. So frustrated did the British become with the amount of frivolous litigation that in 1871 a new stamp tax of fifteen cents for a plaint and five cents per subpoena to call witnesses was imposed. This had a chilling effect on the number of cases brought before the courts for a couple of years, but by 1874 people appeared to become accustomed to the added cost and the number of cases rapidly rose again.[31] Even after the introduction of the stamp, officials were divided about whether the cost of filing criminal cases was still so cheap that it encouraged peasants to regularly engage in them or that it was in fact ruinously expensive for the average peasant. The answer is that it depended on the nature of the case. A simple case of assault could be initiated at little cost, whereas a larger case employing proctors, advocates and subpoenaing many witnesses could cost a lot. A decade later there were still those who worried that stamp duties "resulted in the denial of justice to some of the poorer classes of suitors."[32] It is clear, however, that despite the expense the Ceylonese regularly made a mockery of the court system, appropriating it for their own purposes. Consequently many, especially the young and inexperienced British officials, through frustration failed to see themselves as oppressors, but conversely regarded themselves as victims in an alien world. It is clear, in the case of some of the young British officials, that their experiences in Ceylon had shaken their belief that the rule of law was the greatest gift to the colony.

---

29 *ARC*, 1867, p. 256.
30 AGA Matale, *ARC*, 1872, p. 62.
31 Queen's Advocate, *ARC*, 1871, p. 271; T. Berwick, District Judge Colombo, *ARC*, 1873, p. 3; T. Berwick, District Judge, Colombo, *ARC*, 1874, p. 29.
32 *Weekly Ceylon Observer*, January 9, 1886.

Reluctant witnesses: resistance to the collection of evidence

It seemed paradoxical to the British that although the Ceylonese flooded the courts with cases, they were reluctant to come forward as witnesses and testify in cases brought by the government. However, there is no mystery here; the Ceylonese tended to testify when they believed it to be in their interest to do so and dragged their feet when they did not. Often the government's interest did not coincide with their own and thus the magistrates had great difficulty in collecting reliable evidence. Much of the peasants' resistance stemmed from a lack of commitment to the spirit of English law, but probably also from witness fatigue, as peasants were so often drawn into court cases concerning their neighbours' disputes. The government had great difficulty especially in persuading villagers to testify in murder or gang-related cases. So frustrated were the government officials by this that, as a last resort, they took the highly controversial step of placing a detachment of police for six months in villages that were thought to be withholding evidence and made the locals pay for the service. These "punishment posts" were of questionable legality because they sidestepped the judicial system and punished collectively without establishing guilt. However, it was justified by the government as the only way to deal with what was called the "conspiracy of silence" in certain villages.[33] While this policy angered villagers and deepened their hatred of the police, it failed to make them more cooperative, and villages continued to remain closed to outside surveillance. When witnesses did come forward, much to the frustration of the government, they tended to see their role as supporting whichever side in a dispute they had a stake in, rather than acting as neutral witnesses.[34] Clearly, the liberal ideals underpinning the court system were not well established. Indeed, fear of, or loyalty to, the village headman often discouraged witnesses from coming forward. M. Fisher, the Magistrate at Galagedara, wrote,

> nearly all the cases are compromised by the parties beforehand. If the complainant will not agree to settle the case, the accused go to the headman, who is then bribed; whereupon he calls up the complainant, and acts as a pacificator, and tries his best to settle the case. This he generally manages to do, because the complainant is afraid of offending him. Some of the most severe assault cases, in which both *catties* and knives were used, since I have been here, have all in this way been compromised.[35]

While the British interpreted such compromise as corruption, the Ceylonese saw it as a traditional form of conflict resolution.

In cases where members of the regular police came to a village to investigate a murder, villagers who cooperated risked the wrath of the headman, who was loath to have his authority undermined by the presence of police in his village.

---

33 AGA Kalutara, *ARC*, 1912–13, p. A22.
34 *ARC*, 1885, p. C26.
35 *ARC*, 1869, p. 224.

This acted as a powerful spur to non-cooperation and was often also reinforced by the fear of fellow villagers as well.[36] As the District Judge in Tangalle, F.H. Campbell put it, "fear of the consequences acts as a powerful motive for withholding their own knowledge of the facts, and of suppressing, if they can safely do so, the evidence of others which might operate beneficially in bringing the guilty to punishment."[37] This anxiety was often well founded for, as one GA put it, "when the criminal returns to the village he revenges himself on those who tried unsuccessfully to bring him to justice, he thus intimidates witnesses from appearing against him again and he commits crime with impunity."[38]

F.H. Campbell believed that there was a geography to the likelihood of witnesses coming forward. He said that during his years as coroner in Jaffna most murders were solved because it was hard to hide crimes in a densely settled region. However, he observed that Tangalla, in the Southern Province, was

> a more thinly populated place, where self-preservation can best be secured by maintaining a reticence which it would be dangerous to depart from; a display of knowledge under such circumstances would only stamp the individual as a fitting object of revenge on the part of those whose relation or friend he may have helped to the gallows. The arm of the law becomes weaker the further it is extended, and it cannot be wondered at, that at places where it barely reaches, the people should adopt a system of their own, even if it should be repugnant to the principles of law and order.[39]

While stonewalling as a classic form of resistance by villagers was more pronounced in the more remote villages, it was common everywhere. S. Grenier, the Attorney General, in his directions to magistrates in 1887, warned of "the proneness of villagers to hush up crime" and added that "murders are not infrequent of which the whole neighbourhood is probably aware, and yet the Magistrate, who ought to be the first, is the last to receive information, and this perhaps through an anonymous petition or mere rumour."[40] Such a tactic of leaking information about a crime suggests fear more than solidarity. R. Reid, the District Judge in Ratnapura, wrote that it was almost impossible in his

---

36 The situation was similar in India. Arnold, D. 1985. "Crime and crime control in Madras, 1858–1947," in A.A. Yang (ed.), *Crime and Criminality in British India*. Tucson: University of Arizona Press, p. 81 and Freitag, S.B. 1991. "Crime in the social order of colonial North India." *Modern Asian Studies*, Vol. 25, No. 2, p. 154, document the fear and hatred of the police among villagers in nineteenth-century north India and Madras.

37 F.H. Campbell *ARC*, 1875, p. 29.

38 F.R. Saunders, GA Western Province, *ARC*, 1883, 72A; Mclane, J.R. 1985. "Bengali bandits, police and landlords after the permanent settlement," in A.A. Yang (ed.), *Crime and Criminality in British India*. Tucson: University of Arizona Press, p. 30, reports that in nineteenth-century Bengal witnesses risked death by testifying against local criminals.

39 District Judge, Tangalla, *ARC*, 1871, p. 347.

40 *ARC*, 1887, p. 3C.

district to trace murder cases successfully because the headmen do little in the way of detective work and

> the people are most unwilling to go to Colombo to give evidence. They deny all knowledge of the matter in many cases, or else cunningly tell a little only of what they know, in order to avoid the trouble and expense of going to Colombo.[41]

G.M. Fowler, the GA of the Western Province, wrote in 1902,

> the sense of public duty is not sufficiently developed in the ordinary native to induce him to travel long distances and spend whole days in attending court at his own expense, with the probable result of incurring ill-will in his village.[42]

Again we see the British depoliticising the colonial relation by explaining Ceylonese behaviour as not yet "sufficiently developed." It is surprising though that the British maintained the idea that a sense of duty to a distant bureaucracy might yet be developed over time.

One of the greatest disincentives to testifying in cases where one did not have a direct interest were court postponements. H.W Gilman, the Acting District Judge in Kurunegala stated that he had on average twelve cases a day in his court and, of these, three were decided on evidence, while four were postponed because the parties were not ready and another five were postponed because of lack of time. He continued, "this entailed much needless expense on suitors, because every day they bring witnesses to court, they have to pay their expenses on the road, and otherwise keep them in good humour."[43] His solution was to schedule only half the number of cases daily, which reduced postponements. Of course, it also caused cases to accumulate and in that sense the plan failed to reduce delays. The Inspector General of Police cited a magistrate who reported cases postponed fifteen or twenty times before being struck off. He wrote, "Is it not natural that the misery inflicted on witnesses should make them dread the name of a court, and shrink from being connected in any way with a criminal case; unless they wish to injure a neighbour!!"[44] Under these circumstances it was not uncommon for witnesses to refuse come to court. But, bad as it was for witnesses, it was much worse for the accused; if there appeared to be a valid case, the accused was either forced to pay bail or be locked up.

---

41 *ARC*, 1870, p. 268.
42 *ARC*, 1902, p. B3. Headmen were constantly being called to court as witnesses, but it was not until 1897 that they were paid *batta* (expenses) to attend court in serious cases brought by the Crown. *ARC*, 1897, p. C5. Had *batta* been offered for their attendance as witnesses for normal village disputes, the fear was that they would have come more often whether they were needed or not.
43 *ARC*, 1867, p. 181.
44 *ARC*, 1869, p. 225.

Campbell reported that accused but untried men languished months in jail because of postponements. But even if the accused was able to post bail, he was forced to renew it after each postponement, which meant paying the clerk each time to draw up the bond and paying the village headman as much as a pound to underwrite the bond.[45] One can discern the strategy of a plaintiff who wants to punish an enemy with a false accusation by causing him endless delays and costs. Under these circumstances, even if the plaintiff should lose the case he will have avenged himself. Delays were also routinely used strategically to obstruct justice. Campbell understood this strategy:

> The delay which always takes place in bringing a jury together, especially in thinly populated villages and districts, and the usual formulae which attend all coroners' inquiries, serve in a great measure as an opportunity to persons interested to prepare amongst themselves a tale as will most likely have the effect of concealing the author of the crime about to be investigated. The jurymen themselves are, perforce, selected from among those who may or may not be acquainted more or less with the particulars, and are only too willing to accept the imposition of a duty which at once exonerates them from immediate liability to give evidence, or be charged as principals.[46]

Two decades later the situation was not much improved. C.J.R. le Mesurier, the AGA at Hambantota, reported that, in most cases, people in his district did not report cattle theft, not only because of the distance to the police court, but because of the strong possibility of postponements. He added that consequently the local people were inclined to take the law into their own hands.[47] Vigilantism in such cases can be seen as the strongest form of resistance to the official judicial system.

In response to the slow workings of the courts, the government put pressure on magistrates to do what they could to speed up the process. But this was problematic as well, for the Solicitor General reported that some magistrates

> have cheaply earned a reputation with the government for quick work and no arrears by the questionable process of superficially investigating and recklessly determining cases, to the misfortune of both complainants and defendants. If such vices prevail, the country is really to be pitied, because it has no tangible means of carrying home the charges.[48]

Cases of summary judgements by magistrates continued into the twentieth century, placing the rule of law in Ceylon in jeopardy.[49]

---

45 G.W.R. Campbell, Inspector General of Police, *ARC*, 1869, pp. 226, 230.
46 F.H. Campbell, District Judge, Tangalla, *ARC*, 1871, p. 347.
47 *ARC*, 1892, p. E19.
48 P. Ramanathan, Solicitor General, *ARC*, 1892, p. A1.
49 *Ceylon Observer*, May 12, 1905; January 7, 1910.

A.A. King, the District Judge at Badulla, accurately summed up the multiple reasons for delays in the Ceylon courts:

> The work of the courts is necessarily slower in this country than at home for these reasons. First of all, the unsatisfactory nature of evidence in this country—the proneness on the part of witnesses to falsehood and exaggeration (when even they have right on their side, through a mistaken hope of strengthening the cause they come to support)—renders it necessary that a greater number of witnesses should be examined on one and the same point; and, for the same reason, a much greater freedom and length of examination in chief and cross-examination, is allowed. Besides this, the examination being conducted through interpreters, each question is to be given and answered in a two-fold manner, and then the evidence so gathered has to be carefully committed to writing by the judge. Thus, the length occupied by the trial of a case is often ridiculously out of proportion with the importance of the matter at issue.[50]

Other causes of postponement included the frequent absences of police magistrates and AGAs, who served as district court judges, and were too busy attending to other duties to meet their scheduled court dates. This prompted angry letters in the newspapers about unsatisfactory nature of a system of part-time judges.[51] This use of AGAs, who had many other duties, in order to save money, again reveals the low priority awarded to the criminal justice system. If vigilantism is the strongest form of resistance to the court system from without, the resistance of the colonial government to adequately fund the system can be considered the strongest form of resistance from within.

Due to the very large number of false cases brought before the courts and the habit of tampering with witnesses both for the prosecution and the defence, it became a common practice to require witnesses to post bail.[52] Those who were unable to do so were frequently detained in prison until the trial took place. While the Administration of Justice Ordinance made this legal only in cases where the magistrate thought it absolutely necessary, it was reported in the Legislative Council that the practice had become widespread. Not only were witnesses imprisoned, even if they were willing to post bail, but often the people bringing the cases were imprisoned as well! This move, though illegal, was increasingly adopted by magistrates as a strategy to coerce plaintiffs who refused to prosecute a case that they had begun. Although this practice showed a clear disregard for rule of law and had been criticised by the Solicitor General

---

50 *ARC*, 1869, p. 189.

51 *Colombo Observer*, May 3, 1838; *Ceylon Examiner*, March 17, 1870; *Overland Ceylon Observer*, November 1, 1889.

52 The same policy was followed in India to guarantee the appearance of witnesses at trials. See Mclane, "Bengali Bandits," p. 30.

in 1880, ten years later the new Solicitor General chastised police magistrates for continuing to imprison witnesses in order to force them to give evidence for the Crown.[53] All of this calls into question any easy generalisations about the Ceylonese loving the sport of the courts. Rather it suggests that there must have been considerable distress and tension as litigants dragged their neighbours into court as witnesses in their feuds with fellow villagers.

On the normalisation of perjury

Once in court, witnesses were often reluctant to tell the truth.[54] In large part this was because they were often asked to testify in false cases. Perjury as the major form of resistance had become normalised in Ceylon. It clearly undermined faith in the system on all sides and Giles contends that it was "the most dangerous form of crime in Ceylon, and that which perhaps involves the greatest moral turpitude." He continued

> No man can feel safe while this state of things continues; and the evils are by no means confined to the individual unjustly accused. The prevalence of perjury causes the judiciary to reject evidence which in a purer atmosphere would be unhesitatingly accepted, and criminals benefit by this reluctance.[55]

Giles was certainly aware that perjury had also been normalised in India and was prevalent in England as well, where it was eventually greatly curtailed after mid-century through rigorous cross-examination.[56]

While perjury was believed to be the norm throughout the island, some thought that the Sinhalese engaged in it more than the Tamils or the Moors.[57] Having said that, an editorial in the *Weekly Ceylon Observer* claimed that the Tamils were the more skilful liars; "false testimony is a more scientific matter amongst the Tamils of the northern parts of the island than among the Sinhalese and the witnesses are sometimes exercised at a mock trial before the real one."[58] It was also common to single out village headmen as particularly prone

53 Hon. P. Ramanathan, *Ceylon Hansard (CH)*, 1880, p. 84; *ARC*, 1890, p. A2.
54 In England at the time the law of evidence prevented many potential witnesses from giving evidence. While these barriers were somewhat relaxed after mid-century, the qualifications for being heard as a witness remained much more restricted in England than Ceylon. In England, ownership of property was a major criterion for qualification. This would have been less useful as an exclusionary mechanism in Ceylon where most peasants were property owners. Allen, C.J.W. 1997. *The Law of Evidence in Victorian England*. Cambridge: Cambridge University Press, p. 1; Schneider, W. E. 2015. *Engines of Truth: Producing Veracity in the Victorian Courtroom*. New Haven: Yale University Press, pp. 2–3.
55 Giles, A.H. 1889. *Report on the Administration of Police, Including the Actions of the Courts and the Punishment of Criminals in Ceylon*. Colombo: J.A. Skeen, Government Printer, p. 364.
56 Schneider, *Engines of Truth*.
57 S. Houghton, Acting AGA Mannar, *ARC*, 1882, p. 39A.
58 *Weekly Ceylon Observer*, March 17, 1886.

to giving false testimony. This was considered particularly troubling, not only because headmen were officials of the state, but because they were thought to routinely play all sides, acting as village policemen, informal advisers to one or both sides in criminal cases and witnesses for the prosecution or the defence. Headmen were often the key players in criminal cases as they could to a large extent control the amount and nature of evidence available to the courts.

Well-to-do people were thought to be the most likely to bring frivolous cases, for they could afford to bribe witnesses and headmen and pay court costs.[59] Nevertheless, it appears that people of all classes engaged in this practice. The court records are full of false cases brought by headmen to shield relatives, instances where villagers organised the theft of crops in order to blame their enemies, and stories of people amassing fabricated evidence in order to convict a known village trouble-maker. The records include cases of people committing a crime for the sole purpose of accusing an enemy of having committed it.[60] For example, as a response to the increase in stabbing incidences in the 1880s, Ordinance 16 of 1889 was passed, which allowed whipping to be administered as an additional punishment for causing grievous hurt with a dangerous weapon. Predictably, villagers saw this as a new opportunity to settle scores with enemies. C.M. Lushington the AGA of Trincomalee wrote that

> petty quarrels have led to false charges of stabbing. In every instance the injury has been most trifling, such as might be self-inflicted or caused by scratching. There has not been a single stab wound or a cut more than skin deep.[61]

The Superintendent of Police of the Southern Province agreed, writing,

> I do not consider that the Knife Ordinance has been a success. It has undoubtedly been the means and cause of many false accusations of "causing hurt with a knife." The complainant knows that the accused runs the chance of getting lashes and goes so far as inflicting slight wounds on himself to support the charge.[62]

As a consequence, by 1901 hardly anyone was flogged for knife crime because, as the GA of the Southern Province wrote, "It is very hard if not impossible to prove positively that the wounds are not self-inflicted."[63] Robbery was another favourite accusation. In 1914, the GA of the Southern Province reported that, of

59 Coomaraswamy, *CH*, 1871, p. 72; F. Jayatileke, District Judge, Chilaw, *ARC*, 1871, p. 329.
60 E.F. Tranchell, Acting Inspector General of Police, *ARC*, 1882, p. A26; M. Panabokke, Police Magistrate, Matale, *ARC*, 1898, p. C10; F.R. Saunders, G.A. Western Province, *ARC*, 1883, p. A72; *Weekly Ceylon Observer*, March 27, 1886.
61 *ARC*, 1898, p. F19.
62 *ARC*, 1899, p. E6.
63 G.M. Fowler, *ARC*, 1901, p. E9.

163 cases of robbery reported in that year, 117 cases were judged to be invalid.[64]

Judges, however, were particularly vigilant in cases of cattle stealing for this was the crime most associated with fabricated evidence. In 1886 the Crown Council for the Northern Circuit and Western Province asserted that "the vast majority of cattle stealing cases are proved by false witness." And ten years later, the AGA at Hambantota stated that convictions in cattle stealing cases were difficult as "the so-called eye-witnesses are almost invariably false witnesses."[65] M. Panabokke, the Police Magistrate at Matale in 1898, describes the evidence typically heard in court.

> The usual evidence led was that so-and-so lost his animal and that two or three people at different places, when they went to turn water to their fields, met the accused driving the animals away. The witness speaks to the accused, whereby the witness has an opportunity of recognizing the thieves and the animal with its brand marks. I heard from a judicial officer of great experience that the evidence that he heard in his day of cattle stealing might have been reduced to a stereotyped printed form to be issued by the Government printer to be used by Judges and Magistrates.[66]

The problem of perjury was compounded by the fact that even those who had genuine complaints often fabricated evidence to attempt to bolster their cases.[67] Clearly a structural condition had been created where individuals assumed that others were perjuring themselves and so they felt forced to as well.[68] H.L. Moysey, the AGA at Matale, cites what he claims was a Kandyan proverb "No lie, no victory."[69] There were frequent complaints about homicide cases being spoiled by evidence concocted by relatives of the deceased. E.C. Dumbleton, Crown Council for the Western Circuit gives the following example.

> Numbers of persons are annually killed every year in village quarrels, but the deceased's party in nine cases out of ten deny there was a mutual fight, and by adding a little here and supressing a little there, out of an ordinary case of homicide evolve a cold-blooded murder. The concocted story, rehearsed privately in the village, is reported publicly in court, and as the

---

64 R.B. Hellings, *ARC*, 1914, p. C4.

65 C. Hay, *ARC*, 1886, p. 25C; E.F. Hopkins, *ARC*, 1896, p. E12.

66 *ARC*, 1898, p. C10.

67 A.A. King, District Judge, Badulla, *ARC*, 1869, p. 189. Leonard Woolf was also struck by the prevalence of perjury in true cases. De Silva, P. 2016. *Leonard Woolf as a Judge in Ceylon*. Battaramulla: Neptune Publications, p. 19.

68 For a general discussion of this point in contemporary law see, Northwestern University School of Law. 1974. "Perjury: the forgotten offense." *The Journal of Criminal Law and Criminology*, Vol. 65, No. 3, p. 364.

69 *ARC*, 1883, p. A34. This certainly points to the tactical use of dissimulation.

accused almost always reserve their defence the case has to go to trial as one of murder, though the Crown Prosecutor knows that he is sailing under false colours, and three quarters of his case is false. This is manifest when the Crown witnesses are called: they have forgotten the details of their story, contradict each other on minor points, and though it is perfectly clear that the accused killed the deceased, it is equally clear that he did not do so under the circumstances deposed to. The jurors are then in this difficulty: if they convict upon the complainant's evidence, they must convict of murder, for no extenuating circumstances are disclosed; and if they acquit, the accused, who is in all probability guilty, will get off scot-free.[70]

Another common tactic was to add the names of enemies to the list of those who actually committed a crime. C. Hay, the Acting Solicitor General, cites a typical case.

a villager's house is broken into by a gang; he recognizes three of the burglars, the others he did not recognize. To these three the complainant adds the names of one or two of his enemies and possibly a returned convict, or a village pest; just the same evidence ensures a conviction of those who took no part in the burglary as well as those who did. Thus, three men innocent of that crime are wrongfully convicted. The same will apply to charges of highway robbery, cattle stealing, and other crimes.[71]

And in 1903, the newly arrived Governor Blake wrote to the Colonial Office that, "in the majority of cases tried in Ceylon the evidence is unsatisfactory, as there is not alone an utter disregard for truth, but an extraordinary ingenuity of invention."[72] As an example of bold invention, a 1903 report of a Crown Council suggests that it was a common strategy for the perpetrator of a crime to charge witnesses with having committed it themselves so as to discredit their testimony.[73]

Because of their mistrust of the testimony of all parties, from village officials to the police, magistrates regularly operated under the assumption that cases were false, and, even if the case was credible, that witnesses fabricated or exaggerated what they had seen or heard. Hence, where possible, magistrates sought circumstantial and other forms of material evidence to help decide a case. The

---

70 *ARC*, 1891, p. A6.
71 *ARC*, 1891, p. A2.
72 Governor Henry Blake to Lyttleton, 19 December 1903, CO/54/685, cited in Rogers, *Crime, Justice and Society*, p. 62. Persuasion in order to achieve justice, rather than "truth" as the Governor's believes he can discern it, may be a more relevant issue to the Ceylonese in the colonial courts. See Arendt, H. 1968. "Truth and politics," in *Between Past and Future*. New York: Viking, pp. 227–64.
73 W.S. de Saram, Report of the Crown Council of the Northern and Southern Circuit. *ARC*, 1909, p. A14.

Colonial Secretary in 1859 summed up the problem facing magistrates when he wrote "parties and their witnesses are all, in general, so utterly untrustworthy of credit, that it is in most cases impossible for the courts to pronounce that they are false, though they may not have been satisfactorily proved."[74] While the collection of evidence apparently improved over the next three decades, in 1892 the Solicitor General claimed that "the paucity of convictions in Ceylon is due plainly to the fact that the magistrates are not satisfied with the evidence led in cases, either because the evidence is not strong enough or because it is not credible."[75]

In fairness, the collection of evidence was far from easily accomplished. Some argued that a magistrate, upon hearing of a serious crime, should proceed immediately to the scene to collect evidence before it was destroyed.[76] But the problem with sending a magistrate, as the Inspector General of Police pointed out, is that he is normally too busy to deal with any but local cases.

> At present, many a time, when a case from an outlying district is before the Magistrate, and he knows the witnesses are lying, he is helpless; for he cannot go to the place himself, and he has only the headmen or other village police to send, and he knows that they would lie too.[77]

The magistrates believed that it was sometimes simply not possible to collect satisfactory evidence. In cases of cattle stealing, for example, it was often nearly impossible to identify the stolen animal, for cattle were commonly slaughtered soon after being stolen and the hide cut into pieces and thrown in the jungle or disfigured with burns to obliterate the owner's brand mark.[78] Often the only evidence that the court had was an unreliable eyewitness account. The problem of cattle stealing was compounded by the fact that they could often wander on to other properties or be led by their owner onto a neighbour's property so that the person could be accused of theft.[79]

Disputes over land ownership and use were also very common cases in the court of requests and district courts and were often hard to adjudicate, as identifying boundaries was extremely difficult. As the District Judge in Kegalla wrote,

> It is almost impossible to decide them in Court on the evidence adduced; and the Commissioner cannot find time to go and inspect the lands, many of them being situated at great distance from his Court. Moreover, even if

---

74 Quoted in F.R. Saunders, AGA Sabaragamuva, *ARC*, 1869, p. 24.
75 *ARC*, 1892, p. A9.
76 L. Lee, District Judge, Matara, *ARC*, 1874, p. 49.
77 G.W.R. Campbell, *ARC*, 1869, p. 223.
78 D.E. De Saram, District Judge, Kurunagala, *ARC*, 1869, p. 198.
79 Duncan, J.S. 2016. *In the Shadows of the Tropics: Climate Race and Biopower in Nineteenth Century Ceylon*. London: Routledge, p. 152.

he does go and make an inspection after hearing evidence in Court, it is even then difficult to arrive at a satisfactory decision, for in my experience at least, the boundaries given by witnesses in Court are found to be non-existent, when one proceeds to the spot and endeavours to discover them.[80]

Another judge puts it sarcastically,

so preposterously does verbal evidence sometimes vary with reference to the very same plot of ground, even among witnesses called to support one and the same side, that anyone unaccustomed to native evidence would be often inclined to doubt the very existence of the property disputed.[81]

While an obvious answer would be to use the police and detectives to collect evidence, as I have shown, this would have been problematic as well. Inspector General of Police Knollys summed up the official view when he wrote to the Colonial Secretary,

there is no doubt that a case was seriously prejudiced with a jury for the mere fact of a Detective Officer appearing in it. Indeed, the popular idea of Detective in Ceylon was of a man who can make up evidence where none exists.[82]

The Assistant Superintendent of Police for the Western Province thought the problem lay in the fact that detectives were rewarded for results and that in his experience a detective, "sooner than admit he has been able to gather no information, he will fabricate it."[83] The GA of the Western Province urged that detectives not be rewarded for results because

even men of considerable intelligence and education do not look upon it as a moral offence to fabricate evidence against a person whom they believe to be guilty. Detectives who are too frequently rewarded by results, are generally ready to offer half their receipts to witnesses to falsely testify.[84]

---

80 *ARC*, 1868, p. 566.

81 A.A. King, District Judge, Badulla, *ARC*, 1869, p. 190. Part of the problem stems from the rarity of deeds and the difficulty of using oral evidence. See Kemper, S. 1984. "The Buddhist monkhood, the law, and the state in colonial Sri Lanka." *Comparative Studies in Society and History*, Vol. 26, No. 3, pp. 401–27.

82 Inspector General of Police to the Colonial Secretary, August 14th, 1894, in Dep, A.C. 1969. *A History of the Ceylon Police, Volume 2, 1866–1913*. Colombo: Police Amenities Fund, p. 334.

83 W.E. Thorpe, *ARC*, 1896, p. B10.

84 F.R. Ellis, *ARC*, 1897, p. B11. Freitag, "Collective crime," p. 149, notes that the padding of evidence by the police was very common in North India as well. Dhillon, K.S. 1998. *Defenders of the Establishment: Ruler-Supportive Police Forces of South Asia*. Shimla: Indian Institute of Advanced Study, pp. 118–19, notes that in India the laws of evidence were too complex for untrained police and therefore were ignored.

In the early twentieth century, Leonard Woolf wrote that he found it "extremely difficult to prevent effectively the primitive and illegal methods of the police and headmen in dealing with crime." And that "it was a perpetual—and usually losing—struggle to prevent every kind of pressure, including physical violence and torture, being applied to the accused in order to extract a confession."[85] In his official diary he summed up his view of police methods as follows: "Investigation, nil; methods, obtain a confession; result, acquittal."[86] But even when evidence was successfully collected, the case could still collapse on account of the police. H. Wace, the Acting GA of the Southern Province wrote,

> In the witness box they are utterly unreliable. Having arrested the person charged with an offence, their duty on the side of justice and order ends, in their opinion. In most cases they are influenced by personal interest on one side or the other and suppress or encourage evidence accordingly, their position giving them great facilities in this direction. They are generally corrupt, and when not personally interested are easily bought over. Hence it is not unusual to see the police officer as a witness for the prosecution and the constable arachchi for the defence or vice versa. Consequently the brief inquiry made by these officers on the spot immediately after the commission of the offence, which should be of great service to the Magistrate, is often worthless.[87]

Given the difficulty of collecting reliable in-situ evidence, cross-examination was seen as the best hope against perjury, but witnesses also developed tactics to deal with it.[88] The Kandy Courts reported in 1885 that witnesses who were yet to be cross-examined positioned themselves on verandas just outside the courtroom door so they could overhear the answers of witnesses currently being

85 Woolf, L. 1961. *Growing: An Autobiography of the Years 1904–1911*. New York: Harcourt Brace Jovanovich, pp. 79, 77. The police in India had the same reputation. Dhillon, *Defenders*, p. 117; Sen, S. 2000. *Disciplining Punishment: Colonialism and Convict Society in the Andaman Islands*. New Delhi: Oxford University Press, p. 199; Wiener, M.J. 2009. *An Empire on Trial: Race, Murder and Justice under British Rule, 1870–1935*. Cambridge: Cambridge University Press, p. 154; Singha, R. 1998. *A Despotism of Law: Crime and Justice in Early Colonial India*. New Delhi: Oxford University Press, p. 305.

86 L. Woolf, January 10, 1911. (1963). *Diaries in Ceylon (1908–1911): Records of a Colonial Administrator*. London: The Hogarth Press, p. 214.

87 *ARC*, 1896, p. E3.

88 In England, up until the mid-nineteenth century, perjury was seen as a problem to be dealt with by greatly restricting who could serve as a witness. After mid-century, the problem of perjury in England was increasingly countered by rigorous cross-examination. See Schneider, *Engines of Truth*, pp. 2–3. On the problem of perjury in eighteenth century England, see Oldham, J. 1994. "Truth-telling in the eighteenth-century English courtroom." *Law and History Review*, Vol. 12, No. 1 (Spring), pp. 95–121.

questioned. The court subsequently constructed a witness shed out of earshot from the court to deal with this problem.[89]

The widespread habit of introducing false evidence even into valid cases made magistrates so uncertain about the reliability of any evidence that they were often afraid to convict, for fear that an innocent man might be jailed.[90] They were also reluctant to charge one party with perjury when they felt that it was likely the other party was lying as well. A small step was taken to discourage false testimony when in 1898 the Criminal Procedure Code was amended to allow district courts to prosecute witnesses who "wilfully go back on evidence that they have given before a magistrate."[91] But not all officials even agreed with this step. The AGA of Matara in 1902 argued that, although it might not follow the rule of law, it was pragmatic not to punish the perjurer, for

> If one party leaves the court with a sense of elation, then the unsuccessful party will do all in their power to be even with them. The result is a counter case, and if the previously unsuccessful party succeeds in this, the other party will after a short interval retaliate in a similar manner. The result is a series of cases and counter cases. I have had an actual instance of this from Midigama, where two opposing factions continued bombarding each other with false cases until they were tired of the pastime. The best course is to give neither party cause to exult. Then nothing is left to rankle, and their ill feeling dies a natural death.[92]

In light of these remarks, regularly rehearsed, about the character of the native population, it is hardly surprising that the Ceylonese showed little respect for the criminal justice system, playing it to they own advantage.

The widespread existence of false accusations, as well as the fact that true cases were shot through with perjury, had a corrosive effect on the criminal justice system, undermining magistrates' confidence that justice could be delivered.[93] It further served to delegitimate the system in the eyes of the Ceylonese.[94] For just as the British thought that the Ceylonese were corrupt, so the Ceylonese assumed that the whole administration of British justice was corrupt as well. An editorial in the *Ceylon Observer* stated,

---

89 *ARC*, 1885, p. C26.
90 *ARC*, 1898, p. B16.
91 Attorney General, *CH*, 1905–06, p. 45.
92 *ARC*, 1902, p. E37.
93 For a similar situation in India see Schneider, *Engines of Truth*, pp. 105–07.
94 For a similar delegitimation in eighteenth-century Bengal, see Mann, M. 2004. "Dealing with oriental despotism: British jurisdiction in Bengal, 1772–93," in Fischer-Tine, H. and M. Mann (eds.), *Colonialism as Civilizing Mission: Cultural Ideology in British India*. London: Anthem Press, pp. 29–48.

we know by long experience how ready the natives generally are to impute the worst motives to public officers in the discharge of their duty ... But above all other departments and offices is that of our Criminal Prosecutors liable to suspicion and adverse remark.[95]

This is not surprising, given that it was obvious to all that the law was applied unequally to Europeans and Ceylonese. As E.P. Thompson points out, "The essential precondition for the effectiveness of law, in its function as ideology, is that it display an independence from gross manipulation and shall seem to be just."[96] Neither held sway in nineteenth-century Ceylon. In regard to the treatment of Europeans, the rule of racial and class difference usually won out over the rule of law.

To most colonial officials, perjury was a clear-cut sign of moral failure on the part of the Ceylonese.[97] Yet Rogers suggests that Ceylonese villagers did not hold the idea of truth as an abstract value.[98] While they held that it might be desirable to be truthful, when truth comes in conflict with justice, then truth should give way. Testimony in court, Rogers argues,

> was seen as a formula to manipulate power and was evaluated on the strength of its effectiveness. The morality of court testimony depended not on its truthfulness, but on the intention of the testimony. False testimony in support of a just cause was moral; for an unjust cause it was immoral. In many cases, especially those which were related to long-standing feuds, both sides no doubt viewed themselves as in the right. In such instances court cases were like contests.[99]

An editorial in the *Weekly Ceylon Observer* makes exactly this point;

> it would be very difficult to convince the average villager that he does anything wrong when he suborns a string of false witnesses to convict an enemy of some offence he really suspects him to have perpetrated or commits perjury or forgery to secure some advantage to which he thinks he ought to be entitled.[100]

---

95 *Ceylon Observer*, January 22, 1880.
96 Thompson, E.P. 1975. *Whigs and Hunters: The Origin of the Black Act*. London: Allen Lane, p. 263.
97 But as Schneider, *Engines of Truth*, p. 107 shows, the British conveniently overlooked the prevalence of lying in English courts at the time.
98 Here Rogers, *Crime, Justice and Society*, p. 75 is citing Gombrich, R. 1971. *Precept and Practice: Traditional Buddhism in the Rural Highlands of Ceylon*. Oxford: Clarendon Press, pp. 262–63.
99 Rogers, ibid.
100 *Weekly Ceylon Observer*, March 27, 1886. Herzfeld, M. 1990. "Pride and perjury: time and the oath in the mountain villages of Crete." *Man* (n.s.), Vol. 25, p. 305, found the same attitude among peasants in Crete. He reports "one illiterate old man when asked by a judge whether he knew what perjury was, is said to have replied, 'you get justice [that way].'"

The Superintendent of the Convict Establishment believed that rather than being amoral, "the Sinhalese have a very keen sense of justice, and I know from the names given to some officers, an acute perception of character."[101] L.B. Clarence, Senior Puisne Justice of the Supreme Court, thought that lying, rather than being a character trait of the Ceylonese, was a strategy: "it is unhappily true that the natives are more dishonest in our courts than in their own private life; and this is a fact which we should seriously lay to heart. The inefficiency of our administration of justice promotes dishonesty."[102] The Attorney General went so far as to urge cultural relativism upon his fellow administrators.

> even allowing that every is endowed with a certain sense of right and wrong, I think it cannot be denied that different views as to morals are held by different races and in different countries. These are truths which seem to be frequently forgotten.[103]

Leonard Woolf, reflecting back on why he left the Ceylon Civil Service in 1911, wrote, "I was not prepared to spend my life doing justice to people who thought my justice was injustice."[104] L.B Clarence went further, arguing that "The national propensity for false evidence is fostered by its frequent success; the conviction and sentence lose much of their disgrace, and consequently of their deterrent force, where it is notorious that a large number of convictions are unjust."[105]

The popularity of the British courts and the preference that many Ceylonese showed for using them rather than the village tribunals could be explained as follows: first, the courts had the authority to order greater punishment than the tribunals. Second, villagers sought a non-local source of authority, not only because they feared that the local authorities might be biased against them based on factional grounds,

---

101 M. Mooney, *ARC*, 1887, p. C55.

102 *Weekly Ceylon Observer*, March 27, 1886. A visitor to the courts in Ireland in 1834 noted that lying was rarely an expression of a deep deficiency in character, rather "false oaths are substitutes for weapons." Inglis, H. 1835. *A Journey throughout Ireland during the Spring, Summer and Autumn of 1834*, Volume 1. London: Whittaker and Co., p. 292.

103 F. Fleming, *ARC*, 1883, p. C1.

104 De Silva, *Leonard Woolf as a Judge in Ceylon*, pp. 29–30. Engel, D.M. 2015. "Rights as wrongs: legality and sacrality in Thailand." *Asian Studies Review*, Vol. 39, No. 1, p. 97, argues that in northern Thailand a similar disjuncture occurred between local and national notions of justice. "The Lanna justice concept lived on in the interstices of modern law." It occupied what he terms "a parallel universe outside the formal legal system." Engel found that this alternative notion of justice "shaped the behaviour of litigants in ways that would not be openly acknowledged." Engel, D.M. 2011. "'The spirits were always watching': Buddhism, secular law, and social change, in Thailand," in Sullivan, W. F. et al. (eds.), *After Secular Law*. Palo Alto: Stanford University Press, pp. 242–60. While villagers and local low level bureaucrats often responded to the law using traditional concepts of justice, unlike in Ceylon, the courts were avoided by villagers. See Engel, D.N. 1978. *Code and Custom in a Thai Provincial Court: The Interaction of Formal and Informal Systems of Justice*. Tucson: University of Arizona Press, pp. 1–3.

105 Letter from L.B. Clarence, Senior Puisne Justice on Sir J. Phear's Draft Criminal Procedure Code, to Lieutenant Governor Hon. J. Douglas, 25 May, 1881, *Ceylon Sessional Papers (SP)*, 1882, p. 284.

but also because it would be more difficult to sustain a false case at the local level where the truth of the matter was more likely to be discovered.[106] The great advantage of the British courts, from the point of view of perjurers, was the inability of these courts to discover the facts. What mattered most to court goers was which court was most advantageous to them personally. Essentially, those who lied in court were wagering that the authorities would be unable to discover the truth. Trapped between a wall of silence and a flood of dissimulation, the authorities struggled to control a situation which they found virtually indecipherable.

It might seem paradoxical that villagers who were so resistant to outside intrusion into village affairs would so readily embrace the colonial court system to settle internal disputes. They did so, however, precisely because they were able to do it largely on their own terms. By impeding the collection of evidence, some villagers were able to exert a large measure of control over the proceedings. This strategy, as we have seen, took a variety of forms. Interestingly, headmen employed the structure of the legal system itself to help control the nature of evidence available to magistrates. For example, plaintiffs were required to submit a report on their case by their headman. This not only allowed the headman to shape the evidence, but also to solicit bribes. Furthermore, when the police came to a village to investigate, they were more often than not following up on the line of inquiry suggested to the magistrate by the headman.[107] The position of headmen within the criminal justice bureaucracy, coupled with the reluctance of villagers to cooperate with outsiders, made the villages extraordinarily difficult for the authorities to penetrate.

The lack of a shared cultural notion of what constitutes justice was compounded by a difference in what constitutes an adequate verbal account. A.C. Lawrie, the long-time District Judge of Kandy and an authority on traditional Sinhalese law, noted that "the vagueness and in-exactitude of even truthful witnesses in the Kandyan district, makes parole evidence of less value here than in most countries."[108] It is possible that this "in-exactitude" can be read as flexibility, which the Ceylonese introduced into a system which was in theory quite rigid and impartial and thus lacking in discretionary power. We can see that the formal judicial processes were distorted and appropriated to such an extent that they approximated extra-judicial forms of private restitution.[109]

---

106 For a twentieth-century American example of this same phenomenon, see Merry, *Getting Justice*, p. 176.

107 "Reports by subordinate officers – a dangerous practice." *The Ceylon Examiner*, June 27, 1893; "Crime and bribery in the southern province." *The Ceylon Examiner*, May 1, 1894.

108 *ARC*, 1875, p. 11.

109 There is evidence that courts have often been used in this manner. To give two widely disparate examples, McDonough found that in medieval Marseille, "witnesses made independent judgements of their neighbours' behaviours that often had little to do with the law," and Jakala and Jeffrey argue that in post-conflict Bosnia-Herzegovina individuals had differing interpretations of what constituted justice and challenged the construction of the law. McDonough, S.A. 2013. *Witnesses, Neighbors, and Community in Late Medieval Marseille*. New York: Palgrave Macmillan, p. 4; Jeffrey, A. and M. Jakala. 2015. "Using courts to build states: the competing spaces of citizenship in transitional justice programmes." *Political Geography*, Vol. 47, p. 51.

Conclusion

It should be abundantly clear from the above that uniformity and certainty, the two central pillars of the rule of law as laid out by Macaulay, did not exist in Ceylon. Although the British continued to maintain that the gift of British justice was a key tenet of their right to rule a foreign people who had hitherto been denied justice by their own rulers, this claim was seriously compromised by the "rule of racial difference" that unofficially applied the law unevenly and unfairly. This blatant inequality revealed just how dysfunctional and fragile the colonial court system was. Some failings can be attributed to untrained and incompetent judges, but others stemmed from both active and passive resistance to the judicial system by the Ceylonese. This resistance induced significant adjustments by the government and, one can argue, therefore, that resistance was constitutive of the system of justice. When the formal dictates of law clashed with Ceylonese understandings and methods of achieving justice and restitution, the legal system was compromised and adjusted. By making the court system to some degree "their own," a significant portion of the Ceylonese male population reproduced, and was reproduced by, that system. While they had the power to undermine the system that oppressed them, they could not overthrow it. Consequently, their resistance normalised and perpetuated the system, but in a weakened and more flexible form. Their embrace of the court system, one is tempted to suggest, was closer to the enactment of mockery than mimicry. The British often felt their hands were tied, as the Ceylonese played the system to their advantage, disrupting its authority and undermining it. The Ceylonese use of the court system "domesticated" it, as Judith Butler would say.[110]

While the much-vaunted rule of law offered some protection to Ceylonese life and property, it also oppressed the peasantry by failing to protect common (*chena*) lands from expropriation or to offer much protection against violent attacks by Europeans. Having said that, the court system also unintentionally created a space that could be re-appropriated by the Ceylonese and used to punish their enemies and reward friends. The British deplored the way in which the Ceylonese resisted judicial norms, and they decried their seeming corruption while overlooking their own failure to prosecute European violent crime against the Ceylonese and by all too often making decisions based on expediency rather than justice.

As we have seen, resistance to the courts was a form of negotiation rather than rejection.[111] The system was used heavily, but in ways that were not prescribed by the British. As DeCerteau says the peasants were

---

110 In making similar arguments, Butler refuses the notion of "the subject derived from some classical-humanist formulation whose agency is always and only opposed to power." Butler, *The Psychic Life of Power*, p. 17.

111 On resistance as negotiation rather than negation see Chandra, U. 2015. "Rethinking subaltern resistance." *Journal of Contemporary Asia*, Vol. 45, No. 4, pp. 563–73.

[s]ubmissive and even consenting to their subjection ... [they] ... nevertheless often made of the rituals, representations, and laws imposed on them something quite different from what their conquerors had in mind; they subverted them not only by rejecting or altering them, but by using them with respect to ends and references foreign to the system they had no choice but to accept.[112]

In this sense the system remained intact, while its power was continually blunted. Resistance, as Sharp et al. say, can "reinforce rather than dismantle" a power structure.[113] In their discussion of domination and resistance as always highly entangled and mutually constituted, Sharp et al. point out that whereas "domination is constrained, modified, and conditioned" by the resistance of oppressed groups, "resisting power is constantly in danger of replicating the structures of the dominant." Although the colonial courts were strained financially, and clearly manipulated by the Ceylonese public and lower level bureaucrats, the heavy use of a state structure set up by the British reinforced its place as a primary point of contact between British officialdom and the larger Ceylonese society.

112 De Certeau, M. 1984. *The Practice of Everyday Life.* Berkeley: University of California Press, p. xiii.
113 Sharp, J., P. Routledge, C. Philo and R. Paddison, eds. 2000. "Introduction," in *Entanglements of Power: Geographies of Domination and Resistance.* New York: Routledge, p. 23.

Part IV

The prison and the arts of dark biopower

7 Creating spaces of deterrence

Towards the reformed prison in Britain

Prior to the sixteenth century, prisons in Britain were primarily places to hold people who had been arrested until they were brought to trial. Upon conviction, punishments were normally corporal. Criminals were flogged, branded, put in pillories, transported or executed. In the seventeenth century, however, a number of European countries began to imprison those thought to be guilty of lesser crimes, such as debtors, juvenile offenders and certain types of felons. People were incarcerated for a wide variety of violations of the social order. Socially marginal people were housed together: men and women, the sane and the insane, those awaiting trial and those serving sentences. During the eighteenth century, the impact of enlightenment ideas resulted in gentler visions of deterrence and reform. As O'Brien put it, at the

> core of the political agenda of the modern state was the need to devise new ways of punishing subjects no longer bound by traditional restraints, an impetus which could be found across the globe as old orders collapsed to make place for new modes of governance.[1]

In 1777 the social reformer John Howard published a critical report on the prison system in Britain, advocating that prisoners spend time alone reflecting on their crimes.[2] Howard's ideas were incorporated into the 1779 Penitentiary Act. Jeremy Bentham built upon Howard's ideas, arguing for the separation of different types of prisoners. Debtors and those awaiting trial should be isolated from those sentenced to short terms of imprisonment, who in turn should be separated from those sentenced to longer terms in the interests of avoiding moral contamination. The best known of Bentham's ideas on prison reform is the panopticon, a system whereby the guards would be able to see all the prisoners without the prisoners knowing when they were being watched. The first prison built on Benthamite principles was

1 O'Brien, P. 1996. "Prison reform in France and other European countries in the nineteenth century," in Finzsch, N. and R. Jutte (eds.), *Institutions of Confinement: Hospitals, Asylums and Prisons in Western Europe and North America, 1500–1950*. Cambridge: Cambridge University Press, p. 298.
2 Howard, J. 1784. *The State of Prisons in England and Wales*. London: William Eyres.

Millbank Prison in London, completed in 1821. Much to Bentham's chagrin, it was only his spatial design that was incorporated into prisons and even then, no true panopticon was ever built.[3] For him, the panopticon was meant to be only one part of a broader system of punishments. Drawing upon utilitarian ideas, Bentham outlined a plan for prisons as privately run factories that would employ convicts to efficiently produce goods.[4] While Bentham had campaigned relentlessly for this system, by 1810 he gave up, defeated by other reformers who argued that his system paid insufficient attention to religious instruction and reform.

A major change in the English prison system, however, came in 1835 when the "silent system," developed in Auburn Prison in New York, was introduced. Under this system, prisoners worked together in silence during the day. This was followed in the next decade by the "separate system," which went much further than Howard's original idea. This entailed solitary confinement and had been famously instituted in Pennsylvania at Eastern State Penitentiary in 1829. The system became dominant in Britain in more and less strictly controlled forms during the second half of the nineteenth century. The isolation of prisoners was seen to have the advantage of separating them from the sinful influence of others and providing a therapeutic environment where total control could be exerted by the authorities. In 1842, Pentonville prison was the second major prison in Britain constructed around the ideas of Bentham's panopticon as well as separation and became a model for modern prisons in Britain and the colonies.

The nineteenth-century prison was a veritable monument to instrumental rationality in that everything including the length of sentences, the architectural design and the organisation of daily life was quantified.[5] Arguably, the nineteenth century British prison was the preeminent site of the calculability of space and time in society at the time. The prevailing theory of disease was miasmatic, the now obsolete idea that many diseases are caused by bad air emanating from decaying material. So it was decided that prison cells should be exactly thirteen feet long, seven feet wide and nine feet high, as this would provide the optimum of air quality. Inmates were to work in their cells doing repetitive tasks which would instil a work ethic and cause them to reflect upon their misspent lives. As Ignatieff put it, the penitentiary was conceived as "a machine for the social production of guilt," and the jailers were to be its

---

3 Mitchell argues that the panoptical model of the prison was a colonial invention put in practice in places such as South Asia and the borders of the Ottoman Empire. Mitchell, T. 1991. *Colonising Egypt*. Berkeley: University of California Press, p. 35. I am grateful to David Nally for pointing this out to me.

4 His vision in this regard had to wait for the creation of privately run prisons in twenty-first century America where prisons are profitable thanks to a combination of cheap prison labour and state funding. See Morin, K.M. 2018. *Carceral Spaces, Prisoners and Animals*. London: Routledge.

5 Wener, R.E. 2012. *The Environmental Psychology of Prisons and Jails: Creating Humane Spaces in Secure Settings*. Cambridge: Cambridge University Press, p. 20.

technicians.[6] Through the application of enlightenment rationality, the penitentiary "represented the apotheosis of the idea that a totally controlled environment could produce a reformed and autonomous individual."[7]

Another major innovation was made by Josiah Jebb, an engineer and the Director of Prisons in England. He instituted a three-stage sentence for prisoners: the first stage was solitary confinement, the second was extramural hard labour on public works and the third was conditional release. This model became the basis of the prison system in Ceylon and other colonies. The reformist Jebb was criticised as being too lenient and after his death was replaced by Sir Edward Du Cane who favoured harsh deterrence over rehabilitation, and by his own admission, sought to produce "salutary terror."[8]

There were several reasons for the decline of the reformatory impulse by the early 1860s. The first was the perceived failure of the current methods of reforming prisoners.[9] The second was the passing of the Penal Servitude Acts that ended the transportation of criminals to Australia, thus greatly increasing the number of criminals imprisoned in Britain for serious crimes.[10] And the third was a panic about increased crime rates. In light of these pressures, the Carnarvon Committee of the House of Lords recommended that prisons should be as harsh and deterrent as possible through highly regulated regimes of hard, useless labour, limited food and the spatial separation of prisoners. The latter was put in place, not so much on reformative grounds, but because it was considered more punitive.[11]

Dark biopower in the colonial prison

Foucault posited a discontinuity between eighteenth century, pre-modern forms of punishment based upon physical torture and new nineteenth century, post-enlightenment forms of discipline which sought rational ways to reform criminals. The idea was to shift emphasis from the body to the mind.[12] But, as we

---

6 Ignatieff, M. 1978. *A Just Measure of Pain: The Penitentiary in the Industrial Revolution 1750–1850.* London: Penguin, p. 213.

7 McGowen, R. 1998. "The well-ordered prison: England 1780–1865," in Morris, N. and D. Rothman (eds.), *The Oxford History of the Prison: The Practice of Punishment in Western Society.* New York: Oxford University Press, p. 92.

8 Priestley, P. 1985. *Victorian Prison Lives: English Prison Biography 1830–1914.* New York: Methuen, p. 6.

9 Ignatieff, *A Just Measure of Pain*, p. 200.

10 Wiener, M.J. 1990. *Reconstructing the Criminal: Culture, Law and Policy in England, 1830–1914.* Cambridge: Cambridge University Press, p. 141.

11 McConville, S. 1981. *A History of English Prison Administration. Volume 1. 1750–1877.* London: Routledge and Kegan Paul, pp. 347–54, 406; Evans, R. 1982. *The Fabrication of Virtue: English Prison Architecture, 1750–1840.* Cambridge: Cambridge University Press, pp. 391–93; Davie, N. 2005. *Tracing the Criminal: The Rise of Scientific Criminology in Britain, 1860–1918.* Oxford: Bardwell, p. 52.

12 Foucault was well aware that the reformation of the mind as the Manichean alternative to the mortification of the flesh was difficult to implement in Europe. As Moran and Morin's fine reviews have pointed out, the studies of carceral geographies have been greatly influenced by the theorising of Foucault, Agamben, and Goffman. See Moran, D., N. Gill and D. Conlon, eds. 2013. *Carceral Spaces: Mobility and Agency in Imprisonment and Migrant Detention.* Farnham: Ashgate; Moran,

shall see, from the beginning of British prisons in South Asia, moral reform of prisoners was considered unlikely. For as Arnold expressed it, the British believed that "the body of the 'oriental' might be disciplined, but his 'soul' remained out of reach."[13] And so they set out to discipline the body but, as we shall see, they struggled with the dissonance between modern penal theories and their own practices.

In her discussion of liberalism, Marina Valverde argues that in the practice of governing, technologies are used that don't necessarily express or implement liberal rationality, as much as constitute it. She writes, "although technologies are either logically or as a matter of historical practice, associated with certain rationalities, there is no one-to-one relation between the two."[14] Despite paying lip service to European ideas of penal reform based on liberal theories such as those of Beccaria and Bentham, who argued against cruelty and retribution and in favour of impartial, lawful, proportional punishments, colonial prison officials in practice opted for forms of deterrence considered to be harsh enough to control colonial subjects. Even Beccaria believed that some degree of cruelty was necessary to prevent crime and thus promote freedom for the rest of society, and this may have been amplified in a colonial context.

As Sherman points out, pre-enlightenment corporal punishments like flogging were modified so as to be scientifically administered under the care of a prison doctor, placing officials in a liminal position between penal philosophies.[15] However, once the idea of rehabilitation was effectively abandoned both in Britain and the colonies around the middle of the nineteenth century, the only

---

D. 2015. *Carceral Geography: Spaces and Practices of Incarceration.* Farnham: Ashgate. Morin, K.M. and D. Moran, eds. 2015. "Introduction," in *Historical Geographies of Prisons: Unlocking the Usable Carceral Past.* London: Routledge, pp. 15–32; Foucault, M. 1979. *Discipline and Punish: The Birth of the Prison.* New York: Vintage; Agamben, G. 1998. *Homo Sacer: Sovereign Power and Bare Life.* Stanford: Stanford University Press; Goffman, E. 1961. *Asylums: Essays on the Social Situation of Mental Patients and other Inmates.* Garden City: Anchor. But all three theorists, when applied to the study of prisons, share some common drawbacks. Each create ideal types which are meant to be suggestive of the general contours of institutions, but are less satisfactory guides to a close reading of them. Hence, one would not expect any given prison to show more than a family resemblance to the ideal types implied by these theorists.

13 Arnold, D. 1994. "The colonial prison: power, knowledge and penology in nineteenth century India," in Arnold, D. and D. Hardiman (eds.), *Subaltern Studies VIII: Essays in Honour of Ranajit Guha.* Delhi: Oxford University Press, p. 175.

14 Valverde, M. 1996. "'Despotism' and ethical liberal governance." *International Journal of Human Resource Management,* Vol. 25, No. 3, p. 358.

15 Sherman, T.C. 2009. "Tensions of colonial punishment: perspectives on recent developments in the study of coercive networks in Asia, Africa and the Caribbean." *History Compass,* Vol. 7, No. 3, pp. 659–77; Pete, S. and A. Deven. 2005. "Flogging, fear and food: punishment and race in colonial Natal." *Journal of Southern African Studies,* Vol. 31, No. 1, March, pp. 3–21. Of course, flogging was a form of "officially condoned" torture and whereas its justification was debated, it continued to be used throughout the nineteenth century. While there was a good deal of support for flogging both within and particularly outside government circles, pressure to cut back on it came from Europe where it was rarely used in prisons. Priestley, *Victorian Prison Lives,* p. 217.

alternative considered was the creation of a prison environment so harsh that criminals would be deterred from committing crimes.[16] While the proximate target of the methods of deterrence was the individual prisoner, the whole sub-population of criminals and potential criminals outside the prisons was the main target.[17] While it was hoped that the harsh disciplinary mechanisms of the prison might instil in prisoners a sense of obedience, making them self-disciplined or at the very least docile, the prison authorities thought that it was much more realistic to hope that the sub-population of the criminally inclined, upon hearing the details of harsh prison life, would think twice before committing crimes.[18]

And so, as the Ceylonese, as a race, were thought to love ease, prisoners were made to work long hours; as they were thought to love sociability, they were made to remain silent, and as they were unaccustomed to a regimented life, they were made to follow the strictest of daily schedules. Prison authorities devised highly calibrated, often cruel, punishments based on the utilitarian principle of the denial of pleasure as an object lesson.

Foucault identified the key principles underpinning the nineteenth-century penitentiary as: total control over time and space,[19] the organisation of space to control visibility, the removal of the individual from the environment of crime, and the creation of systems of behavioural accountancy directed by jailers and doctors, who produced records of individuals' labour, diet and health.[20] Despite highly asymmetrical power relations between prisoners and jailers, the former were both able to systematically undermine such goals. As Valverde points out, Foucault treated governing practices in a radically anti-functionalist manner, as "ever-changing, contingent, site specific, pragmatically put together collections of governing techniques whose success or failure depends on their usefulness

---

16 McConville, *A History of English Prison Administration*, p. 347.

17 Nally, D. 2011. *Human Encumbrances: Political Violence and the Great Irish Famine*. South Bend: University of Notre Dame Press has similarly argued that the poor laws in Ireland, while ostensibly targeting paupers, sought to rehabilitate an entire class of small farmers.

18 Garland, D. 1985. *Punishment and Welfare: A History of Penal Strategies*. Aldershot: Gower, p. 12. Anderson et al. argue that transportation likewise targeted populations through fear. Anderson, C., C.M. Crockett, C.G. de Vito, T. Miyamoto, K. Moss, K. Roscoe and M. Sakata. 2015. "Punishment space and place, c. 1750–1900," in Morin, K.M. and D. Moran (eds.), *Historical Geographies of Prisons: Unlocking the Usable Carceral Past*. London: Routledge, pp. 206–33.

19 As Foucault put it, "In the first instance, discipline proceeds from the distribution of individuals in space." Foucault, M. 1979. *Discipline and Punish: The Birth of the Prison*, New York: Vintage, p. 141.

20 Foucault, *Discipline and Punish*, pp. 184–93. Also see Driver, F. 1993. *Power and Pauperism: The Workhouse System, 1834–1884*. Cambridge: Cambridge University Press. For an early critique of Foucault's view of prisons see Garland, *Punishment and Welfare*. For an examination of Garland's and Foucault's views in light of the Irish penal system see Carroll-Burke, P. 2000. *Colonial Discipline: The Making of the Irish Convict System*. Dublin: Four Courts Press. For a discussion of Foucault's temporal assumptions regarding types of discipline in light of Indian data see Kaplan, M. 1995. "Panopticon in Poona: an essay on Foucault and colonialism." *Cultural Anthropology*, Vol. 10, pp. 85–98.

not to society, but rather to contenders in particular battles or struggles."[21] This is an apt description of the workings of the prison system in colonial Ceylon, where micro-battles over time and space flared continually.

The rule of law, to the extent that it informed penal practice, required that prison punishments for any individual should not exceed what was handed down in his sentence.[22] For example, if a sentence was a year's imprisonment with hard labour and little food, the punishment was not to be of such intensity that it could kill a prisoner or ruin his health.[23] In practice this meant that hard labour and diet were used to punish the body as severely as possible without doing permanent damage. There were no studies on the long term health of ex-prisoners, instead, the metric became merely to minimise the number of prisoners who fell ill or died while incarcerated. This criterion for monitoring punishment was seen as both lawful and modern, based as it was on the new scientific model of the body as a machine. Science was used to guide prison officials in quantifying exactly how much punishment a body could withstand. And so, prison officials tapped into burgeoning research on energy and the human body.[24] Based upon the recently formulated theory of thermodynamics, techniques were sought to measure the expenditure of energy by the body with the goal of making workers more productive. Such a research programme drew upon new theories of nutrition and physiology. The tools of science were brought to bear on these questions and, beginning in the 1870s, increasingly sophisticated studies of motion and fatigue were conducted in Europe. The results were quantitative measures of energy expended in foot-tons, which

21 Valverde, M. 2008. "Police, sovereignty and law. Foucauldian reflections," in Dubber, M. and M. Valverde (eds.), *Police and the Liberal State*. Stanford: Stanford University Press, p. 17.

22 Sherman ("Tensions of colonial punishment") argues that rule of law was not followed in prisons. I will show, as I have with the police and the courts, that while rule of law was often violated, it served to somewhat constrain and set limits upon the free exercise of arbitrary violence. The degree to which it did so varied from place to place, but it was always present, if only as a potential threat to those who violated it.

23 In practice, however, the Government accepted a higher death rate in prisons like the Mahara Quarry because they needed stone for government building projects. Likewise, the Government of India tolerated very high death rates in the penal colony of the Andamans because prisoners were needed to clear malarial jungles and swamps. Sen, S. 2000. *Disciplining Punishment: Colonialism and Convict Society in the Andaman Islands*. New Delhi: Oxford University Press, p. 123. In both cases the governments were making biopolitical choices that prison labour was more expendable than free labour. But some, like Inspector General F.R. Ellis, believed that the rule of law was secondary to the goal of deterrence. He wrote that it is "absolutely necessary that at some risk to health, and even life, jail should be made distasteful to all without exception, and as long as this is done, the danger of high sick rates must remain." He continues, a "cause of the high sick rate is the necessity of enforcing the performance of the appointed task." *Administration Reports, Ceylon (ARC)*, 1893, C10.

24 Kamminga, H. and A. Cunningham. 1995. "Introduction," in Kamminga, H. and A. Cunningham (eds.), *The Science and Culture of Nutrition, 1840–1940*. Amsterdam: Rodopi, p. 1. For a discussion of the role of science in informing nineteenth-century Irish penological practice, see Carroll-Burke, *Colonial Discipline*, pp. 179–231.

allowed employers there to calculate the body's precise capacity for labour. These ideas were then picked up by prison administrators in Britain and Ceylon who used them in devising regimes of punishment.

Surveys of the food habits of the poor in Europe had begun earlier in the nineteenth century. After mid-century scientists began studying factory workers in Europe to systematically establish how much food was necessary for survival and how it could be provided cheaply. To this end, scientific standards of bare subsistence were established.[25] This growing body of information was used not only to make free labour more productive, but also to provide a rational basis for the design of deterrent prison and workhouse environments. The strategies used to make workers more productive through scientifically managing workloads and nutrition can be understood as a biopolitical technology, or what Dean describes as the dark side of biopower.[26] In prisons, such finely calibrated knowledge of the body was drawn on to increase the production of suffering.

By the last decades of the nineteenth century, a large body of knowledge about the relationship between energy and the body had been gained, but it was mainly based on studies of European bodies and so it was not clear how well it applied to "native" bodies in tropical environments. Colonial jailers and prison doctors from the mid-nineteenth century on sought to answer this question, albeit in an amateur and dangerous manner.[27] As a total institution, the prison was deemed a perfect place to test new nutritional and workload theories. Firstly, test conditions should in theory be controlled so that investigators could determine exactly how much energy prisoners expended and how much food they consumed.[28] Secondly, prisoners were considered appropriate subjects for such experiments; although their lives were seen as more than "bare life," in that unintended sickness or death of prisoners were to be officially accounted for, prisoners were considered neither deserving nor valuable to society.[29]

Prison officials, as we shall see, wished to discover how far "native" bodies could be pushed before they became too unwell to work. Prison doctors helped to design experiments and then monitored them to ensure that they were not too dangerous. Prison doctors became technicians of the body expected to know how to administer

---

25 Rabinbach, A. 1992. *The Human Motor: Energy, Fatigue and the Origins of Modernity*. Berkeley: University of California Press; Milles, D. 1995. "Working capacity and calorie consumption: the history of rational physical economy," in Kamminga, H. and A. Cunningham (eds.), *The Science and Culture of Nutrition, 1840–1940*. Amsterdam: Rodopi, p. 82.

26 Dean, M. 2010. *Governmentality: Power and Rule in Modern Society*. 2nd edition. Los Angeles: Sage, p. 164.

27 Research was conducted in Indian prisons on the amount of food needed to keep prisoners in health. It was estimated that one and a half pounds of grain per day was sufficient if supplemented by vegetables and some fish or meat. Although these figures were based upon sound medical research at the time they were often disputed and undercut by officials who worried about pampering prisoners, encouraging the hungry poor to flock to prisons, especially during times of famine, or simply the cost of feeding prisoners. See Arnold, D. 1994. "The 'discovery' of malnutrition and diet in colonial India." *The Indian Economic and Social History Review*, Vol. 31, No. 1, pp. 1–26.

28 On the notion of a total institution see Goffman, *Asylums*.

29 Agamben, *Homo Sacer*.

lawful suffering and also how to prevent unlawful degrees of cruelty. It was here that dark biopower strove to bring the formerly "private concerns of the body to public control."[30] There was an increasing tension as the century progressed between penal practices designed to humiliate and punish the body, and medical imperatives to manage the care of the body.[31] This dynamic between state-sanctioned violence and pastoral biopower played out slightly differently in India than Ceylon, for in the former the norm was for a medical officer to be head of prisons, and in the latter there was continual tension between administrators and prison doctors. The struggles around scientifically calculated punishment were compounded by pressures from evangelical and humanitarian groups in Britain. However, the most effective forms of resistance to harsh penal technologies were from the prisoners themselves who employed classic practices of resistance such as foot dragging to reduce labour quotas, scrounging food to supplement their diet, threatening and bribing guards not to enforce regulations, and enlisting the support of doctors when punishments were especially onerous. As Arnold put it, the spaces of the prison were colonised by "other, unofficial networks of power and knowledge than those represented by formal prison authority."[32]

Although doctors routinely protected the prisoners from being worked or starved to death, in some instances prisoners were so weakened by the harsh regime that they died from these and other causes. When the death rates became too high, prison authorities had to explain themselves to the government. The acceptable death rate in a colonial prison was expected to lie within a statistically determined normal range; what Foucault has termed "a bandwidth of the acceptable that must not be exceeded."[33] While senior prison officials usually followed their prescriptions, subordinate jailers flouted them on a routine basis.

The order to make prisons in Ceylon more deterrent

In light of the reformed prison movement in England, in 1863 the Duke of Newcastle sent circular dispatches to the governors of all colonies presenting them with the findings of both the Committee of the House of Lords on the State of Discipline in Gaols, and the Report and Evidence presented by the Royal Commission on Penal Servitude. However, within two years, it was clear that the report had been widely ignored across the empire. Consequently, it was followed in 1865 by another dispatch urging that all colonial prisons be brought in line with prisons in Britain.[34] This was influenced in part by the liberal tenet that the criminal justice system should be rational and uniform throughout the

---

30 Rejali, D. 1994. *Torture and Modernity: Self, Society and State in Modern Iran.* Boulder: Westview Press, p. 63.
31 Sherman, T.C. 2010. *State Violence and Punishment in India.* London: Routledge, p. 6.
32 Arnold, "The colonial prison," p. 155.
33 Foucault, 2007. *Security, Territory, Population,* p. 21.
34 Circular Dispatch from the Right Hon'ble the Secretary of State for the Colonies, 16th January 1865, *SP,* 1866, p. 73.

Empire.[35] The circular requested information on the state of prisons in Ceylon and provided a set of guidelines for their reform. Underpinning these guidelines was the belief that suffering was character building and, following the utilitarian ideology, that severe punishment would teach prisoners impulse control, the value of deferred gratification and long-run consequences of one's actions.[36] The separate system was thought to stop the "moral contamination" of new prisoners by hardened habitual criminals. Prison sentences were also to be divided into three stages calculated to further produce self-control and a consideration for long-run consequences. The first stage was meant to ensure "severe suffering in the early portion of the sentence" as it was believed that this would have the greatest deterrent effect on "the class of persons by whom offences are generally committed [as they] do not look far forward, and they are governed by what is presently, and not by what is distantly, within their view."[37] Kyla Schuller's study of nineteenth-century biopower through "technologies of feeling" is interesting in this regard. She sees the dark side of biopower as the management of bodies, which at that time were seen to be differentially malleable and differentially impressible depending upon race and degree of civilisation. Tropical races were believed to be less impressible, or mouldable, and thus less able to respond to discipline except in the harshest form. She also argues that much of nineteenth-century theory was Lamarkian in its concern with bad habits acquired through contagion and improved habits as the bodily response to harsh deterrent environments.[38] All of these ideas are evident in these directives from the Secretary of State for the Colonies.

The Committee of the House of Lords laid out specific guidelines on how to implement the separate system and the three stages of sentences in colonial prisons.[39] The British were concerned that non-Europeans might not value freedom as much as did Europeans and that imprisonment without corporal punishment would be insufficient.[40] Consequently, spatial separation, rigorous hard labour and semi-starvation diet were to be the new biopolitical rationale targeting the criminal sub-population.[41] Below I outline the struggles of the Ceylon

---

35 Wiener, *Reconstructing the Criminal*, p. 61.

36 *Ibid.*, p. 111.

37 *Ceylon Sessional Papers (SP)*, 1866, pp. 73–74.

38 Schuller, K. 2018. *The Biopolitics of Feeling: Race, Sex and Science in the Nineteenth Century.* Durham: Duke University Press.

39 Committee of the House of Lords on the State of Discipline in Gaols, and the Report and Evidence presented by the Royal Commission on Penal Servitude, in Digest and Summary of Information Respecting Prisons in the Colonies, supplied by the Governors of Her Majesty's Colonial Possessions in Answer to Mr. Secretary Cardwell's Circular Despatches of the 16th and 17th January 1865. Presented to both Houses of Parliament by Command of Her Majesty. Vol. LVII, Pt. 2, p. 84. London: George Edward Eyre and William Spottiswoode, 1867.

40 The same concern was expressed in Africa. Bernault, F. 2007. "The shadow of rule: colonial power and modern punishment in Africa," in Dikotter, F. and I. Brown (eds.), *Cultures of Confinement: A History of the Prison in Africa, Asia and Latin America.* London: Hurst, p. 74.

41 DuCane, E. 1885. *The Punishment and Prevention of Crime.* London: Macmillan, pp. 1–2.

Government to put these guidelines in place and the manner in which both the guards and prisoners persistently resisted all new penal practices. But, first, I will describe the state of Ceylon prisons before the reforms.

Ceylon prisons before the reforms

In response to the Duke of Newcastle's letter to the colonies in 1863 urging an inquiry into prison conditions, the Government of Ceylon conducted a survey on the state of jails covering everything from the spatial layout of jails, separation of prisoners, their dietary, type and extent of labour, illness and punishment for prison offences. There were at the time twenty-eight jails in Ceylon, including the central penitentiary, Welikada in Colombo.[42]

As directed by the Secretary of State for the Colonies in January 1866, Governor Robinson appointed a Committee on Prison Discipline to suggest reforms to the Ceylon prisons. Drawing on the 1863 survey, the Committee summarised their conclusions as follows: "[t]he great and glaring evil of the present state of things is the deficiency of classification ... mere lads and grown men, criminals of every shade of guilt herd together at night and work or loiter about together by day."[43] The Committee added that separation and silence were not even attempted during the day and thus the system, as it was actually practiced, was a "source of moral contamination."[44]

The Committee found that the food provided in jail was "too good and plentiful ... at present the superiority in quality and quantity of the food in the Jail over what many of them get out of Jail is a positive inducement to the commission of offences."[45] By way of example, the Committee reported that at Caltura Jail the authorities went so far as to give prisoners money daily and allowed them to go to the local bazaar to purchase their own food.[46] The prison diets were clearly healthier before the reforms of the 1860s. In particular, vegetables which had been a regular part of the prison diet were excluded after the reforms.[47]

---

42 *Ceylon Blue Book*, 1863. Return the Gaols and Houses of Correction. For India see Anderson, C. and D. Arnold. 2007. "Envisioning the colonial prison," in Dikotter, F. and I. Brown (eds.), *Cultures of Confinement: A History of the Prison in Africa, Asia and Latin America*. London: Hurst, p. 311.

43 *SP*, 1866, p. 66.

44 *Ibid.*

45 *Ibid.*

46 *Ibid.* This was the custom in Indian prisons until the early 1840. When the British stopped it, there were riots both among prisoners and in some cases among towns people as well over the caste implications of common messing. Singha, R. 1998. *A Despotism of Law: Crime and Justice in Early Colonial India*. New Delhi: Oxford University Press, pp. 229–84; Yang, A.A. 1987. "Disciplining 'natives': prisons and prisoners in early nineteenth century India." *South Asia*, Vol. 10, No. 2, p. 34.

47 It is interesting to note that troops stationed in Ceylon at the time were also given virtually no vegetables, although it was known at the time that they were important to the maintenance of health. Military Sanitary Reports, Ceylon 1864. *SP*, 1864, p. 207.

The Committee on Prison Discipline reported that the existing system of discipline in Ceylon prisons was lax and applied unevenly. Most prisoners were employed outside the prison walls doing light labour such as sweeping walks and weeding official residences.[48] Furthermore, bribery was rife; "We fear there is too much reason for the general belief, that a Criminal who is wealthy or who has wealthy and influential friends, never does a hard day's work, whatever be the term for which he is sentenced."[49] A part of the problem, they reported, were guards whose pay was so low that they were forced to take bribes.[50]

In terms of health, the Committee admitted that sewerage and drainage were inadequate, as were the latrines, bathing facilities and hospital facilities. The Committee also recorded that in their view ventilation was poor, which would have been especially worrying to the authorities given the prevailing miasmatic conception of disease.[51] Improvement in sanitation was emerging as a primary security concern in the management of the poor in Britain and the colonies, and represents the widening exercise of biopower in the mid-nineteenth century.

Having catalogued this litany of shortcomings, the Committee agreed to follow the instructions of the Secretary of State and implement the British model.[52] But they made it clear that the colony must give

> due heed to questions of expense and to local points of importance as to climate, as to diet, as to physical capacity for work, and as to the moral and intellectual characteristics of the mass of the inhabitants of this Island.[53]

In short, the Committee thought that the English model of penology should be modified, not only with reference to Ceylon's financial resources, but because of the tropical climate and racial differences.[54] The new sciences of race and

---

48 During the 1860s all labour both inside and outside of prisons was either industrial or public works. *CBB*, 1862, "Return of the Gaols and Houses of Correction."

49 *SP*, 1866, p. 66.

50 *Ibid.*, p. 62.

51 *Ibid.*, p. 66.

52 By the mid-nineteenth century most colonial officials accepted the principles of Benthamite reform (Arnold, "The colonial prison," p. 164).

53 *SP*, 1866, p. 63.

54 In India cultural, climatic and financial reasons were also given for why English penal practices need not be strictly followed. Arnold, D. 2007. "India: the contested prison," in Dikotter, F. and I. Brown (eds.), *Cultures of Confinement: A History of the Prison in Africa, Asia and Latin America*. London: Hurst, pp. 147–84. Similarly, in Latin America, the penitentiary system was introduced on a tight budget, more with a view to deterrence rather than reform. Aguirre, C. 2007. "Prisons and prisoners in modernising Latin America," in Dikotter, F. and I. Brown (eds.), *Cultures of Confinement: A History of the Prison in Africa, Asia and Latin America*. London: Hurst, pp. 14–54. During the nineteenth century French colonial Vietnamese prisons made little attempt to institute European theories of prison reform. They were notoriously violent and corrupt institutions. See Zinoman, P. 2001. *The Colonial Bastille: A History of Imprisonment in Vietnam, 1862–1940*. Berkeley: University of California Press.

climate should, in their view, take precedence over theories of incarceration solely based upon the English experience. For example, the Committee took issue with the recommendation that long sentences were more efficacious than short ones. It argued that, unlike in England, in Ceylon a short, sharp sentence would normally suffice, because the Ceylonese are "far more timid, far less recalcitrant against coercion, and far less obdurate than Europeans are."[55] No doubt a second unacknowledged reason for short sentences was economic. As I will show, the government tried to spend as little as possible on prisons, and when pressed by the Colonial Office to increase expenditure on improvements, they were often resistant, albeit on grounds that were not explicitly financial.

The quest to shame prisoners

A basic problem that worried prison administrators was their lack of knowledge concerning the cultural attitudes of the Ceylonese to incarceration. Was freedom valued by "natives?" Was prison a welcome relief from family responsibilities for the very poor? Was jail food an attractive alternative to hunger? There was a persistent anxiety around the assumption that prisoners appeared to feel no shame, guilt and importantly no social stigma at having served time in prison.[56] Was it because so many innocent men were imprisoned on false charges? As L. Lee, the Acting District Judge at Matara wrote,

> the uncertainty of result—the effect of inquiry at a distance from the scene —has led to the absence of that shame and social degradation which else- where accompanies a convicted criminal. Punishment for crime is of little deterrent effect, if the conviction does not carry with it social debasement. The released criminal in Ceylon returns to his previous social position and privileges, and is not thought the less of that he was convicted and pun- ished for theft. I attribute this principally to the want of education, but in no small degree to the large element of uncertainty which is present in every case, whether civil or criminal. This element can only be eliminated by local investigation.[57]

The British administrators in Ceylon decided that if incarceration itself was not enough to produce shame, then they must find other methods of shaming pris- oners. One measure adopted in 1871 was to cut short the hair of all prisoners upon admission. This horrified Sinhalese prisoners, who traditionally wore their

---

55 *SP*, 1866, p. 66.
56 Ahmed, S. 2004. *The Cultural Politics of Emotion*. Edinburgh: Edinburgh University Press argues that emotions are cultural practices, which if understood, can be manipulated to the advantage of those in power within an institution such as a prison. The British administrators often expressed frustration at their lack of knowledge about how to predict and gauge Ceylonese emotional reac- tions to their disciplinary measures.
57 *ARC*, 1874, 50.

hair long.[58] Another was to purposely ignore caste so that there was common messing for all prisoners and everyone without exception had to do latrine duty.[59] This policy was purposely put in place in the hope of humiliating prisoners from the higher castes.[60] While these policies upset the prisoners, especially high-caste Tamils, it appears that ex-prisoners suffered from little shame from having served time and this was undoubtedly due to the widespread belief that the judicial system to consistently failed to produce justice.[61]

The spatial-temporal re-organisation of prison life

A central tenet of the British reformed prison movement was that a rational temporal-spatial organisation at a range of scales was necessary to discipline prisoners. Within the prison, space-time was highly rationalised according to penal theory.[62] The prison itself excluded prisoners from their home communities for specified periods of time. It separated prisoners from each other and strictly regulated their daily activities. Different types of crime were punished by different lengths of sentence and each sentence was divided chronologically into stages, with each stage based on a specific space-time configuration. In Ceylon, as soon as it was considered economically feasible, separate prisons were built to house prisoners at different stages of their sentence. The prisoners were thus enveloped in a kind of Russian egg of nested space-time organisation,

58 Regulations were subsequently put in place that shorn hair had to be cut into three-inch lengths to prevent guards fighting over who got to sell it in the market. As the number of regulations grew, each new one spawned responses from prisoners and guards requiring further regulations. *ARC*, 1880, p. C139.

59 What the British did know was that caste was fundamental to the prisoners' identities and that shame was brought on by failure to abide by caste sanctions. And so they manipulated them into feeling shame and extreme disgust at work that they saw as ritually contaminating to their bodies and their very caste identities, such as cleaning toilets or being forced to eat food prepared by lower caste peons. On the loss of caste through imprisonment see "Abolition of imprisonment for debt of kanganies and labourers," *Ceylon Hansard (CH)*, 1909, p. 15. On the point more generally, see Ahmed, *The Cultural Politics of Emotion*, on the political uses of emotions.

60 *ARC*, 1871, p. 355; 1887, p. C53.

61 Emotions can appear personal and ahistorical and thus difficult not to naturalise. From the point of view of the British the very fact of incarceration should be shameful and stigmatising and should thus act as a deterrent to crime. Incarceration, as they saw it, should act as what Ahmed calls a sign that can "stick" to an identity and "impress" emotionally upon one. Kyla Schuller, *The Biopolitics of Feeling*, also uses the term "impress" or "impressibility" in a related way, but sees it as a specifically nineteenth-century idea that conceptualized different races as differentially "impressible" such that the more modern and civilised, the more impressible. Impressibility was seen as evidence of fitness to participate in liberal democracy. Thus, according to Schuller, the British would have seen the lack of shame as further evidence of not being culturally evolved enough for self-government.

62 See Moran, D. 2012. "Doing time in carceral space: timespace and carceral geography." *Geografiska Annaler B Human Geography*, Vol. 94, No. 4, pp. 305–16 on "timespace" and avoiding the dualism of space and time.

all of which was intended to produce a scientifically managed harsh but lawful penal regime.[63]

Following guidance from England in 1866, the Committee on Prison Discipline agreed to increase the amount of spatial separation within the prisons. As a start, debtors and those awaiting trial were separated from those who were convicted.[64] The Committee, however, didn't accept the recommended figure of 1,200 cubic feet of air per cell based upon the Indian experience, instead initially calling for 1,000 cubic feet[65] and reducing it three years later to 900 cubic feet, arguing

> that the climate of Ceylon makes a smaller cell sufficient for health than is required in India or even in England. We have not the extremely high temperature that prevails for the greater part of the year in India; and above all, we do not have the oppressively hot winds, which are felt at some seasons in that country. Nor is there any season of wintry cold such as that which is experienced in the larger part of India for some months of the year, and which causes buildings there to be constructed with a less amount of opening and ventilation, than is usual in a more equitable climate.[66]

But in fact even the 900 cubic foot standard was largely ignored.

The Committee agreed that hard labour during the first six months, referred to as the penal labour stage, was to be conducted in isolation when possible, and promised that, even when face-to-face interactions could not avoided, prisoners would be prevented from communicating.[67] Shot drill in the prison yard was the form of hard labour favoured.[68] While acknowledging that shot drill had the disadvantage of allowing prisoners to see each other, they promised that no prisoner would be allowed to communicate or come within six yards of another while performing the task.[69]

The Commissioners cautioned that while separation was possible during the six month penal phase of a sentence, it was impossible during the latter stages

---

63 On the importance of British conceptions of time to the imperial project, see Mawani, R. 2014. "Law as temporality: colonial politics and Indian settlers." *University of California Irvine Law Review*, Vol. 4, No. 1, pp. 70–95.

64 *SP*, 1866, p. 64.

65 *Ibid.*, p. 67. While this was held to be a general rule, allowances were made for climate. For example, it was argued that in Nuwara Eliya, as it was cooler and hence less miasmatic, cells could be smaller than at lower-lying locations.

66 Report of the Prison Discipline Commission of 8th July 1869, p. 37. However, the Military Sanitary Reports for Ceylon in 1864 argued that each soldier needed at least 1,500 cubic feet of air (Military Sanitary Reports, Ceylon 1864. *SP*, 1864, p. 207).

67 In England the penal stage remained nine months until 1899 when it was reduced to six months. Priestley, *Victorian Prison Lives*, p. 38.

68 Shot drill was the carrying of a heavy iron ball back and forth between two stands for hours.

69 *SP*, 1866, p. 69.

of a sentence, due to the nature of the labour that prisoners were required to do. However, the Commissioners offered assurances that all prisoners would be separated at night throughout their entire sentences.[70] These pledges of nocturnal separation and silence while they laboured were in fact rarely honoured.[71]

While the Committee's response to the directives from London in 1866 was to accept the full reformist agenda in theory, it was clear by their subsequent request for a reduction of cell size that they were primarily concerned about the costs of implementing it. The truth of the matter is that the physical state of the prisons in Ceylon was so dismal that there was no possibility of the total separation of prisoners. The Government of Ceylon, which was investing in building an infrastructure to support the lucrative coffee industry, was reluctant to divert money to a major prison building programme. Preventing and punishing crime was a low priority as it appeared to pose no real threat to British rule and violent crime against Europeans was rare.

James Fitzmaurice, the Inspector General of Prisons in 1867, wrote that Welikada Prison in Colombo was the only facility "approaching to what a prison building ought to be"; but even there, he noted, a portion of the building remained unfinished for years.[72] Two years later, £26,000 was allocated to transform it into the country's "central penitentiary," and only in 1873 did it approach the level of spatial organisation the committee had recommended.[73] The prison was divided into six divisions around a main quadrangle: the first and largest section containing 270 cells with shot drill yards and coir-beating compartments for Ceylonese penal stage prisoners.[74] The second section was composed of thirty-eight cells and one small ward for European and Burgher prisoners, who from the beginning were segregated from the other prisoners.[75] Second stage prisoners slept in the third section, which comprised three large wards that could hold sixty-five men. During the day they worked in prison services. The fourth section contained kitchens and the fifth, offices, workshops and an overflow ward for thirty prisoners, and the sixth, the hospital, dead house. After serious attacks by prisoners on guards in 1876, the main quadrangle was subdivided into four yards separated by thirteen-foot walls to

---

70 *Ibid.*, p. 70.

71 In England and Wales many smaller jails were closed because they were unable to afford to implement the separate system. Ogborn, M. 1995. "Discipline, government and law: separate confinement in the prisons of England and Wales, 1830–1877." *Transactions of the Institute of British Geographers*, Vol. 20, pp. 295–311.

72 *ARC*, 1867, p. 259.

73 *ARC*, 1869, p. 146.

74 Coir beating is the pounding of coconut husk into strands of fibre.

75 The same system was followed in India where European prisoners were given separate and better living conditions. Arnold, *The Contested Prison*, p. 174. For a similar system in late-nineteenth century Africa, see Bernault, *The Shadow of Rule*, p. 73. Colonial Office officials often reminded prison officials that there were not supposed to be separate rules for European prisoners, but such warnings had little effect. Dikotter, F. 2004. "'A paradise for rascals': colonialism, punishment and the prison in Hong Kong (1841–1898)." *Crime, History and Societies*, Vol. 8, No. 1, pp. 49–63.

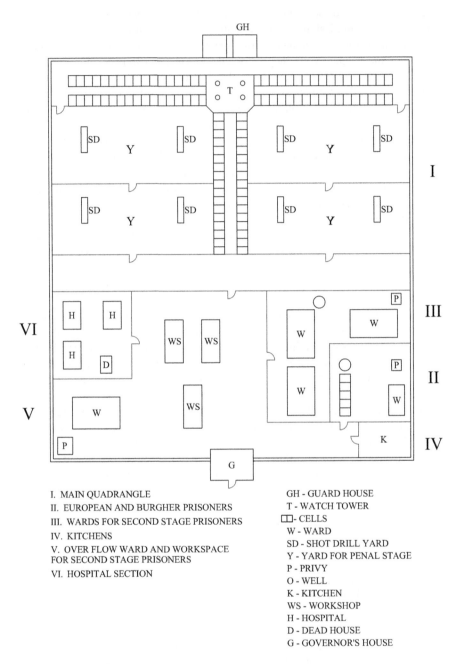

GH

T

SD · Y · SD · SD · Y · SD

I

SD · Y · SD · SD · Y · SD

H · H

WS · WS

W · P

W

III

VI

H

D

W

P

W

II

V

W

WS

W

P

K

IV

G

I. MAIN QUADRANGLE
II. EUROPEAN AND BURGHER PRISONERS
III. WARDS FOR SECOND STAGE PRISONERS
IV. KITCHENS
V. OVER FLOW WARD AND WORKSPACE
FOR SECOND STAGE PRISONERS
VI. HOSPITAL SECTION

GH - GUARD HOUSE
T - WATCH TOWER
⊓⊔- CELLS
W - WARD
SD - SHOT DRILL YARD
Y - YARD FOR PENAL STAGE
P - PRIVY
O - WELL
K - KITCHEN
WS - WORKSHOP
H - HOSPITAL
D - DEAD HOUSE
G - GOVERNOR'S HOUSE

Figure 7.1 Schematic diagram of Welikada Prison in the mid-1870s.
Source: Adapted from diagram in *Ceylon Blue Book*, 1863.

discourage attacks on guards by gangs of prisoners, and in the case of a general uprising, all of the prisoners would be unable to congregate in one place.[76]

Hulfsdorf, Colombo's other prison, was divided into two sections. The first had thirty cells and a ward for 170 male prisoners, and was used exclusively for un-convicted prisoners and debtors. Another section was used for women prisoners, who represented between only one and two percent of the prison population.[77] They were housed separately from men and the only internal separation was between prostitutes and the others, as the prostitutes were considered morally and medically dangerous.[78] The small number of female prisoners and the fact that they gave little trouble to the authorities, as well as the belief that they were not habitual criminals, may account for their near total absence from the archive.[79]

The other major silence in the prison reports concerns relations between Sinhalese, Tamil and Moor prisoners. One would assume that there was little communication between them as few Tamils and Moors spoke Sinhalese; however, if there had been much inter-ethnic conflict, it would presumably have appeared at least once in the reports between 1867 and 1914. This absence is in line with annual police reports showing little inter-ethnic conflict during the nineteenth century.

There is no question that some significant progress in meeting Whitehall's goals had been made by the mid-1870s. F.R. Saunders, the Inspector General, wrote in his annual report that

> no one who has seen our jails, and who can remember what they were, can fail to be struck with the wonderful change that has taken place in them. Even as late as 1872, many of our jails were hovels not fit for human beings. Even our best jails were overcrowded and filthy; two, three, four, five prisoners being shut up in a cell fit only for one.[80]

76 *ARC*, 1876, p. C21.

77 *ARC*, 1873, pp. 8–9.

78 *ARC*, 1875, p. 6. Prostitution in Ceylon was legal, but prostitutes were required to be registered and they were regulated through contagious diseases acts. Therefore, not many prostitutes were incarcerated, except for violations of the regulations or other crimes. The regulation of nineteenth-century colonial prostitution is another biopolitical project whose story is admirably told by both Philip Howell and Philippa Levine. Howell, P. 2009. *Geographies of Regulation: Policing Prostitution in Nineteenth Century Britain and the Empire*. Cambridge: Cambridge University Press; Levine, P. 2003. *Prostitution, Race and Politics: Policing Venereal Disease in the British Empire*. New York: Routledge. And for a later period see Legg, S. 2015. *Prostitution and the Ends of Empire: Scale, Governmentalities and Interwar India*. Durham: Duke University Press.

79 On a trip to India in 1867, the English reformer Mary Carpenter found that the condition of female convicts was even worse than that of males and that they regularly occupied the worst part of the prisons and were allowed no visitors. See Carpenter, M. 1867. *Suggestions on Prison Discipline and Female Education in India*. London: Longmans, p. 17. However, I was not able to find any information on the treatment of female convicts in Ceylon. It does not appear to have been an important concern to prison officials at the time.

80 *ARC*, 1875, p. C20.

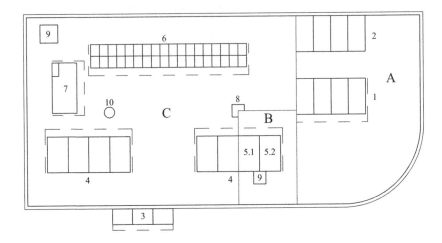

A. JAILER'S SECTION
B. WOMEN PRISONER'S SECTION
C. MALE PRISONER'S SECTION
---- COVERED VERANDA

1 - JAILER'S HOUSE
2 - GODOWNS
3 - GUARD HOUSE AND ENTRANCE
4 - UNCONVICTED PRISONERS AND DEBTORS' WARD
5.1 - PROSTITUTES' WARD
5.2 - OTHER WOMEN PRISONERS' WARD
6 - CELLS
7 - COOK HOUSE
8 - STORE ROOM
9 - PRIVIES
10 - WELL

Figure 7.2 Schematic diagram of Hulfsdorp Jail in the mid-1870s.
Source: Adapted from diagram in *Ceylon Blue Book*, 1863.

In fact, the physical condition of outstation jails had been much worse. Among the worst in the Central Province in the early 1870s was Matale, where it was said that "under a midday sun the heat is almost beyond the power of human endurance," and Nuwara Eliya, where the jail was "simply the end room of a dilapidated shed."[81]

More typical of outstations jails was the one in Avishawelle which was organised around a secure central courtyard. Such jails held prisoners guilty of minor offences and those waiting to be transferred to the convict establishment.

The progress cited by the Inspector General was not enough for the Colonial Office and in 1875, Lord Carnarvon, then Secretary of State for the Colonies, became impatient with the slow implementation of prison reform in Ceylon. He

81 *ARC*, 1871, p. 356.

Table 7.1 Daily average men and
women in prison 1871–1911

Year	Men	Women
1871	1559	39 = 2.4%
1881	2395	27 = 1.1%
1891	3530	42 = 1.2%
1901	2231	32 = 1.4%
1911	2770	54 = 1.9%

Table 7.2 Ethnicity of prisoners 1875–1908

	1875	*1888*	*1898*	*1908*
European	98	72	33	18
Burgher	34	39	44	33
Sinhalese	5,035	9,587	5,147	6,267
Tamil	2,513	1,409	1,559	1,977
Moors	577	647	516	530
Malays	27	50	46	50
Others	17	43	85	125
Total	8,605	11,847	7,430	9,000

Table 7.3 Occupation of prisoners 1870

Soldiers	20
Clerks	10
Merchants	1
Seamen	4
Carpenters	35
Cultivators	692
Coolies	219
Blacksmiths	7
Kanganies	12
Fishers	42
Traders	106
Domestic Servants	20
Cart drivers	43
Toddy drawers	12
Miscellaneous	132
Occupation not given	30
Total number confined on Dec 31	1875

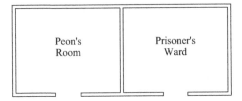

Figure 7.3 Schematic diagram of Nuwara Eliya Lockup in the mid-1870s.
Source: Adapted from diagram in *Ceylon Blue Book*, 1863.

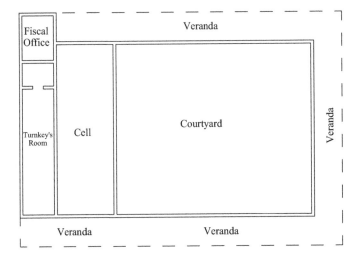

Figure 7.4 Schematic diagram of Avishawelle Common Jail in the mid-1870s.
Source: Adapted from diagram in *Ceylon Blue Book*, 1863.

ordered that all prisoners, both convicted and un-convicted, have separate cells. Governor Gregory responded ambiguously claiming that the island would do so "as fast as financial considerations permit."[82]

The Ceylon Government's response to pressure from Whitehall to spend more money on prisons was to try to make them as remunerative as possible. This policy undercut what was considered best penological practice at the time,

---

82 *CH*, 1876–77, p. 10. Members of the Legislative Council were less ambiguous about priorities. The planting member thought it more important to fund the railway extension and roads to benefit the European community on the island. *Ibid.*, p. 158. The Ceylon government's foot dragging on separation was no doubt encouraged by the fact that India gave up on the ideal of separate cells for every prisoner in 1868 (Arnold, "India: the contested prison," pp. 159–61).

where hard, useless labour was seen to be more effective than productive labour, because it was thought to be less satisfying to prisoners.[83] As a result of this policy, Slave Island and Breakwater Prisons were constructed specifically to house prisoners working on the construction of the breakwater in Colombo. In 1875 over 600 of the worst offenders in Ceylon were housed in association wards in these prisons. When Lord Carnarvon wrote that it was unacceptable that such convicts be housed in association wards, Governor Gregory defended the ward system in the Slave Island Prison, not on economic grounds, which would have been unpersuasive, but on moral, medical, and environmental ones. He wrote that while in Britain "the separate system is necessary on account of the vices prevalent among bad characters when confined in association wards. Those vices are, I believe, absolutely unknown among the native occupants of the Ceylon gaols."[84] The governor supported his views with letters from the Principal Civil Medical Officer, who argued that cellular imprisonment on Slave Island would produce more sickness than found in the well-ventilated association wards and from J.L. Vanderstraaten, the Medical Officer of Welikada Jail, who claimed that

> cellular confinement is not suited to either European or Native prisoners in this Island. In the hot season both classes suffer from bilious derangement while in cellular confinement ... With reference to the subject of unnatural crime, I beg to say that during the period of two-and-a-half years that I have had medical charge of convicts, I have only heard of two cases, and both these were crimes committed in the cells at Welikada.[85]

However, Carnarvon was unyielding on the issue, arguing that, "the conclusions respecting prison discipline arrived at by Her Majesty's Government about ten years ago were founded upon a laborious and extensive enquiry into European, American and Asiatic experience ... [and is] ... indispensable to a well-regulated prison."[86] The result of this standoff between the Colonial Office and the Government of Ceylon was the following compromise: the wards in the Slave Island and Breakwater prisons were to be divided into cubicles with seven-foot-high wooden partitions, with each prisoner shackled in his cubicle at

---

83 DuCane was against productive labour on the grounds that it was less punishing than the non-productive kind. Priestley, *Victorian Prison Lives*, p. 127.

84 Governor the Right Hon. W.H. Gregory to The Earl of Carnarvon, May 8, 1875, in Papers Relating to the Improvement of Prison Discipline in the Colonies. *Command Papers*, August 1875, Vol. LI, p. 210.

85 *Ibid.*, p. 212. This exchange reveals the concern at the time about homosexual behaviour. Such behaviour was punished by flogging throughout the nineteenth century. In contrast, in French prisons there was a much more relaxed attitude toward homosexuality. See O'Brien, P. 1982. *The Promise of Punishment: Prisons in Nineteenth-Century France*. Princeton: Princeton University Press, p. 92.

86 The Earl of Carnarvon to Governor the Right Hon. W.H. Gregory, October 7, 1875, in Papers Relating to the Improvement of Prison Discipline in the Colonies. *Command Papers*, August 1875, Vol. LI, p. 215.

night. This prevented prisoners from what was considered "unnatural" sexual contact, but they could talk freely, as the partitions did not reach the ceiling.[87] The same inexpensive, open cubicle system was inserted into the wards in the Breakwater Prison. This arrangement violated one of the central tenets of separation; that there be no "moral contamination" through communication.[88] The cubicle system was disingenuously justified to Whitehall as temporary, although it lasted for decades. The Ceylon Government successfully argued that cells were unnecessary at the Mahara quarry prison, because it also was a temporary facility, lasting only as long as the stone was needed for public works. The government continued to refer to Mahara as a temporary prison for decades, even though it held a daily average of over 300 of those incarcerated for the most serious offences. As late as 1903, prisoners in Mahara slept in wards on cots covered with coir nets so that they couldn't leave their beds. The wards were supervised by other prisoners known as orderlies.[89]

Over the years, the foot dragging resistance of local authorities to cellular separation had become so normalised that the Inspector General of Prisons could in all seriousness argue in his annual report that, "there is no overcrowding in 1888. Some of the jails had a few more men over their proper strength, but these were wholesomely lodged in verandas, corridors, and other suitable places."[90] Such a casual attitude towards separation, as if it were a non-issue, astonished R.E. Firminger, the newly arrived Superintendent of the Convict Establishment in 1889. At the end of his first year in charge he produced a scathing report. "In comparison with an English convict prison," he wrote, "the disciplinary arrangement of the Convict Establishment here appears crude in the extreme. The term Convict Establishment is a misnomer."[91] The Slave Island and Breakwater prisons, central pieces of the penitentiary system, he dismissed as "without any pretention to proper prison accommodation" and the ward system in Mahara he wrote off as "pernicious."[92] While he reluctantly accepted that wards would continue to exist in the minor prisons, he felt that they were unacceptable within the Convict Establishment, as there could be no proper surveillance of the wards.[93] He wrote,

> it is in these wards that unnatural offences are committed and rarely discovered, conversation is carried on, tobacco smuggled into jail is consumed with impunity, and recalcitrant prisoners gain fresh courage to resist authority from the praise, admiration and advice of their fellows.[94]

87 Inspector General of Prisons, *ARC*, 1876, pp. C30–31.
88 F. Keyt, Medical Officer of the Convict Establishment. *ARC*, 1883, p. C66.
89 R.E. Firminger, *ARC*, 1903, p. C13.
90 G.W.R. Campbell, *ARC*, 1888, p. C49.
91 *ARC*, 1889, p. C7.
92 *Ibid*.
93 This objection lay at the heart of opposition to the ward system. For when no guard was present the prisoners ruled the ward, an unacceptable condition for a penal institution.
94 *ARC*, 1890, p. C3.

Within three years Firminger had managed to transform penal stage punishment by greatly rationalising "space-time." Five prisons were designated to house those convicted of serious crimes. All prisoners sentenced to over a month served their three-month penal stage in Welikada, Kandy, Galle, Jaffna or Kuru-negala. These jails were especially set up to confine prisoners to their cells when they were not working. For hard labour they were marched from their cells to individual stalls in single file, three paces apart, and were not allowed to speak or communicate in any way.[95] But this rigorous approach only applied to the Convict Establishment and only so long as strict senior officials were in charge. When they went on extended leave or were replaced, much of the rigor disappeared. Furthermore, this strict treatment applied only to "native" prison-ers. On the grounds of health and their supposed lower tolerance to oppressive climatic conditions, European and Burgher prisoners were given the best cells, better food and clothing, and worked lightly, if at all.[96] The only way the Cey-lonese could avoid hard labour was if they spoke English. These prisoners were moved out of the penal phase to serve as clerks and orderlies in the prison hos-pital. In 1892 the practice was stopped in the name of deterrence, which then resulted in a dearth of clerks.[97]

The Annual Report on Prisons in 1875 reported that while the expansion of jails was anticipated, Ceylon's finances were such that cellular separation was unlikely to be fully implemented across the island. The Director of Prisons went on to note that in India and even in England, "it is found impossible in practice to at once give effect to theoretical principles."[98] The term "theoretical" here was used disparagingly, as empirically minded Eng-lish criminologists tended to be disdainful of theory. Having said that, they uncritically accepted prevailing environmental determinism as common sense because it appeared to be confirmed by their own empirical experi-ence. So from the beginning of British rule in Ceylon, officials had some doubts, in some cases genuine, in others strategic, as to whether the English ideas of punishment were fully transferable to the tropics. Environmental theories of the tropics conveniently suggested qualifications to that expen-sive model.

As I have noted, officials were forced to choose between adequately funding the criminal justice system or investing in European capitalist development, and it was the latter that was given priority. Nevertheless, under pressure from Whitehall, improvements were gradually made and the percentage of prisoners kept in separate cells grew relative to those kept in wards. However, by the

---

95 *Ceylon Blue Books (CBB)*, 1892, p. 302.
96 *ARC*, 1876, p. 18c. The same rule of racial difference prevailed in Bengal, Singapore and Africa. Sen, M. 2007. *Prisons in Colonial Bengal 1838–1919*. Kolkata: Thema, p. 41; Pieris, *Hidden Hands*; Pete and Deven, "Flogging, fear and food," pp. 3–21.
97 *ARC*, 1892, p. C1.
98 *ARC*, 1875, p. 10.

early twentieth century smaller jails continued to be overcrowded. Leonard Woolf, reflecting back on his days as an AGA in the Southern Province in the early twentieth century, wrote,

> in those days the prison system was more barbarous even than the law. The prisoners were confined in cages like those in the lion house in the Regent Park Zoo, two, three, or even four men sometimes in a cage. The buildings were horrible.[99]

Although progress towards separation had been made in the three decades since the prison system was reformed, in 1901 the Inspector General still pointed to better separation as the primary need of the prison system.[100] In the following chapter, I will examine the English model of punitive hard labour that was introduced in Ceylon and show that resistance to this model arose from within the prison system itself.

99 Woolf, L. 1961. *Growing: An Autobiography of the Years 1904–1911*. New York: Harcourt Brace Jovanovich, p. 169.
100 L.F. Knollys, *ARC*, 1901, p. C7.

8　Experiments in the production of bodily suffering

In a letter to Governor Robinson dated 29 July, 1866, the Prison Commissioners stated that deterrence should be the policy in Ceylon, because "the listless, unenergetic natives are comparatively little open to … [reform]"; and "it would be unfair to the honest part of the community, who pay the taxes out of which prisoners are maintained, to keep offenders under mild and instructive treatment for any protracted periods of time."[1] Nevertheless, for the first couple of years hard labour was supplemented on Sundays by classes in Christian religious instruction and learning to read English. However, by 1870 such attempts at the "reform" of prisoners had ended due to lack of interest among the prisoners, and officials reiterated that "a system enforcing hard, distasteful labour [was] better calculated to repress crime in Ceylon and keep its jails tolerably empty."[2]

Undergirding the penal philosophy of deterrence, in fact, was an implicit reformatory goal, to instil in prisoners a rudimentary utilitarian consciousness through physical pain. This, however, was very much less ambitious than the Foucauldian notion of governmentality as the "conduct of conduct" whereby the state or other governing institution leads a population to self-government such that individuals following their perceived self-interest do as the state thinks they ought.[3] As Sen points out, such self-identification with the state was unlikely in the colonies because the people were subjects rather than citizens. Perhaps the most that the British could have hoped for was to teach subordination.[4] However, the British were unable to decide if hard labour would teach prisoners the value of work or if they would become repelled by the experience, forever

---

1 *Ceylon Sessional Papers (SP)*, 1866, p. 67.
2 *Administration Reports, Ceylon (ARC)*, 1870, p. 390. By 1905, in response to shifting views on imprisonment in Britain, juveniles within the system were receiving two hours of schooling per day and docile prisoners after completing the penal phase of their sentence were given the opportunity of learning a trade. Both were done in an attempt to curtail recidivism. *Ceylon Sessional Papers (SP)*, 1908–09, lxiii; *ARC*, 1905, C3.
3 Scott, D. 1995. "Colonial governmentality." *Social Text*, Vol. 43, p. 202.
4 Sen, S. 2000. *Disciplining Punishment: Colonialism and Convict Society in the Andaman Islands*. New Delhi: Oxford University Press, pp. 2, 14.

associating labour and regimentation with their time of suffering. Thus, they did not hold out high hopes of reforming prisoners.

The scientific production of official violence was organised rationally through space-time management, allowing an opportunity for the prisoner to reduce the pain through self-discipline.[5] The latter necessitated goal-oriented behaviour and the deferment of gratification, all of which were thought to be the antithesis of the aimless life of leisure that was considered the mark of a criminal. The principal technology for the production of self-control was the three-stage system which allowed well-behaved prisoners to move smoothly from stage to stage, each one being less onerous than the one before. Those who failed to be obedient would remain throughout their sentence in the harshest environment.

The production of self-discipline: the three-stage system

Following the British model, new prisoners were to undergo a six-month intramural penal stage doing as much hard labour as they could endure. In theory this stage was to be monitored by doctors so that prisoners would suffer no permanent bodily damage. It was therefore adjusted for women, children and older men. Women had their own walled-in area where they performed their labour. When not doing shot drill, male prisoners were to be employed in their cells beating coir, or breaking stone.[6] Every prisoner was to be kept in a separate cell at night and fed a highly reduced penal diet. If the prisoner conducted himself well during the first six months, he was raised to the second stage of industrial hard labour. In this stage, prisoners were employed at extramural labours such as stone breaking, quarrying, scavenging and road making. While at work prisoners were to be kept as much apart from each other as was practicable, and silence was to be strictly enforced. In order to be promoted to the third stage, a prisoner had to acquire a certain number of marks for good conduct and hard work. The third stage consisted of light labour such as cooking or serving as a warder or hospital orderly. If prisoners obtained the equivalent of nine marks a week for three-quarters of their terms of imprisonment, exclusive of the period passed in the penal stage, they were recommended for remission of their sentences.[7]

The calculation of hard labour

The ordering of everyday life during the initial six-month penal phase was calculated to be as harsh as possible. In theory, the regime was to be strictly

---

5 On this see Weindling, P., ed. 2017. "Introduction," in *From Clinic to Concentration Camp: Reassessing Nazi Medical and Racial Research, 1933–45*. London: Routledge, p. 4.
6 Shot drill was the carrying of a twenty-four pound weight between tripod stands. Coir is coconut husk.
7 Rules and Regulations of the Wellicadde Prison, 12 June 1867, reproduced in *Committee of the House of Lords*, p. 77.

enforced in all prisons, but in fact it was only in the Convict Establishment in Colombo and later in the new jails in Kandy and Jaffna that it was followed with any regularity. According to penological theory, hard labour during the six-month penal phase was supposed to be intramural and entail severe physical exertion, but should be sustainable for a full nine-hour day. It was also meant to be monotonous and uninteresting, and able to be done without the assistance of another prisoner. Importantly, it should be possible to accurately measure the amount of effort a prisoners must exert to accomplish particular tasks.[8] The creation of such a highly regulated labour regime required detailed knowledge of the physical abilities and psychology of prisoners as well as the ability to rapidly and accurately measure the output of hard labour. We will explore how the authorities in Ceylon prisons experimented with different types of labour in search of work that filled as many of these criteria as possible.

The controlled production of suffering necessitated a knowledge of the intimate connection between physical exertion and long-term health, knowledge which would have been rudimentary at best during this time. Tasks had to be designed so that prisoners could endure them, but only just. Advised by doctors, administrators believed that there was a razor-thin difference between effective methods of deterrence and bodily harm. Tasks were to be sufficiently onerous, able to be performed for long enough to be boring, and medically approved. As it was difficult to know exactly where this threshold lay, experimentation was undertaken.

Furthermore, prisoners often arrived to begin their sentences in poor physical health. But up until the mid-1870s newly arrived prisoners in the Convict Establishment were admitted with, at best, a cursory medical inspection and at the smaller outstation jails where there was no resident doctor, the examination would have been even more superficial. As it turned out, the lack of a proper examination in the rural jails was usually not a problem, as hard labour was rarely enforced at outstation jails. After the 1870s the Convict Establishment began to develop a more rigorous admissions schedule of evaluating the bodies of incoming convicts to ascertain how much punishment a prisoner could withstand. But the method, being based on aggregate data, was not well individualised. Prisoners were judged as to whether or not they fell below an artificially constructed norm. As Foucault points out, "disciplining normalisation consists first of all in positing a model—and trying to get people, movements and all actions to conform to this model."[9] The sick or weak prisoners who fell below the standard were temporarily excused from hard labour. However, all others were treated as "normal," the norm having been based on a strong, 130-pound man, and all punishments were calibrated to that artificial standard. Of course, real differences in strength went undetected, as did decreases in strength as prisoners became ill or weak from disease, prison diet and/or over work. Doctors

8 *ARC*, 1890, p. C7.
9 Foucault, M. 2007. *Security, Territory, Population: Lectures at the College De France, 1977–78.* Edited by M. Senellart. New York: Palgrave Macmillan, p. 85.

were instructed to oversee bodily changes, but they rarely intervened before prisoners became ill. Ironically, because their elaborate experiments and precise calculations were based on an artificial norm, the uniformity in punishments that prison administrators sought was not possible. The danger of such biopolitical strategies was that although they were administratively efficient, they could be deadly, especially in an institution such as the prison where lives were often pushed to the edge, thereby exposing the dark side of biopolitics.

Upon arrival, prisoners who had been sentenced to three months or more were told to wash and were given uniforms to distinguish them by race; Europeans and Burghers a suit of blue cotton with a shirt and shoes, and others a much less expensive piece of cloth six feet long and three feet wide.[10] The second morning, they were vaccinated against small pox and held under medical observation for ten to fifteen days. During this time experiments were conducted to establish their ability to endure hard labour. They were then divided into those capable of hard labour and those suited to only light labour until their bodies were capable of withstanding greater punishment. If seriously ill, they were sent to one of the prison sanitaria at Negombo or Kalutara. During this initial sorting period, all prisoners were under the control of the prison doctors, whose decisions about fitness were final.[11] Doctors and jailers were on the lookout for malingerers during this period, especially men who had been imprisoned before and understood the benefits of failing to meet the standard of health and strength expected of an average man. These inspections served another important function as well. They provided prison authorities with scientific legitimation for the harsh, sometimes deadly, hard labour experiments conducted in the prisons. Jailers could point out that from the start of a sentence, hard labour was overseen by medical professionals. When Whitehall argued that the disciplinary regime was endangering prisoners' health and life by exceeding sentences handed down by the courts, the official answer from prison administrators was that the sickness and death rates must be normal for such men, as prison doctors had been monitoring the health of prisoners throughout. In this way prison authorities were able to evade the rule of law concerning just sentencing.

Those judged capable of hard labour faced a daunting schedule. The organisation of space-time in prison was designed much like a factory, albeit one whose product was bodily fatigue and hunger. In order to fill the prisoner's day, the authorities devised a system of alternating strenuous and non-strenuous punishments. The physically demanding ones were shot drill, pingo carrying and stone breaking, and the less-demanding one was coir beating.[12] When shot drill, the lynchpin in the disciplinary system, was put into effect in Welikada Prison in 1867, the chief medical officer

---

10 *Ceylon Blue Books (CBB)*, 1862, Return of the Gaols and Houses of Correction. India prisons followed the same policy Anderson, C. 2004. *Legible Bodies: Race, Criminality and Colonialism in South Asia*. Oxford: Berg, p. 119.

11 F. Keyt, Medical Officer of the Convict Establishment. *ARC*, 1883, p. C64.

12 A pingo is two baskets containing twenty pounds of stone each on a pole. The corresponding punishments in England were the treadmill, stone breaking and oakum picking. Priestley, P. 1985. *Victorian Prison Lives: English Prison Biography 1830–1914*. New York: Methuen, p. 121.

for prisons noted with satisfaction that it was "highly obnoxious to prisoners."[13] During the penal phase of their sentence, prisoners faced what was, on paper at least, a crushing daily workload. They were woken at 5 a.m. and given a light meal in their cells at 5.30. They then did shot drill in the prison yard from 6 to 8 am. From 8 to 11 am they broke stone or beat coir in their cells or carried two twenty-pound pingos in the yard. They then ate lunch at 11 and were allowed to rest in their cells. From 1.30 to 2.30 they beat coir in their cells, and from 2.30 to 4.30 they did shot drill again in the yard.[14] At 5.00 they ate their last meal of the day in their cells and at 6.00 were shut in for the night. The silence gong was sounded at 8.00 and the corridors were patrolled by three guards every half hour throughout the night in an attempt of prevent conversation and what were termed "unnatural acts." In the smaller jails, corridors might be patrolled only several times per night if at all.[15]

As the women prisoners were excused from shot drill, they were required to beat coir. And as the norm for women was determined to be ninety pounds, they were made to do two-thirds the amount expected of a man.[16] Boys under sixteen and male prisoners who were judged by the prison medical officer to be unfit did the drill with a fourteen-pound shot.[17] Prisoners were allowed to bathe twice a week and daily if they broke stone.[18] By the end of the century, shot drill had been abandoned and prisoners spent six hours daily on a treadmill or carrying pingos. Those who were too weak for these tasks beat coir for six hours daily. Those in the second stage of punishment worked six hours daily breaking rock or building the breakwaters in Colombo Harbour.[19]

The Committee on Prison Discipline's injunction that punishment be rationalised and so measured out exactly to fit the crime was taken quite literally. Uniform and precise punishment for each type of crime was thought of as a contractual obligation. The exactitude of the design of penal technologies provided, at the very least, a veneer of science and liberal justice. Beginning in the 1870s, Dr Haughton's formula for calculating the dynamic value of work in foot-tons (the amount of effort necessary for raising one ton one foot in height) was employed. As we can see, human labour was treated as a "mechanical system which could be decomposed into energy transfers, motion, and the physics of work."[20]

13 *ARC, 1867*, p. 265. The *CBB* for 1867 shows that this was the only prison in Ceylon to introduce shot drill in that year.
14 *ARC, 1870*, p. 389.
15 *CBB*, 1871, p. 591.
16 Throughout the nineteenth century, British officials were surprised by how little female criminality there was in comparison to Britain.
17 *ARC, 1872*, pp. 436–38.
18 F. Keyt, Medical Officer of the Convict Establishment, *ARC*, 1883, p. C64.
19 *CBB*, 1876, p. 655; A.W. De Wilton, Inspector General of Prisons, *CBB*, 1902, p. Y5; F.R. Ellis, Inspector General of Prisons, *ARC*, 1893, p. C2.
20 Scott, J.C. 1998. *Seeing like a State: How Certain Schemes to Improve the Human Condition Have Failed*. New Haven: Yale University Press, p. 98.

Figure 8.1 Prisoners working on the Breakwater, 1875.
(Source: Manuscripts Room, University of Cambridge).

$$\frac{(W + Wt) \times D}{20 \times 2,240}$$

Where: W= weight of person; WT = weight carried; D = distance walked in feet; 20 = coefficient of traction; 2,240 = number of pounds in a ton. Haughton estimated that for the average European male, a limit of 150 foot-tons was necessary to maintain an adult male in health. 300 foot-tons represented an average day's work, 400 foot-tons a hard day's work and 500 foot-tons a very hard day's work. Having estimated the average weight of a Ceylonese male to be 130 pounds, an average day's work was calculated at 210 foot-tons.[21] This became the measure, in theory at least, of the minimum that prisoners should be worked and tasks were set to only modestly exceed this measure. By tailoring hard labour to a norm, some of the smaller and weaker prisoners suffered. Once they were judged fit, all were required to perform the same

21 *ARC*, 1876, p. C116. Foot pounds was the standard measure of energy into the 1870s. After this time, calories and Joules became the more common measure. Hargrove, J.L. 2006. "History of the calorie in nutrition." *The Journal of Nutrition*, Vol. 138, p. 2960.

amount of work irrespective of how strong they were unless they could convince a doctor to temporarily reduce the task. If they failed to fulfil their task, they were punished, often by reducing their food, which of course made them weaker still. As we shall see, it was very common for prisoners to fail to fulfil their quota. Although the jailers couldn't be certain whether this was because they were physically incapable of doing so or because they were shirking, they usually assumed the latter and of course, this attitude worked against those who were physically weak.

Taking the example of shot drill, the principal form of penal labour in the 1870s and 1880s, we can see how theories of labour, physics, physiology and punishment become entwined. The following is the instruction given to prison administrators on how shot drill should be organised:

A man stoops down and lifts a 24-pound shot from a tripod stand 18 inches high,[22] erects himself, steps 13 feet, and lowers the shot to another similar stand; he returns empty to the first stand, lifts another shot, carries it to the second stand, and so on for two hours at a time. Four double journeys are performed per minute; he therefore walks 26 feet 240 times per hour or 6,240 feet in that period of time, carrying his own weight (130 lbs) and also for half the distance, a shot of 24 lbs. He, in addition, lifts and puts down this weight eight times per minute, a height of 18 inches or 720 feet per hour. Assuming the average weight of an ordinary prisoner to be 130 lbs., I find that the work is equal to 109 tons raised one foot.[23]

All of this was in theory only, for it assumed vigilant guards who obeyed prison rules and compliant prisoners. In fact, neither proved to be the case. A decade after its introduction, shot drill, the mainstay of the penal phase, was widely considered a failure due to widespread resistance by prisoners and guards.[24] Prisoners protected themselves from the crushing workload by routinely moving at a slower pace than prescribed and lightening the load by hugging the iron ball to their body rather than holding it with their arms outstretched. Prisoners often risked punishment by not doing shot drill as prescribed, although they tended to injure themselves when they did it correctly. While Houghton's formula calculated the energy expended in lifting and carrying the ball, it failed to consider whether prisoners had the strength to repeatedly lift the ball from the low stands without getting hernias or damaging their backs. After a decade of witnessing injuries among the newly arrived prisoners and

22 The recommendations from London in 1866 were that the tripods be two feet in height on the grounds that repetitive bending lower than that with a twenty-four pound weight might result in the rupturing of a man's spleen. *Ceylon Sessional Papers (SP)*, 1866, p. 69.
23 *ARC*, 1876, p. C116. A slightly different variant of this drill was employed in Mauritius Anderson, C. 2008. "The politics of punishment in colonial Mauritius, 1766–1887." *Cultural and Social History*, Vol. 5, No. 4, pp. 411–22.
24 In England as well, guards were regularly bribed by prisoners to ease off on the task required. Priestley, *Victorian Prison Lives*, p. 140.

widespread evasion by old hands, the Inspector General concluded that "if done properly, it sent men to hospital; if not done properly it was not hard labour."[25]

In 1887 Superintendent Mooney sought to save shot drill by calling for a new, more scientifically designed form using a fourteen-pound shot and four-inch stands placed eighteen feet apart. He argued that, according to his calculations, this combination of weight, depth and distance was perfectly suited to the non-European body. However, hard labour using the shot drill was seen to be so difficult to implement, due to prisoners' tactics of evasion and assaults on guards, that his suggestion was not taken up and finally, in 1890, its use was prohibited by the prison doctors.[26]

Shot drill was normally interleaved with five and a half hours a day of coir beating or coir-twisting, which were monotonous but not exhausting tasks and consequently used to fill in the time between sessions of hard labour. Interspersing hard labour with light labour was thought to maximise the amount of punishment that the body could withstand without breaking down. Some officials objected to the fact that this routine failed to exhaust a prisoner, but that was to miss the point. Its function was to be unpleasantly boring and to allow prisoners to continue working without collapsing. The other advantage of coir beating was that prison doctors rarely forbade its use. As with other punishments, they sometimes simply reduced the task for weak prisoners to one-half, one-quarter, or one-eighth of full task.[27] Instead of hard labour, prisoners under sixteen years old did 5/7 of full task of coir beating.[28] Women and road defaulters likewise were only required to beat coir.[29] Adult males were tasked to produce five pounds of wet coir or thirty fathoms of yarn per day. It was estimated that this

25 *ARC*, 1890, p. C7. Though designed in England, shot drill was virtually never used there as it was considered dangerous to the health of prisoners. Priestley, *Victorian Prison Lives*, p. 131. The twenty-four pound ball was designed for larger framed English males, rather than the slighter Ceylonese.

26 Evidence given by Mr. Assistant Superintendent Mooney, Report of the Prison Committee, *SP* 1887, p. 228; *CBB*, 1887, p. 524; *ARC*, 1890, p. C7.

27 F.R. Ellis, Inspector General of Prisons, *ARC*, 1893, p. C 10.

28 A.B. Santiago, Medical Officer of the Convict Establishment, *ARC*, 1900, p. C9.

29 A. Kalenberg, Medical Officer of the Kandy Jails, *ARC*, 1900, p. C11. Ordinance 10 of 1848 required that every male between the ages of eighteen and fifty-five years of age labour for six days per year on the improvement of the thoroughfares on the island. Alternatively, he could commute his liability by paying a fee of Rs. 2. Those who refused to work and defaulted on the fee were sentenced to one month in prison. The police, judges and jailers were ambivalent about the defaulters, because if a man was willing to spend a month in prison rather than pay a Rs. 2 fee, it could be argued that his only "crime" was being poor. Consequently, road defaulters were given lenient treatment in prison. For years, the number of defaulters in prison per year numbered under three thousand. But in 1878, as hard times hit with the collapse of coffee, the number rose to 8,000, only to rise again three years later to 14,000. By 1884 when the numbers were averaging 15,000 per year the Government passed Ordinance 31, changing the penalty from Simple Imprisonment (no hard labour) to one month's hard labour on bread and water. The effect of this strategy was dramatic. By 1888 the number of defaulters had dropped to 4,902; by 1893 to 1,127, and by 1906 only 575. *ARC*, 1893, p. C3; *ARC* 1906, p. C4.

work corresponded to the equivalent of oakum picking in English prisons and therefore equalled 150.3 tons lifted one foot. So, the daily work performed on shot drill and coir beating during the penal stage was equal to 259.3 tons lifted one foot, safely above the 210 ton minimum.[30]

The problem with coir beating, however, was that unlike oakum picking

> each pound of coir did not equal a similar quantity of labour; some was more easily beaten than others, owing to the wetness of the husk or ripeness of the fruit. Furthermore, evaporation caused significant differences in the weight and rendered it difficult to discover any accurate check.[31]

Consequently, prisoners tried to wet the beaten fibre to increase its weight. Measurement caused continual disputes between prisoners and the guards. Nevertheless, coir beating continued to be used throughout the nineteenth century as no other punishment could be devised to fit this particular bio-disciplinary niche in this highly quantified penal regime.[32] By the late nineteenth century the quota of wet coir to be produced was reduced to three and a half pounds, as few prisoners could achieve the required five. But in 1907 the Jaffna prison quota was raised to five pounds in order to more severely punish the habitual criminals who were imprisoned there. The prisoners initially refused to produce that amount, but were flogged until they were seen to try.[33] This is but one example of the uneven application of the rules.

Those in the second stage of hard labour at Welikada worked constructing a breakwater in the Port of Colombo. A day's labour moving stone was calculated as the equivalent of lifting 259.3 tons one foot. The walk from the prison to the breakwater and back was calculated at another 17.67 foot-tons. The total of 277 foot-tons was judged to be not excessive, "even for a native."[34] Stone breaking was considered by many prison administrators to be the best form of labour, as it was hard work, was capable of being accurately measured, and had the added advantage of producing crushed stone to sell to the government for fill on the breakwaters and metal (stone) for roads.[35] This form of labour was found to be so useful that a special prison was built at the Mahara quarry to supply rock for the breakwater.

Following Dr. Haughton's formula for the "Dynamic Value of Work," a man was tasked to produce a ton of broken stone every three days.[36] As shot drill was phased out, prisoners increasingly broke stone during the penal phase of their sentence. However, when there was a lag in government building projects, as there was in the early 1890s, it was difficult to dispose of the crushed stone.

---

30 *ARC*, 1876, p. C116.
31 *ARC*, 1890, p. C7.
32 *ARC*, 1894, p. C13.
33 *ARC*, 1907, p. C7.
34 *ARC*, 1876, p. C116.
35 Metal in road building is broken stone. The term comes from the Latin *metallum* for quarry.
36 *ARC*, 1889, p. C6.

The Slave Island Prison in Colombo, where prisoners regularly broke stone that was shipped to them in blocks from the Mahara Quarry Prison, ran out of space to put the broken stone. The Inspector General of Prisons wrote in his annual report that it required a hundred men to furnish penal stage prisoners at Slave Island with enough stone to break and that "there was not left … a yard of land on which to heap metal."[37] As a result, the prison was abandoned and the prisoners sent to Welikada where they carried pingos and beat coir instead. Elsewhere, due to reduced demand, stone breaking was only used as a punishment for certain prison offences rather than as part of regular hard labour.[38] When public works projects resumed in the early twentieth century, stone breaking again became widespread in the prisons.

From the 1870s on, pingo carrying was a common form of unproductive hard labour. A prisoner was required to carry two twenty-pound weights on a pole balanced on the shoulders for eight hours walking at two miles per hour for a total of sixteen miles per day. This was calculated as falling within the acceptable level of foot-tons per day.[39] Newly arrived prisoners walked only eight miles per day for the first three days after their arrival, because otherwise too many were unable to continue due to shoulder abrasions.[40] Pingo carrying was simpler to administer than other measured labour, but prisoners could evade the full effect by stopping frequently or not walking at the required pace, and there was continual conflict with guards over correct procedure. There was also a concern that prisoners preferred this to breaking stone or working on the breakwaters.[41]

Strategies of enforcement and tactics of resistance

The great problem with all tasked systems of labour was that if prisoners were not compliant, then guards would have had to be sufficiently honest and in control to be able force prisoners to obey, and this was often not the case. As we have seen, prison authorities were unable to enforce regulation shot drill without sending prisoners to the hospital and so it was eventually abandoned. Likewise, prisoners systematically resisted their quotas of stone breaking and coir beating. So concerned were authorities about the scale of non-completion of tasks, referred to as "short-task," that in 1892 they reviewed the records of 300 recent admissions to see how many actually performed tasks. Only one-fifth consistently completed their tasks, two-fifths failed less than ten times, one-fifth failed ten to thirty times, and one-fifth failed more than thirty times.[42] The authorities concluded from this that the tasks were doable, but that there was systematic

---

37 *ARC*, 1891, p. C7.
38 *ARC*, 1892, p. C1.
39 H.G. Tomasz, Medical Officer of the Kandy Jail, *ARC*, 1895, p. C17.
40 *Ibid.*
41 *ARC*, 1896, p. C2.
42 *ARC*, 1892, p. C12.

resistance on the part of longer-serving prisoners. And so, punishments for non-completion were increased. A decade later, resistance had increased to the point that over 50% of all prisoners doing stone breaking or coir beating were being punished at any given time.[43] It is unclear to what extent this was due to prisoner resistance or their sheer physical inability to meet the quota.

Table 8.1 sheds light on types of prisoner resistance to regulations. The first type of violation reveals a significant amount of non-violent aggression, primarily directed towards fellow prisoners as they fought over scarce resources, but also towards guards.[44] The next most common type is a classic illustration of the weapons of the weak—partial or inadequate compliance sometimes disguised as misunderstanding orders. Only with "Disorderly Conduct and Idleness" and "Refusing Work" do we find instances of open and unambiguous challenges to authority, and these represent only a small percentage of the violations. This is hardly surprising, as direct challenges to prison regulations were met with severe penalties.[45] Punishments for mild but effective forms of resistance were a severely restricted diet lasting a few days, increased labour or loss of marks.[46] Only those who blatantly refused to work or continued to do less than they were assigned were put in solitary confinement, flogged or caned.[47] But even those who openly refused orders remained in the words of the Superintendent of the Convict Establishment "perfectly submissive and subservient" in their demeanour.[48] In this way they hoped to soften their challenge to authority and somewhat lessen the punishments they would receive. By the early 1880s, loss of marks had become the most common form of punishment reducing reliance on other types.[49]

While there was constant low-level resistance in the form of short-task[50] and evasive tactics to reduce the number of foot-tons required, there were in some rare instances prisoners who organised collective resistance. When, in August 1890, prisoners at the Mahara Quarry Prison had their quota raised from

---

43 R.E. Firminger, Superintendent of the Convict Establishment, *ARC*, 1903, p. C13.
44 This figure suggests that there was a good deal of prisoner on prisoner violence as well. Given that little was reported, it probably took place out of sight of the jailer's gaze. The rare murder of a prisoner by another was given sensational coverage in the press. "A fatal conflict between Welikada prisoners," *The Ceylon Times*, January 3, 1891.
45 *ARC*, 1879, p. C112.
46 F.R. Ellis, Director and Inspector General of Prisons, *ARC*, 1895, p. C10; A.R. Santiago, Medical Officer of the Convict Establishment *ARC*, 1901, p. C8. In England refractory diet was the standard punishment for failure to meet one's quota of work. Priestley, *Victorian Prison Lives*, p. 135.
47 F.R. Ellis, Director and Inspector General of Prisons, *ARC*, 1894, p. C9. Corporal punishment was to be reserved for grave offences, such as persistent refusal to do hard labour, attacks on prison officers, escapes, and the commission of "unnatural acts." *Committee of the House of Lords*, p. 104.
48 *ARC*, 1891, p. C10.
49 E. Elliott, Inspector General of Prisons, *ARC*, 1882, p. C41; W. Thompson, Superintendent, Convict Establishment. *ARC*, 1882, p. C48.
50 Short task is the failure to complete the amount of labour assigned. This was usually measured in foot-tons.

Table 8.1 Prison offences in 1879.

Shouting, Quarreling & Using Abusive Language	209
Not Performing Task Completely	189
Disobedience of Orders	117
Possessing Prohibited Articles	95
Quitting Rank without Permission	66
Disorderly Conduct	33
Idleness and Refusing Work	32
All Others Under 20 Each	150
Total	891

one ton of crushed rock every three days to one and a half tons, there was an organised opposition that entailed going slow and failing to meet even the former quotas. During the autumn of that year, punishments for "short task" rose to ninety per week from the normal average of thirty-five. In response to what the authorities took to be organised resistance, they increased the level of punishment for short task. By the end of the year the resistance was broken through a combination of caning and punishment diets and the normal rate of punishments was re-established.[51] The idea of a normal rate of punishment was important, for it marked the level of resistance that jailers were willing to reluctantly tolerate. Such norms, I would argue, were created through a persistent level of resistance by prisoners. Norm manipulation was one of the most important long-term tactics of resistance by the cumulative actions of prisoners, for it served to re-set norms to a lower level than had been established by the authorities. In this sense it was similar to the low expectations set by the Ceylonese police and court officials, as I discussed above.

In the autumn of 1890, word of the organised resistance at Mahara spread throughout the prison system and convicts elsewhere were emboldened to resist. The Inspector General of Prisons wrote

> The difficulty experienced in getting the men to do their tasks has been almost incredible. Many men, especially those previously convicted, resolutely refused to contemplate their tasks. Refractory diet, solitary confinement, caning, every species of punishment, was tried, but in some cases without effect. That the task was one within the power of all is shown by the fact that a large number of first admissions did it, but it was found almost impossible to deal with men who had made up their minds that, come what might, they would not do the work. The example set by these men was very injurious to prison discipline … The problem has not yet

---

51 R.E. Firminger, Superintendent of the Convict Establishment, *ARC*, 1890, p. C3.

been solved, but experiments are still being conducted which will, I have no doubt, produce in the end the wished for result."[52]

Punishment options were narrowing, though. By the early 1890s there was pressure on the prison authorities from Whitehall to cut back on flogging and from prison doctors to stop using severely restricted diets.[53] Solitary confinement, often in a dark cell, was rarely the punishment of choice in Ceylon, in part because of shortages of space within prisons, and also because curiously there was no agreement on its effectiveness. In 1880, the Inspector General said that he believed prisoners were not bothered by the idleness or the darkness of the solitary cell. In contrast, however, Assistant Superintendent of Prisons Mooney in his evidence to the Prison Committee in 1887 said that solitary confinement was little used because it was "a severe punishment for natives." Ten years later, the new Inspector General argued that, while solitary confinement was dreaded in England, in Ceylon it only gave a prisoner "the rest and sleep he desires."[54] It is interesting that even in a controlled environment, such as the prison, the administrators did not have enough local knowledge to determine if prisoners dreaded or appreciated solitary confinement. Of course, it is quite possible that the reaction to isolation varied from individual to individual and that different jailers generalised from single cases to the population. Given the European tendency to homogenise the psychology of "natives," that is very likely the case.

With restrictions placed on corporal punishments by Whitehall and on the use of the punishment diet by doctors, prisons began to import treadmills from England to use in addition to pingo carrying. And by the end of the century these had largely replaced dietary punishments.[55] The treadmill was a machine with large paddle-wheels twenty feet in diameter with twenty-four steps around a six-foot cylinder. The attraction of the treadmill was that the machine itself forced prisoners to do the required work. Each treadmill could accommodate twenty-four prisoners, who were tied onto the wheel, compelling them to keep moving at the pace set by the machine.[56] But prisoners soon discovered that

---

52 *ARC*, 1891, p. C7.
53 *ARC*, 1892, p. C11; *ARC* 1895, p. C2.
54 *ARC*, 1880, p. C139; *SP* 1887, p. 228; *SP*, 1898, pp. C5–6.
55 The treadmill was invented in 1818 and along with stone breaking was the principal form of hard labour in England and Wales. It gained in popularity in Britain in the 1820s as solitary confinement went out of fashion. It was discovered that weak prisoners suffered more from it than strong prisoners and so uniformity was a problem. It, along with all other forms of unproductive hard labour, were abolished in England 1898. It is telling of attitudes towards racial difference that it was abandoned in Britain while being introduced to the colonies. Evans, R. 1982. *The Fabrication of Virtue: English Prison Architecture, 1750–1840*. Cambridge: Cambridge University Press, pp. 301–09; Priestley, *Victorian Prison Lives*, pp. 125, 129; Bhushan, V. 1970. *Prison Administration in India*. Delhi: S. Chand, p. 13; *ARC*, 1985, p. C15.
56 The machines were calibrated to average fifty-seven steps per minute. Priestley, *Victorian Prison Lives*, p. 127.

although the machine forced them to lift their bodies up three feet with each step, by stepping lightly on the edge of each step, the force of the machine would help to raise them up. Prisoners also learned that it was less tiring if they leaned against the wooden cross-pieces on the machines. By 1913, under pressure from Whitehall, the use of the treadmill was abandoned in the colonies as it was considered a relic of the nineteenth century. In fact, by then the notion of purely punitive and non-remunerative hard labour had largely fallen out of favour in the colonies as it had done earlier in Britain.[57]

Another experiment suggested by the Inspector General in 1890 entailed tweaking the pingo system so as to remove any possibility of evasion.

> Each man has to hold a short string or trace; this is fastened to a rope, which runs through two pulleys fastened to each end of a shed. The rope goes round with the prisoners, and no man, without falling out of line, can avoid doing the same distance in the same time as the others. If he attempts to walk slower than the others the trace slackens, and he falls back on the man behind him. The pace is regulated by minute glasses and indicators, the latter enabling a supervising official to check at any time the number of revolutions performed. Given the pace and the hour of starting, it is of course easy to discover how many revolutions should have been done at any particular time of day.[58]

By 1903 this type of pingo carrying had become the most used form of penal hard labour as it countered any type of prisoner resistance. Once a prisoner was roped into the pingo line, he had to do the full-allotted labour. As we have seen, penal policies were continually re-evaluated in response to prisoner resistance and new strategies were devised. However, three years after it had been celebrated as the ideal form of hard labour, pingo carrying was almost completely abolished, because it was non-remunerative and jailers reported that prisoners appeared to prefer it to other forms of hard labour.[59] In its place prisoners were again made to break stone and beat coir and pingos were only used when prisoners were physically unable to break rock. Because authorities deduced that prisoners found it too easy, the weight of the pingos was increased, as was the number of miles they were to be carried.[60]

There was, however, the rare prisoner who simply refused to do any hard labour. Superintendent of Prisons Ellis pointed out in frustration that there were only two ways to deal with such men: the punishment diet and corporal punishment, but he complained that doctors interfered with punishment diet, halting it when a prisoner became weak, and that the Ceylon government had by then

57 *ARC*, 1912–1913, p. C4.
58 FR Ellis, Superintendent of Prisons, *ARC*, 1895, p. C2.
59 *ARC*, 1906, p. C8.
60 *ARC*, 1906, p. C13.

placed severe restrictions on corporal punishment.[61] Regardless, he encouraged the liberal use of the cane. He wrote of one prisoner who

> refused to get on the treadmill, laid down if a pingo were given to him, and refused to lift a mallet to beat husk. The struggle could have but one termination: the Prison authorities dare not give in; the prospect of a possible success would immediately induce many to enter into a similar contest.

The prisoner was caned on twenty-five separate occasions before he finally began to work.[62]

The guards: liminal figures

The prison guards in the Convict Establishment, like the convicts, faced a crushing daily schedule. They were required to be at the jail by 5.15 a.m. in order to eat their breakfast ration before going on duty at 5.30. Once on duty they had to remain walking or standing until their forty-minute lunch break at 11 a.m. They returned to duty at 11.40 and remained until 5.20 p.m. They were punished if they sat down or even leaned against a wall during the eleven hours that they were on duty. It is little wonder that the prisons department found it difficult to hire and retain guards, given the heavy workload and low pay.[63] In spite of all the exact calculations of dynamic expenditure of energy that existed on paper, the actual practice of penal labour was far from the ideal, due to a lack of commitment to the system on the part of the guards who were often lax in managing the prisoners.

There were only a small handful of trained prison administrators in the country and the majority of the prison heads in district jails were AGAs who paid infrequent visits to their jails. As was the case with police outstations, the small prisons were run by untrained, uncommitted and poorly paid jailers with little or no supervision. The Acting Superintendant of Welikada Prison in 1873 wrote that even here, "the overseers are, with but few exceptions, totally unfit for

---

61 *ARC*, 1893, p. C9. A similar situation pertained in Bengal (Sen, M. 2007. *Prisons in Colonial Bengal 1838–1919*. Kolkata: Thema, p. 50) and Burma where flogging was heavily relied upon to discipline prisoners. Under pressure from London it was cut back somewhat after 1880. Brown, I. 2007. "South East Asia: reform and the colonial prison," in F. Dikotter and I. Brown (eds.), *Cultures of Confinement: A History of the Prison in Africa, Asia and Latin America*. London: Hurst, p. 239.

62 *ARC*, 1895, p. C11.

63 R.E. Firminger, Superintendent, Convict Establishment. *ARC*, 1899, p. C9. The subordinate ranks of the prison staff was composed of overseers, sub-overseers, guards and prison constables. The latter were drawn from well-behaved third stage prisoners. They were in charge of up to six prisoners and were paid one rupee per month. They were seen as the weakest link in the prison system. *ARC*, 1872, p. 436.

their posts, and are not to be trusted in the slightest degree, and bribery and corruption have poisoned the whole establishment."[64] When the authorities attempted to enforce more discipline in the jails in 1874, they identified the guards as a key problem.[65] The Inspector General wrote in his annual report of 1875 that a guard "ought not to be paid, as at present, less than an ordinary cooly." In fact, the separate system itself was called into question as unworkable, for "guards and overseers are so untrustworthy that there are far more opportunities for guards to favour or spite a prisoner."[66] Prisoners regularly complained that guards refused to let them out of their cells at night to use the urinals, knowing that the prisoners would be punished in the morning for soiling their cells. Consequently in 1879 night pans were put in the cells to stop this practice, as well as to stop prisoners from harassing guards by making multiple requests to go to the toilet.[67] By the late 1880s, although guards in the convict establishment were paid a labourer's salary of fifteen rupees a month, guards in outstations were only paid nine rupees.[68] The Inspector General argued that it was unsurprising that guards at outstation jails were untrustworthy and inefficient and, until their pay was increased by at least 50%, there was little chance of any improvement.[69]

In the mid-1880s the AGA of Matale wrote in his official diary that the guards at his jail were "starving on the crumbs that fall from the prisoners' tables" and that their salaries must be doubled to stamp out corruption.[70] The Inspector General reported that guards regularly exchanged favours with prisoners for a share of the latter's food.[71] The Report of the Prison Committee on the condition of the jails stated that "a large number of under-guards in the Colombo jails, and especially at the Convict Establishment are very unfit from every point of view to enforce discipline."[72] In 1890 the Inspector General admitted that the problem of corrupt guards was insoluble as long as they could earn as much smuggling contraband in a week from prisoners' friends as the government paid them in a month.[73] The authorities attempted to combat this by randomly searching guards for contraband and while this reduced its supply they continued to risk getting caught in order to maintain this profitable trade.

64 *ARC*, 1873, p. 30.
65 The same problem of underpaid, untrustworthy guards was found throughout India. Arnold, D. 2007. "India: the contested prison," in F. Dikotter and I. Brown (eds.), *Cultures of Confinement: A History of the Prison in Africa, Asia and Latin America*. London: Hurst, p. 17; Chakrabarti, R. 2009. *Terror, Crime and Punishment: Order and Disorder in Early Colonial Bengal, 1800–1860*. Kolkata: Readers Service.
66 *ARC*, 1875, p. C3.
67 *ARC*, 1875, p. C3; 1879, p. C112.
68 Inspector General of Prisons, *ARC*, 1880, p. C121; *ARC*, 1889, p. C5.
69 *ARC*, 1881, p. C41.
70 Diary of AGA Matale, September 2, 1887. *ARC*, 1887, p. A101.
71 *ARC*, 1881, p. C30.
72 *SP*, 1887, p. 218.
73 *ARC*, 1890, p. C3.

Prisoners colluded with guards by refusing to reveal their source and saying simply that they found it lying on the ground.[74]

The supervision of the task system provided another important source of illicit income for guards, as they were required to weigh or measure and log each man's task at the end of the day. In the 1880s steps had to be taken to stop overseers from accepting bribes to award prisoners a full task when they had failed to finish. A superior officer was required to measure the guards' totals before handing out punishment or marks.[75] While prison authorities worried primarily about prisoners bribing or threatening guards, upon enquiry it was discovered that "many of the subordinate officers, especially the Malays ... did not hesitate to bring false or exaggerated charges against the prisoners for the purpose of keeping up their character as smart disciplinarians." The Inspector General continued, "taken as a lot, the Malay subordinates are not reliable or desirable men ... and while ... perhaps superior to our other races as disciplinarians, are just as inferior as supervisors of work."[76]

It is interesting to see how guards used the British anxieties about their lack of local knowledge and rapport with the Ceylonese to their own advantage. As in the police and court bureaucracies, British prison officials were unsure whether to believe their own subordinate officers. As the Inspector General wrote in 1906, "it often happens that the man with the fewest reports is the man in whom you can place least trust, simply because he is artful."[77] Consequently, jail rules were changed to deal with this uncertainty. In 1879 both prisoners and guards were placed under increased surveillance by a "superior who is above all suspicion of personal feeling ... [to allow] ... the truth to be recorded before evidence can be concocted on either side." Prisoners were allowed to call witnesses when accused for the same reason. The following year, a second guard was added to groups of prisoners working outside the prison, not only to discourage escapes, but "to prove charges of idleness or insubordination when all prisoners claim it wasn't so."[78]

It is predictable that the guards showed little commitment to an institution that treated them so badly. Only those desperate for work turned to the prison service, and those who did expected to profit illegally from the job. In the early

74 *Report of the Prisons*, p. 93.

75 *ARC*, 1891, p. C7.

76 *ARC*, 1876, p. 21C. Reservations were also expressed about the abilities of second-generation Europeans who grew up in the tropics due to degeneration and/or lack of exposure to European values. For example, in 1877 the Inspector General of Prisons argued that while jailers born in England should be hired at the top level of the prison service, that the service should consider hiring "country-born or bred" men as "heads of second class jails, and valuable assistants to the English trained jailers. The latter will keep the discipline, regularity, and method up to a high standard, while the former, with their knowledge of the language, and native ways, will be a medium of communication, and able to keep their superior warned against the tricks of the native subordinates." *ARC*, 1877, pp. C66, C97.

77 A.W. de Wilton, *ARC*, 1906, p. C12.

78 *ARC*, 1879, pp. C89, C107; 1880, p. C119.

twentieth century there were reports of guards letting prisoners know that their time in prison could be happier if their families offered bribes.[79] The guards often found themselves in league with the prisoners against senior jail officials. The extent of this collusion is revealed by the fact that in 1901 guards in Ceylon were punished 1,404 times. As the Inspector General pointed out, this meant that more guards were punished per capita than prisoners. By 1914, the number of punishments for guards had been reduced; however, guards continued to commit more violations than prisoners.[80] Assuming that guards were often not caught, there were probably many thousands of violations.[81] Thus guards can be considered liminal figures, representatives of the administration who were little more trusted by the prison officials than the prisoners they guarded. Their ambivalent relationship with the higher administrators and the prisoners complicated the oppressed/oppressor relationship, and there was never a clear-cut dichotomy in any of the branches of the criminal justice system.

Open resistance: violence against guards

In 1869, before the new directives from London to make prison life harsher had been implemented, James T. Fitzmaurice, the Inspector General of Prisons, claimed "that there is no such thing as prison offences among the native prisoners." He reasoned that prisoners have "neither sufficient strength of character or physique, to be disposed to rebel against a well-defined authority … [and] if the Sinhalese have no other specialty, they certainly have that of making the best prisoners imaginable."[82] A more plausible explanation for why prisoners appeared docile during the early years was that the combination of lenient prison conditions and draconian punishment, such as flogging for attacking guards or attempting to escape, led prisoners to conclude that open resistance would be irrational.[83] However, as I have discussed, by the early 1870s, the new reformed penal philosophy dictated that prisons become more oppressive. In addition, the termination of transportation of prisoners to the Straits

---

79 *Report of the Prisons*, p. 93.
80 L.F. Knollys, *ARC*, 1901, p. C6; A.W. De Wilton, *ARC*, 1914, p. C3; *ARC*, 1902, p. C7. Over the decades the punishments were given for neglect of duty, absence without leave, late for duty and trafficking. *ARC*, 1884, p. C26.
81 Research on contemporary prisons in North America reveals a similar pattern of resistance to prison rules by both prisoners and guards. See Arford, T. 2016. "Prisons as sites of power/resistance," in Courpasson, D. and S. Vallas (eds.), *The Sage Handbook of Resistance*. Los Angeles: Sage, pp. 385–417; Ross, J.I. 2009. "Resisting the carceral state: prisoner resistance from the bottom up." *Social Justice*, Vol. 36, No. 3, pp. 28–45; Ross, J.I. 2013. "Deconstructing correctional officer deviance: towards typologies of actions and controls." *Criminal Justice Review*, Vol. 38, No. 1, pp. 110–26.
82 Report on Prisons, *ARC*, 1869, p. 207.
83 The standard punishment for escape was fifty lashes with the cat o' nine tails and an additional year of hard labour. Dickman, C. 1872. *The Ceylon Civil Service Manual*. Colombo: F. Fonseka, Government Printer, p. 128.

Settlements led to the Ceylon jails housing prisoners convicted of more serious crimes with longer jail terms.[84] It was assumed that these long-term prisoners had less to lose by open resistance than did short-sentenced men. In light of all this, it is hardly surprising that there were some instances of open resistance on the part of prisoners.[85] In the 1870s, new head jailors were brought from England to tighten discipline, and throughout the early 1870s there was sporadic violence on the part of inmates against guards as a reaction to the increased discipline. By 1875 there was "constant punishment of prisoners for having betel and tobacco in their possession and for small acts of insubordination." The screws were tightened further at this time by weeding out some of the most corrupt guards.[86] While the administrators believed that the violence was primarily due to resistance to increasingly arduous prison conditions, they also admitted that tension was caused by their own "ignorance of the language and native ways and misconception of orders, both by subordinates as well as convicts."[87] In fact, prisoners often complained that they were unaware of prison rules as they were not posted. The standard practice for new inmates was to ask other inmates to explain the rules, but often they were misinformed, or they would ask guards, but the guards were not to be trusted as they were known to deliberately give wrong answers. A prison inquiry supported the latter view as "the large mass of prison rules and regulations afford an officer ill-disposed towards a prisoner ample opportunity to persecute a prisoner with no risk to himself."[88]

Violence against guards and other personnel began to escalate dramatically during the mid-1870s in response to the arrival of three strict disciplinarians from England. In 1876 a Burgher medical officer was assaulted in Welikada and an old Ceylonese officer in the same prison was beaten to death.[89] The prison authorities treated these two serious assaults as random events, lamentable, but within "normal" levels of violent resistance to strict prison conditions. They were, however, deeply shocked by the attacks on European jailers the following year. The first of these attacks was on Mr Castle, the recently arrived strict English jailor at the Breakwater Prison in October 1877. Two prisoners in a work gang on the site attacked him with crow bars, striking him first on the

---

84 Several thousand convicts were transported to Mauritius and South East Asia between 1815 and 1868. Anderson, C. 2007. "Sepoys, servants and settlers: convict transportation in the Indian Ocean 1787–1945," in Dikotter, F. and I. Brown (eds.), *Cultures of Confinement: A History of the Prison in Africa, Asia and Latin America*. London: Hurst, p. 188.

85 It would appear that there was a higher level of violence in Indian jails than in Ceylonese ones. See Anderson, C. 2007. *The Indian Uprising of 1857–58. Prisons, Prisoners and Rebellion*. London: Anthem, pp. 36–39.

86 *ARC*, 1875, p. 2.

87 Inspector General of Prisons, *ARC*, 1881, p. C41.

88 *Report of the Prisons*, pp. 88, 93.

89 Inspector General of Prisons, *ARC*, 1881, p. C 41; Inspector General of Prisons, *ARC*, 1877, p. C65.

arm so he could not draw his revolver, and then beating him on the head and knees to incapacitate him. While it was not uncommon for prisoners to come to the aid of a jailer who was being attacked, as extra marks were given for protecting jailers, as well as giving information on plots, preventing escapes, and preserving order during disturbances, in this case none of the working party including the overseer and guard made any move to help Mr Castle, who was rescued by a European engine driver working nearby.[90] It is clear that the assailants intended to remove the guard from the prison rather than kill him, for they certainly had the time and the means to do the latter. The assailants were sentenced to forty-five and fifty lashes apiece; the other prisoners who simply watched lost ninety good conduct marks apiece and the overseer and guard were dismissed, while Mr Castle returned to England to recover.[91] The following January, a foreman by the name of Hobbs was attacked with a crowbar by a prisoner at the Breakwater Prison. Another prisoner rushed at Hobbs with a crowbar from the other side but was knocked down by a mason before he reached him. The prisoner claimed he attacked Hobbes for reporting him the day before for negligence. He was sentenced to twenty-five lashes. The Inspector General concluded that both of these attacks were premeditated, and an inquiry revealed the intention was to incapacitate jailers who were considered strict disciplinarians. More troubling still, the Inspector General concluded that:

> There were indications too that an organised system of intimidation was in existence, and that with a few exceptions, most of our non-European officers were afraid of our convicts, while the assaults on Messrs. Castle and Hobbs were attempts to try to coerce the few who were not afraid to do their duty.[92]

The Inspector General believed furthermore that attacks on European officers were one of the ways older convicts "intimidated the new comers and prevented them from doing as much work as they might."[93] This resetting of expectations, along with violence towards newly arrived prisoners who were unwilling to collude, were important ways in which "go-slow" was enforced as an evasive strategy on work projects. The great increase in violence in the mid-1870s can be seen as an attempt by prisoners to re-establish the norm in terms of allowable deviation from prison rules which had existed up to that time. Violence was a high-risk negotiating tactic, but it could be effective.

---

90 E. Elliott, Inspector General of Prisons, *ARC*, 1882, p. C41; W. Thompson, Superintendent, Convict Establishment. *ARC*, 1882, p. C48.

91 Inspector General of the Convict Establishment, *ARC*, 1878, pp. C64–65.

92 *Ibid.*, p. C65. The intimidation of guards by some prisoners was considered a problem in England as well. Priestley, *Victorian Prison Lives*, p. 257.

93 *ARC*, 1878, p. C 65.

Two strategies were put in place in an attempt to forestall future violence. Firstly, an armed European guard was placed in the Breakwater yard with orders to shoot any convict who attacked a European jailer with a crowbar or hammer.[94] Secondly, in 1877, a separate section for violent prisoners supervised by European guards was created at Welikada.[95] These unruly prisoners were made to work in the same long shed where European prisoners were doing light labour. The authorities reasoned that should the prisoners attack the European guards, the European prisoners would show racial solidarity and take the side of the guards.[96] For the first few years, the twenty or so convicts in the special section gave little trouble. But in early 1883 trouble began to flare up there as well. Soon there were many acts of insubordination and prisoners fighting among themselves.[97] The authorities responded by making prisoners work in stalls with barred doors where they were unable to see or communicate with one another. Discipline was further enforced by the rule that prisoners could not return to the general prison population until three months after their last infraction of the rules.[98]

In spite of these measures, assaults by prisoners continued, although at a fairly low level, throughout the 1880s. During the mid-1880s, it became clear to the authorities that some of the incidents of violence were organised by fellow inmates, but more worryingly, prisoners were communicating between prisons. Within less than a month in the summer of 1884 there were four major disturbances in prisons around the island. The most serious took place in Galle where prisoners mutinied, attacked guards, broke into the tool stores arming themselves with pickaxes, crowbars, and so on, and then took over the jail until armed police were brought in to put down the uprising. This was followed a week later by assaults on guards at the Banksall-Street works in Colombo. A few days later there was an uprising at Mahara, followed a week later by another disturbance in Galle jail. All of the prisoners involved in these uprisings were punished and the ringleaders flogged and isolated in the Special Unit in Welikada.

Guards who made prisoners work harder or faster than the norm on outdoor work sites or during shot drill were more often assaulted.[99] In 1888 there were two separate escapes by prisoners working on land reclamation along the shore in Colombo. In both cases they attacked their guards, crippling one for life,

---

94 *ARC*, 1881, p. C 41.

95 *ARC*, 1877, p. C64.

96 Inspector General of Prisons, *ARC*, 1880, p. C119.

97 This is one of the few mentions in the prison reports of prisoner on prisoner violence. One can assume, however, from reports about intimidation of new prisoners that violence, sexual and otherwise, especially in the wards at night existed.

98 W. Thompson, Superintendent of the Convict Establishment, *ARC*, 1883, p. C62; *ARC*, 1885, p. C42.

99 W. Thompson, Superintendent of the Convict Establishment, *ARC*, 1886, p. C55. Assaults on guards supervising labour were common in nineteenth century India as well. Anderson, *The Indian Uprising*, p. 32.

before fleeing. When recaptured they gave as their reasons unhappiness with the severe treatment by guards.[100] In that same year there were four serious outbreaks in provincial jails around the country where guards were savagely beaten or murdered.[101] Assaults on officers were significantly diminished by the early 1890s due to the shift from shot drill to pingo carrying as the most common form of hard labour. As we have seen, it was difficult for prisoners to evade hard labour with the pingo system and so there was much less intervention by guards and correspondingly less hostility towards them.[102]

It was also believed that disputes between prisoners and guards over goods smuggled into prison were responsible for a large number of the reported assaults on guards by prisoners.[103] The authorities took a much more tolerant view of assaults on Ceylonese than on European guards, especially when they thought the reason was a dispute over contraband, thinking that such assaults might help curtail the practice. Overall, the reduction of violence by the early 1890s was such that the special section in Welikada was virtually empty by 1893.[104]

At times, attacks on guards were followed by attempted escapes. The number of escapes shot up from the mid-1870s to the mid-1880s as a reaction to harsher prison conditions, but dropped down again in subsequent decades as prisoners adapted their tactics to the new conditions. In the five-year period ending in 1890, of the 175 prisoners who escaped, 170 were recaptured. During the early twentieth century, the number of escapes declined, and the recapture rate remained high. Between 1897 and 1914, 167 escaped and 162 were recaptured.[105] After 1884, the chances of successful escape were very slim.

Given the high odds of being caught and flogged, the decision to escape would appear to have been based on purely emotional reactions to harsh treatment or utterly unrealistic views of one's chances. As ever, open resistance to power was high-risk and entailed a willingness to sacrifice everything for one's freedom. As the Superintendent of the Convict Establishment claimed, "a man who escapes must begin a perfectly fresh life, relinquishing his property and abandoning his family."[106] But this did not mean that escapes or even the threat of escapes didn't impact on the organisation of prison life, because clearly they did. As we have seen, guards were afraid of being punished for allowing prisoners to escape and they responded by making life easier for prisoners in extramural working parties. Thus, the mere threat of escape opened up spaces of resistance for prisoners to do somewhat less work or receive contraband and other favoured treatment. Even the Inspector General felt his

100 G.W.R. Campbell, Inspector General of Prisons, *ARC*, 1888, p. C48.
101 G.W.R. Campbell, Inspector General of Prisons, *ARC*, 1888, p. C49.
102 L.F. Knollys, Acting Inspector General of Prisons, *ARC*, 1896, p. C8.
103 M. Mooney, Acting Superintendent of the Convict Establishment, *ARC*, 1887, p. C53; *Report of the Prisons*, p. 95.
104 R.E. Firminger, Superintendent Convict establishment. *ARC*, 1893, p. C14.
105 F.R. Ellis, Director and Inspector General of Prisons, *ARC*, 1891, p. C3; A.W. DeWilton, Inspector General of Prisons, *ARC*, 1914, p. C2.
106 R.E. Firminger, *ARC*, 1890, p. C1.

hands were tied in that knowing that if he tried to make jails more deterrent by increasing hard labour and decreasing food, there would be more escapes.[107]

The doctors: the normalisation of sickness and death rates

Like guards, doctors also occupied an ambivalent role vis-à-vis prisoners in that, on the one hand, they were involved in the periodic reframing of the diet and labour requirements of the harsh penal regime, and on the other hand, often took seriously their role of protecting prisoners when punishments exceeded the prisoner's sentence or endanger their health. It would not be too strong to say that the different remits of the doctors and prison administrators put them on a collision course. It is possible that this conflict in Ceylon was exacerbated by racism, as the senior jailers in the convict establishment were British and the majority of the prison doctors were Burgher.[108] In addition to inspecting the sanitary state of prisons every three months, doctors were to inspect every prisoner once a week, and those confined in solitary punishment cells, once a day. The doctors regularly ordered reduced tasks for weak prisoners and admitted men to hospital over the objections of the jailers.[109] The fact that prison administrators had the authority to suspend any officer other than the medical officer and the matron who was in charge of the welfare of women prisoners was testament to the importance assigned to their role.

Prisoners routinely feigned sickness, trying to play the doctors against the jailers. Some prisoners even made themselves ill by eating soap to cause diarrhoea and scratching themselves until their wounds became infected.[110] In 1878, the Inspector General of Prisons warned the guards not to conduct tests that were "certain to cause great bodily harm" to prisoners who claimed to be ill,

---

107 G.W.R. Campbell, Inspector General of Prisons, *ARC*, 1884, p. C24. Anderson in her study of Indian prisons also found that various forms of prisoner resistance influenced the management of prisons. Anderson, *The Indian Uprising*, p. 36.

108 In England there were similar conflicts between doctors and jailers over prisoners' labour and diet. Priestley, *Victorian Prison Lives*, p. 169. Conflict was reduced somewhat in India by having the position of Inspector General of Prisons and the superintendents of most central jails filled by members of the Indian Medical Service. Arnold, *The Contested Prison*, p. 167. Until the last two decades of the nineteenth century, a career in the imperial service was seen as low status for British doctors. This began to change due to crowding in the medical field in Britain and the rise in prestige of tropical medicine at the end of the century. Haynes, D.M. 1996. "Social status and imperial service: tropical medicine and the British medical profession in the nineteenth century," in Arnold, D. (ed.), *Warm Climates and Western Medicine: The Emergence of Tropical Medicine, 1500–1900.* Amsterdam: Rodopi, pp. 208–26. On the prestige of tropical medicine beginning at the end of the nineteenth century see Harrison, M. 1992. "Tropical medicine in nineteenth century India." *British Journal of the History of Science*, Vol. 25, pp. 299–318. It is interesting to note that during the last decade of the nineteenth century, just as tropical medicine as a specialty was becoming prestigious, Burghers were losing prestige in Ceylon, for as I have noted above the hardening of racism produced a shift in the view of mixed race Burghers.

109 *CBB*, 1887, 524.

110 In India, prisoners used poisons like arsenic that was kept to kill rats to enlarge sores so that they could escape hard labour. Anderson, *The Indian Uprising*, p. 36.

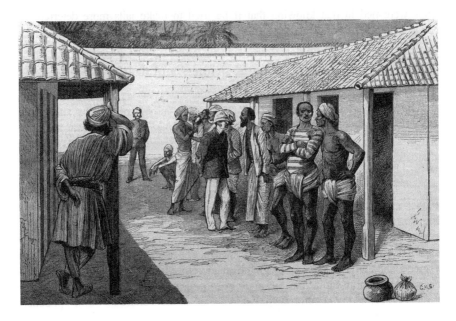

Figure 8.2 The Doctor's Prison Parade.
(Source: *The Graphic*, October 6, 1883).

but, if no evidence of the alleged illness could be found by doctors, then the prisoner was to be sent back to hard labour. If the prisoner continued to refuse to work, then he would be put in solitary confinement at two-thirds rations. If after thirty days he still refused to work, then he should be re-examined by the prison doctor and, if found healthy, be flogged, put in irons, or have a further six months added to his sentence.[111] Feigning insanity also became more common at the end of the century as prisoners discovered that doctors had begun to consider madness a disease. To discourage this deceit it was made a flogging offence.[112] We can see that the regimes of health and punishment set off, as Sen puts it, "guerrilla wars" between prisoners and doctors and between doctors and jailers.[113]

When the government commissioned a review of prisons in 1889, one of its briefs was to resolve this conflict. The head of the commission, a policeman,

111 *ARC*, 1878, p. C83. Persistent "malingering" continued to be punished by corporal punishment well into the twentieth century. *Report of the Prisons Inquiry Commission, 1931*. 1932. Colombo: Ceylon Government Press, p. 58.
112 *Report of the Prisons Inquiry Commission, 1931*. 1932. Colombo: Ceylon Government Press, p. 108.
113 Sen, *Disciplining Punishment*, pp. 133, 19.

predictably came down on the side of harsh methods of deterrence and recommended that

> the medical officer should be looked upon as an adviser and not as a dictator, the advice of a second medical officer or of a visitor being taken where there was a difference of opinion as to the fitness of the prisoner for work.[114]

In the opinion of the inspector general of prisons, it was "very difficult to get the two authorities to view matters in the same light or work towards the same end."[115] It was clear from the various government commissions and annual reports that monitored both the effectiveness of penal regimes and sickness and death rates of prisoners, that this structural contradiction between the remit of the administrators and the medical officers had caused continual conflict over the decades. In the mid-1870s a Ceylonese member of the Legislative Council felt it necessary to remind jailers that "so long as a man has not been sentenced to death, Government was bound to take reasonable precautions to keep him alive." Six years later, jailers again had to be reminded that the old guideline that "no permanent injury to a prisoner's health" was still in force.[116] By accusing doctors of overfeeding and underworking prisoners, "in order that the sick rate and death rate may be kept as favourable as possible," Inspector-General Campbell reveals the dehumanising attitudes of administrators towards prisoners as mere statistical objects. And the following year the medical officer of the Convict Establishment argued in reply that doctors did only the minimum to maintain prisoners "in ordinary health and to save life," and that "to endanger their lives would be both unlawful and inhuman." Some superintendents, such as the strict disciplinarian F.R. Ellis in the mid-1890s, despaired saying, "the day may come when Government will have to choose between a deterrent system of imprisonment and a low sick rate."[117] The struggle within the prison system was not so much over whether the disastrous effects of imprisonment were in themselves inhumane and unlawful, but over what were considered "acceptable" rates of sickness and death.

Outstation jails: a carceral shadowland

In spite of the principle of uniformity underpinning penal reform theory, there was, throughout the century, significant geographical variation in penal practice. While Welikada Prison in Colombo adopted the most systematic and rigorous approach, followed by Kandy and Jaffna prisons, the smaller jails were more

---

114 Giles, *SP*, 1889, p. 373.
115 *ARC*, 1896, p. C7.
116 *Ceylon Hansard (CH)*, 1875–76, p. 75; *ARC*, 1881, p. C44.
117 *ARC*, 1886, p. C55; J. Attygalle, *ARC*, 1887, p. C58; *ARC*, 1893, p. C2.

lax.[118] The Inspector General of Prisons in 1871 offered the following in his damning report on prison discipline in the smaller jails:

> Prison discipline in any form is very little known or understood at any of the minor gaols, which term includes three-fourths of the whole. Shot drill, as a mode of enforcing hard labour, has, it is true, been introduced at most of them, and is professedly put in practice; but on inspection it is only too clear, by the slovenly manner in which it is performed, and occasionally by the entire ignorance displayed both by gaoler and prisoners, that the practice is in reality little more than the profession.[119]

The Inspector General continued,

> the hard labour of a prisoner at these gaols is reduced to mere desultory work, very rarely making or repairing roads, and if so, under no proper supervision. His life is one of comparative ease and comfort; he works but little, and in most cases fares very much better than he would do at home.[120]

He concluded by saying that the new system of penal discipline had not, in fact, been given a fair trial at any of the small prisons. Prison inspectors argued that the failure of discipline was largely due to the lack of supervision and enforcement of prison routine. P.A. Templer, the Fiscal of Kandy Prison, claimed that the guarding of prisoners in outstation jails was "utterly defective" throughout the island, because guards, and especially the peons, were paid only a pittance so that they were forced to become corrupt.[121] It had, he wrote, "become notorious beyond question that no jail peon can be trusted to withstand the temptation of a bribe."[122] In one case described in the press as particularly "brazen," a jail peon in Ratnapura regularly released two prisoners at night to commit robberies and then, after sharing the proceeds, locked them up again before dawn.[123]

118 In Welikada Prison there was a prison officer to every fifteen prisoners. Other prisons had higher ratios of prisoners per guard. Arnold, "The colonial prison," p. 153 notes that in India as well, administrators worried that they were unable to effectively enforce hard labour.

119 *ARC*, 1871, pp. 356–57.

120 *ARC*, 1871, pp. 356–57.

121 Precisely the same complaints about corrupt, inefficient guards were made in Britain and in India. In fact, such was the concern about guards, who it must be remembered were drawn from much the same classes as the prisoners, that Bentham's panopticon was set up not only to be able to survey prisoners but guards as well. Ignatieff, M. 1978. *A Just Measure of Pain: The Penitentiary in the Industrial Revolution 1750–1850*. London: Penguin, pp. 77–78, 105, 191; Arnold, "The colonial prison," pp. 154, 186.

122 *ARC*, 1873, p. 36.

123 *Colombo Observer*, July 10, 1863.

In 1874 a new manual was circulated to all jails in the hope of standardising procedures across the island. However, prisoners in the outstation jails, away from the administrative centre, continued to have a relatively easy time compared to those in the convict establishment in Colombo. They were given small tasks such as cleaning the jail and the town streets and occasionally light road repair.[124] This was defended on several grounds: first that guards capable of enforcing hard labour could not be hired for the pay, second that most prisoners in outstation jails had not been convicted of serious crimes, and third, that the towns relied on cheap prison labour to keep their streets clean and to help with minor public works.[125]

The rule of racial difference: European prisoners

Although the calculations of labour capacity were originally based upon the larger European body, ironically European prisoners were rarely worked hard in Ceylon.[126] The issue of hard labour for Europeans is interesting, because it lies at the intersection of a number of British colonial anxieties concerning, on the one hand, the negative impact of tropical climate on temperate constitutions, and on the other, that it should not appear obvious that there were different rules of justice for different races. Moreover, there was a more general fear that Europeans might collectively lose face if some among them were shown to be criminal. The European criminal was a profoundly troubling social type for the British in Ceylon because his very presence was thought to belie the claims to moral and intellectual superiority that underpinned imperial rule and thus, British criminality itself struck at the very heart of the legitimacy of racialised rule.[127] Anderson points out that European prisoners were considered "first as British rather than as criminals" and thus were treated as racially superior and more deserving of respectful treatment.[128] Ten years after the Committee on Prison Discipline published its recommendation for the reform of prisons, it was clear that European prisoners were not expected to do as much of the hard labour specified in the recommendations as non-European prisoners. This negated the principle of uniformity undergirding the liberal penal ideal. The

---

124 G.W.R. Campbell, Inspector General of Prisons, *ARC*, 1884, p. C24.

125 *ARC*, 1879, p. C111; *ARC*, 1883, p. C57.

126 The same policy was followed in India and throughout the Middle East. Gorman, A. 2007. "Regulation, reform and resistance in the Middle Eastern prison," in Dikotter, F. and I. Brown (eds.), *Cultures of Confinement: A History of the Prison in Africa, Asia and Latin America*. London: Hurst, pp. 95–146; Sen, *Prisons*, p. 225.

127 The very same concerns were expressed in India. Arnold, "The colonial prison," p. 170; Arnold, "India: The contested prison," p. 150.

128 Anderson, *Legible Bodies*, p. 121. The French in their penal colony on Devil's Island also used race as a basis for the allocation of food, space and labour. Redfield, P. 2005. "Foucault in the tropics: displacing the panopticon," in Inda, J.X. (ed.), *Anthropologies of Modernity: Foucault, Governmentality and Life Politics*. Oxford: Blackwell, pp. 56–57.

Inspector General of Prisons justified this position by arguing that the same sentence would be harsher for the European than the non-European races if the former were required to do hard labour in the tropics. He wrote,

> whatever may be its effects on the native prisoners, the penal stage is most severe on the Europeans, so much so that the penal stage (though reduced to three months) is in few instances carried out in its entirety. After a few weeks of penal diet with four hours a day shot drill and six hours of beating coir, &c., in a tropical climate, the European prisoner is either a patient in hospital or is declared by the medical officer to be fit only for 'light labour,' the sentence passed on him becoming to a certain extent cancelled … My opinion is that (both to further the ends of justice and for his own sake) when a European prisoner is sentenced to hard labour, he be at once transferred to a climate where the sentence can be properly and completely carried out without injury to his health.[129]

This statement captures one of the central dilemmas of liberal, penal ideology. Should prisoners of different races should be given different levels of punishment on the grounds of equity? This view was premised on the assumption that the European-born were physically weaker than the native born because of the effects of the tropical climate which paradoxically challenged the commonly held view that South Asian men were inherently weak and more feminine than European men. The issue of European prisoners and hard labour provoked heated debate among members of the Legislative Council when the government sought the permission of the Crown to send long-term European prisoners sentenced to hard labour to prisons in Gibraltar. The Colonial Secretary, in making his case before Council, cited two recent cases of Europeans sentenced to hard labour. The first was sentenced to ten years, but pardoned after two on the advice of medical officers, who claimed he could die. The second was moved after six months of his seven-year sentence to a seaside jail where he would experience fresh sea air and thus would survive his sentence. As recorded in the *Ceylon Hansard*, prisoners were often removed from Ceylon entirely because:

> there could be no doubt the European prisoner could not in this climate do the hard labour to which he was sentenced in common with natives, and consequently the invidious distinctions were made, where one prisoner was, according to the medical officer's opinion, unable to do anything but sit idly all day, while another, sentenced, perhaps, at the same time, and maybe for a similar offence was strictly undergoing the hard labour which the sentence of the Court required.[130]

129 *ARC*, 1876, p. C20.
130 *CH*, 1875–76, p. 171.

The Sinhalese member of Council asked if this might also apply to Burghers and those of European ancestry born in the Island.[131] He was told by the Queen's Advocate that this proposal was solely for European-born prisoners. The Governor added that he saw this removal as necessary for otherwise the government would be forced to choose between sanctioning the death of Europeans or giving the impression that there was one law for Europeans and another for the Ceylonese.[132] The Tamil member of Council argued against the motion, saying that his "native friends did not look with favour on this motion. When told it was done in a spirit of mercy, they were by no means contented with this explanation." Rather, he said people assumed that the policy was corrupt and that "once away from here, pardons and remissions of sentence might be the rule and European offenders thus let go scot free. Such are the fears of the people."[133] In spite of objections by the two non-European members, the motion was passed. Not all European prisoners sentenced to hard labour were thereafter sent to Gibraltar, but those who remained in Ceylon were in fact not subjected to hard labour.

The issue arose again in 1883 when the Secretary of State for the Colonies introduced into the House of Lords a Bill which gave the Colonial authorities power to remove to England any prisoner "who has been born and brought up in a temperate climate," on the grounds that their health was likely to be affected by imprisonment in a tropical climate. W. Thompson, the Acting Inspector General of Prisons supported this move as

> it can hardly be said that there is a single European prisoner who could undergo a lengthened imprisonment in this country without his health being affected by it ... certainly not if the sentence is carried out with any approach to due severity.[134]

G.W.R. Campbell, the acting Inspector General of Prisons, was less sympathetic when he wrote in 1884 that European prisoners "have always had an easy time of it on account of the dangers of the climate, real or alleged." What was needed for Europeans, he argued, was to create a programme of "such really hard and irksome labour as can be safely exacted in this climate."[135] In 1887 a Prisons Committee appointed by the governor concluded that Europeans in practice were given preferential treatment by guards and their "so-called hard labour" consisted of an hour's walk from seven to eight in the morning and from four to five in the afternoon.[136] One of the recommendations of the committee was that European prisoners "do the same kind and amount of work as

---

131 *Ibid.*
132 *Ibid.*, pp. 171–72.
133 *Ibid.*, p. 172.
134 *ARC*, 1883, pp. C56–57.
135 *ARC*, 1884, p. C24.
136 *SP*, 1887, p. 228.

the native prisoners, care being taken to keep Europeans and natives separate from one another."[137] But this was never carried through. These biopolitical debates about the impress of the tropics on European constitutions call into question the extent to which the science of climatic determinism was believed and the extent to which it was used strategically to favour prisoners who were thought of as Europeans first and criminals only secondarily. The answer probably is complex. The medical and the political rationalities were thoroughly interpenetrated in that the imprimatur of science, whether truly believed or not, lent credibility to that which was politically desirable and feasible.

Making prisons remunerative

As I have outlined, in the mid-1860s Whitehall ordered that all prisoners do hard, monotonous, un-remunerative, intramural labour during the first six months of their sentence, as such work was thought to be most deterrent. However, as I have shown, the British government was unwilling to adequately fund the new reformed prison system, so the Ceylon government sought to help defray the costs of prison by having prisoners do some useful labour as well. From the start, the location and size of prisons was largely influenced by economic considerations. There was a market for prison labour in Colombo, which was the site of the main penitentiary and 100 convicts were assigned to the salt works at Hambantota because labour could not be recruited to do such back breaking and unpleasant work. In the early years, only enough men were imprisoned in Kandy, Galle, Jaffna, and Trincomalee to fill the local labour needs on public works.[138] The Ceylon Government signalled from the outset that, in spite of Whitehall's dictates, their primary concern was that prisons be run as economically as possible. In 1874 the Ceylon government reduced the penal phase of intramural labour from six to three months in order to release more labour to work on the new Colombo breakwater and established a large sprawling prison which included a quarry at Mahara ten miles from Colombo so that prisoners could quarry stone intramurally for the breakwaters.[139]

A decade after the reduction in intramural hard labour took place, the Inspector-General of Prisons wrote that while, "labour entirely intramural is the most deterrent ... I doubt we can afford to get the highest good in this direction at the cost of losing the convict's labour on works valuable for the public good, which could not otherwise be afforded."[140] But there were prison officials who argued against this.

---

137 *Ibid.*, p. 218.
138 Report of the Prison Discipline Commission of 8th July 1869, *SP*, 1869–70, p. 37.
139 Report of J.L. Vanderstraaten, Assistant Colonial Surgeon, Colombo Convict Hospital for 1874. *ARC*, 1874, p. 20.
140 *ARC*, 1886, p. C55. This was also justified as "following the example of the English convict prisons, every effort has been made to put as many convicts as possible on public works." *ARC*, 1881, p. C42. Likewise French prisons concentrated on manufacturing rather than hard labour. O'Brien, P. 1982. *The Promise of Punishment: Prisons in Nineteenth-Century France*. Princeton: Princeton University Press, p. 92.

In 1887 Marcus Mooney, the newly arrived acting superintendent of the convict establishment, condemned the disciplinary regime in Ceylon prisons stating that "by no stretch of language can it be called a system." In some countries, he wrote, prison discipline is "almost a science" while in Ceylon it had been "ignored or misunderstood." Rather, he said, a uniform system of discipline "is sacrificed to an idea that money must be saved to the public by the employment of prison labour."[141] However, even those who supported the idea of prisoners doing remunerative extramural labour admitted that the quality of such work was often unsatisfactory as the prisoners tended to be uncooperative. For example, the Municipal Council of Kandy in 1890 routinely paid the wages of guards supervising prisoners who cleared the silt traps on Kandy Lake, but it reported that the work done by prisoners was "very unsatisfactory, as it was difficult to task them, and even if tasked they did not appear to be liable to any punishment for non-fulfilment of work."[142] The principal evasive tactic used by prisoners was foot-dragging.[143] The British often mistook such resistance for the natural laziness of the "natives" and, as Scott points out, such racialised judgements tended to work in favour of subaltern groups by reducing expectations of the amount of work that could be accomplished.[144]

The AGA of the Central Province blamed the guards, writing that the

> overseers and guards as a rule take no interest whatever in the work as to how it is done, or how much is done, the chief object being apparently to make it as pleasant as possible to the prisoners and to prevent escape.[145]

It is not surprising that this sort of extramural labour was highly desired by prisoners as it tended to be lightly supervised and therefore a ready source of contraband. Friends and relatives would hand prisoners tobacco on the street, or in an enclosed worksite throw it over the fence when the guard was distracted.[146] Such was the concern in the Convict Establishment about prisoners procuring contraband that, wherever possible, prisoners working on the Breakwater projects in Colombo were housed at the projects themselves rather than being marched through the city.[147] Likewise in 1891, "habitual thieves,"

141 Evidence given by Assistant Superintendent of the Convict Establishment Marcus Mooney, *SP*, 1887, p. 228.
142 *ARC*, 1890, p. N8.
143 On the same strategies in Mauritius see Anderson, "The politics of punishment."
144 Scott, J.C. 1990. *Domination and the Arts of Resistance*. New Haven: Yale University Press, pp. 34–35.
145 *ARC*, 1888, p. A312.
146 Report of the Prisons, p. 92. Anderson rightly points out that the desire to make prisons productive made them much more porous. Anderson, C. 2014. "The power of words in nineteenth century prisons: colonial Mauritius, 1835–1887," in Paisley, F. and K. Reid (eds.), *Critical Perspectives on Colonialism: Writing the Empire from Below*. New York: Routledge, p. 201.
147 Ellis, *ARC*, 1891, p. C7.

who were thought to be predominantly Sinhalese, were sent to Jaffna Prison, as it was assumed that they would have no friends or family in this Tamil-speaking area who could smuggle contraband to them.[148] This became the toughest prison on the island, as the intramural penal stage was extended there to twelve months, the maximum time that "native" bodies were calculated to be able to withstand hard labour. Conversely, Tamil habitual criminals served their time in Colombo, also in an attempt to isolate them from their friends outside of prison.[149]

Although, in the early 1890s, prison officials such as F.R. Ellis and R.E. Firminger tried to reinstate the importance of intramural punitive labour and the isolation of prisoners from each other, this position could not be maintained against the tight budget of the colony and the demands of government for cheap labour.[150] By the end of the century, work at the Mahara Quarry Prison increased once again as public works projects came on line and the majority of prisoners in the Convict Establishment were employed extramurally building the Colombo Harbour extension and docks.[151]

The move away from unproductive hard labour was in line with changes in contemporary English practice in the late 1890s. English prison practices, under the harsh, autocratic governance of Edmund Du Cane, had been insulated from newer, more pragmatic, but also more humane, European penological policy and practice.[152] England had not officially participated in the international penological congresses where new ideas were debated from the mid-1870s until Du Cane's retirement in the mid-1890s. In fact, Du Cane dismissed penal reformers as well-intentioned theorists who sentimentalised the criminal.[153] Ceylon, as a colonial backwater, was conservative and less enlightened, even by English penological standards, which helps explain why, into the twentieth century, penal policy remained steeped in utilitarianism and the deterrence principle of "less eligibility," while the newer turn of the century theories of crime, its causes, and prison administration had little impact.

However, the combined effect of the move in English penal circles away from purely punitive labour and the demand that prisons be run as cheaply as possible did in fact deal a crippling blow to the old system of intramural hard labour. The penal stage of intramural labour was reduced to two months and then in 1911 to one month so that prisoners could be employed on public works sooner. In 1909 a system of portable prisons was approved to move prisoners

---

148 *Ibid.*, p. C3.
149 *ARC*, 1892, p. C3; 1894, p. C1; 1893, p. C1.
150 *ARC*, 1890, pp. C6, 7.
151 *ARC*, 1899, pp. C7-8; *ARC*, 1900, p. C7; *ARC*, 1901, p. C6.
152 Wiener, M.J. 1990. *Reconstructing the Criminal: Culture, Law and Policy in England, 1830–1914*. Cambridge: Cambridge University Press, p. 337.
153 Garland, D. 1985. *Punishment and Welfare: A History of Penal Strategies*. Aldershot: Gower, p. 108; McConville, S. 2018. *English Local Prisons: 1860–1900. Next Only to Death*. London: Routledge.

around the country to work on public works projects in isolated locations. Under the circumstances, questions of the separation of prisoners and worries over moral contamination became irrelevant. Those who continued to work intramurally, after the first month of coir beating, were moved to higher value craft work within the prison walls. Such prisoners had largely become a source of slave labour for the government.[154]

This chapter has explored the first part of a two-part programme of dark bio-power in the prisons, informal experiments undertaken to calculate the maximum amount of hard labour that could be demanded in order to push the bodies of native prisoners to their absolute physical limits. These experiments were resisted in numerous ways by prisoners in order to diminish their impact on their bodies. Even in a highly controlled "total institution," convicts were able to routinely enrol prison doctors and guards into their strategies of resistance through dissimulation, intimidation and bribery. Finally, the most systematic violation of rule of law was racially differential punishments allegedly based on climatic science.[155]

154 *SP*, 1908–09, p. lxiii; *ARC*, 1910–11, p. C4; *ARC*, 1911–12, p. C4; *ARC*, 1914, p. C4.
155 In Chapter 5 we saw how medical theories about diseased native bodies were routinely used to avoid Europeans being charged with murder.

9   Determining the limits of bare life

Bare life as defined by Agamben is life unprotected by law.[1] There is a threshold beyond which life ceases to be politically relevant and can be eliminated without legal sanctions. It was this threshold, prescribed by the rule of law, which prison officials, prisoners and doctors negotiated continuously over the decades. Prison administrators sought to calculate the knife-edge between health and long-term bodily damage. Prisoner lives were considered to be neither valuable nor bare life, i.e. expendable.[2] The colonial prison administrators, as a general rule, took seriously their responsibility to punish bodily and severely, but not to let die. And as long as the rate of death and sickness fell within an acceptable, normalised range, then they could feel they were fulfilling their bureaucratic duty and observing the rule of law. Thus, it was politics as much as the actual prisoners' bodies that established this norm.

The science and politics of nutrition

Diet was an "emergent site of political power" during the nineteenth century.[3] In fact, the regulation of food consumption in the management of criminal populations became almost an obsession and, as I will show, a source of anxiety

---

1 Agamben, G. 1998. *Homo Sacer: Sovereign Power and Bare Life*. Stanford: Stanford University Press.
2 Fassin claims that bio-politics is more than power over life. "Contemporary societies are characterised by the legitimacy they attach to life … intervention in lives is a production of inequalities." Fassin, D. 2009. "Another politics of life is possible." *Theory, Culture and Society*, Vol. 26, No. 5, p. 44.
3 Schuller, K. 2016. "Biopower before and below the individual." *GLQ: A Journal of Lesbian and Gay Studies*, Vol. 22, No. 4, p. 635. Fassin uses the terms bio-legitimacy and bio-inequality to conceptualise the issue of differential value of biological life based on citizenship, race, criminality, and other social categories. On the shifting politics of provisioning see, Bohstedt, J. 2010. *The Politics of Provisioning, Food Riots, Moral Economy, and Market Transition in England, c 1550–1850*. Farnham: Ashgate; Nally, D. 2011. "The biopolitics of food provisioning." *Transactions. I.B.G.*, Vol. 36, pp. 37–53. Nally argues that a moral economy of hunger was gradually replaced by the emergence of liberalism and laissez faire, shaping how hunger and scarcity were viewed. With the spread of market society to the colonies, paternalism declined and there was a reworking of the "laws of necessity." Food provisioning was increasingly left to market mechanisms. This harsher, less paternalistic view was reflected even in prisons where the state acknowledged its responsibility to ensure minimal conditions of life.

and conflict between prison administrators and doctors. Food, like everything else in the prison, was managed using a classificatory grid and was allocated on the basis of stage of sentence, race, climate and prisoner health.[4] Jailers sought to calibrate the amount of each type of food allowed in order to ensure that prisoners did not become ill while doing hard labour.[5] For such calculations to be successful, a knowledge of nutritional theory and its relationship to labour was required. By 1848, analyses of the nutrient content of most common foods in Britain had been published, making it possible to more effectively analyse prisoners' diets there. By the 1860s, prison administrators in Ceylon could draw upon several decades of British research to help them develop a regulatory apparatus governing food and labour. Prisons had been shown to be particularly useful sites for the fledgling science of human nutrition because prisoners were fed the same types and quantities of food over long periods of time.[6] Prisoners' bodies thus became sites of what Arnold terms "the medical reconnaissance of society," for knowledge extracted from experiments on captive bodies was seen as applicable to calculating the minimum needed to feed the free poor during famine times.[7] Experiments were constructed in an attempt to ascertain how long prisoners could continue to do physical labour on a diet of bread and water, how much protein a body needs, to what extent vegetables are necessary to maintain health, and if these answers might vary by race, age, sex and initial state of health. By the mid-1860s, doctors in Europe were certain that prisoners

4 On the uses of classificatory grids see Anderson, B. 1991. *Imagined Communities: Reflections on the Origins and Spread of Nationalism*. London: Verso, p. 184.

5 In the early 1840s experiments were conducted in England on prisoners to see how little they needed to eat in order to avoid severely damaging their health. Priestley, P. 1985. *Victorian Prison Lives: English Prison Biography 1830–1914*. New York: Methuen, p. 151.

6 Such experiments could be conducted also because prisoners were considered people of low value. who were under the control of prison doctors (Carpenter, K.J. 2006. "Nutritional studies in Victorian prisons." *The Journal of Nutrition*, Vol. 136, No. 1, pp. 1–8; Tomlinson, M. 1978. "Not an instrument of punishment: prison diet in the mid-nineteenth century." *Consumer Studies*, Vol. 2, No. 1, pp. 15–26). Callous as these early experiments were, they anticipated the even worse Nazi clinics and concentration camps which were the sites of experiments in the name of science to establish the limits of human endurance (Weindling, P. 2017. "Introduction: a new historiography of the Nazi medical experiments and coerced research," in P. Weindling(ed.), *From Clinic to Concentration Camp: Reassessing Nazi Medical and Racial Research, 1933–45*. London: Routledge, pp. 3–33). From the 1940s through the 1970s prisoners in American jails were infected with malaria, typhoid, cholera, syphilis, etc. in the name of science. By the early 1970s, 90% of pharmaceutical testing was conducted on prisoners (Morin, K.M. 2018. *Carceral Spaces, Prisoners and Animals*. London: Routledge, p. 58). These experiments are clear examples of what Fassin refers to as bio-inequality. Some people's lives could be risked for the sake of saving others (Fassin, "Another politics of life is possible").

7 Arnold, D. 1994. "The colonial prison: power, knowledge and penology in nineteenth century India," in D. Arnold and D. Hardiman (eds.), *Subaltern Studies VIII: Essays in Honour of Ranajit Guha*. Delhi: Oxford University Press, pp. 181–83. Also see Miller, I. 2015. *Reforming Food in Post-famine Ireland: Medicine, Science and Improvement, 1845–1922*. Manchester: Manchester University Press, pp. 76–78 on post-famine Ireland.

could not survive long on bread and water alone and that both protein and vegetables were necessary to maintain health.[8]

In addition to experiments on prisoners, inquiries were made into the diet of the free poor, both in England and South Asia, in order to help frame prison dietaries. On the basis of what was known in Britain as "less eligibility," an idea originating in the English poor law, such information was considered necessary in order to ensure that prisoners ate less than the free poor.[9] Debates about prison food among administrators and parliamentarians, as well as in the press from the 1860s on, were often over the credibility of nutritional science. Humanitarians, who thought that prisoners should be fed sufficiently well in order to be strong enough upon release to obtain honest work, accepted the new science of nutrition, while those who argued that prisons would not be deterrent unless food was greatly restricted questioned it. Nowhere was this battle clearer than in the Report of the Committee on Dietaries for English jails, which established policy for penal reform in Britain and the colonies in the mid-1860s. The Report recommended a harsh approach largely disregarding nutritional science. It concluded, "the facts which have been placed at our disposal, have satisfied us that scientific data have but a limited and uncertain practical application … It is to experience, therefore, that we must turn for guidance."[10]

In 1864, in the interest of bringing the reformed prison movement to the colonies, the Secretary of State for India requested that information on the diet of the free and incarcerated populations of India be collected.[11] One such report by Cornish, on the diet of peasants and institutional dietaries in the Presidency of Madras, demonstrated that by the mid-1860s the government was in possession of the most up-to-date information on nutrition.[12] In another report on prison diet in India at the time, Cornish argued that prisoners needed one and a half pounds of grain

---

8 Guy, W.A. 1863. "On sufficient and insufficient dietaries, with special reference to the dietaries of prisoners." *Journal of the Statistical Society of London*, Vol. 26, No. 3, pp. 239–80; Carpenter, "Nutritional studies," pp. 1–8.; Priestley, *Victorian Prison Lives*, p. 251; Rabinbach, A. 1992. *The Human Motor: Energy, Fatigue and the Origins of Modernity*. Berkeley: University of California Press, p. 129.

9 Priestley, *Victorian Prison Lives*, p. 154. For a discussion of the workhouse system in nineteenth-century Britain see Driver, F. 1993. *Power and Pauperism: The Workhouse System, 1834–1884*. Cambridge: Cambridge University Press.

10 Guy, "On sufficient and insufficient dietaries;" Guy, W.A., J. Maitland and V.C. Clarke. 1864. "Report of the committee on the dietaries of county and borough gaols." *Parliamentary Papers*, Vol. XLIX, pp. 563–719. The official prison dietaries were attacked in 1868 by the English surgeon Edwin Lankester in the *British Medical Journal* for ignoring what was known at the time about the negative impacts of semi-starvation on long-term health (see Miller, *Reforming Food*, p. 77).

11 Cornish, W.R. 1864. *Observations on the Nature of the Food of the Inhabitants of Southern India and on Dietaries in the Madras Presidency*. Madras: Gantz, p. 1. Reprinted from the *Madras Quarterly Journal of Medical Science*, No. 15, 1864; Priestley, *Victorian Prison Lives*, p. 154.

12 Cornish, *Observations*. It stated that all foods are formed of four elements: carbon, hydrogen, oxygen and nitrogen, and that a number of minerals such as sodium, potassium and iron are also needed for health. It went on to state that albuminous or nitrogenous elements are used to build up the muscles and nerves and that non-nitrogenous elements such as starches, oils, fats and sugar are used as fuel.

per day, supplemented by vegetables, fish or meat to remain in good health. These recommendations became the basis of famine relief in India in the mid-1870s and represented in Cornish's judgement a base line for the minimum needed to sustain health.[13] The point of gathering such material on Indian diet was two-fold; first to calculate the nutritive value of local foods and second to establish the peasant diet as a norm or benchmark to ensure that prisoners ate less well than the free poor.[14] Lying behind this was a utilitarian conception of prison as a market option that the poor might be tempted to choose if they could eat better in prison than they would as free men.[15] Hence, part of the deterrent strategy was to have prisoners constantly feel "the bite of hunger."

The largely unfounded fear that the poor might choose to be fed by the state rather than work had also haunted Britain throughout the nineteenth century. The New Poor Law of 1834, based on utilitarianism and the writings of Bentham and his disciples, sought to make the workhouse deterrent to paupers by ensuring that conditions there were less attractive than those of the poorest labourer. The problem with devising a metric of less eligibility was that in the mid-nineteenth century many agricultural labourers in Britain had an unhealthy diet consisting mostly of bread with a bit of meat, butter and cheese from time to time, and often went hungry.[16] In Ceylon there were also persistent anxieties about prisons being insufficiently harsh. But the idea that the poor would choose prison was a myth, for while the peasantry were often hungry, starvation was rare and there is no evidence that the poor wished to be incarcerated in order to eat even in the early days of lenient prisons. Such a view of peasant hunger contradicted the oft-repeated assertion that Ceylonese peasants were lazy because they enjoyed a life of tropical ease and abundance. At times, even as harsh a disciplinarian as Inspector General Campbell became exasperated by the claims of certain administrators and members of the non-official European community that life was so good in the jails that it encouraged crime. He wrote, "if his life is so luxurious, whence, I would ask, the crowd of piteous petitioners for even a little remission of sentence, which I am daily sending from convicts to the Government."[17]

While the techniques of dark biopower were developed with a close eye on the adherence to the rule of law, there were ongoing disputes between prison

13 Arnold, D. 1994. "The 'discovery' of malnutrition and diet in colonial India." *The Indian Economic and Social History Review*, Vol. 31, No. 1, p. 6.

14 Yang discusses this policy in early nineteenth century India in Yang, A.A. 1987. "Disciplining 'natives': prisons and prisoners in early nineteenth century India." *South Asia*, Vol. 10, No. 2, p. 31. Also see Vernon, J. 2007. *Hunger: A Modern History*. Cambridge: Harvard University Press.

15 Garland, D. 1985. *Punishment and Welfare: A History of Penal Strategies*. Aldershot: Gower, p. 46.

16 Griffin, E. 2018. "Diets, hunger and living standards during the British industrial revolution." *Past and Present*, No. 239, May, p. 101; Eyler, J.M. 1992. "The sick poor and the state: Arthur Newsholme on poverty, disease, and responsibility," in Rosenberg, C.E. and J. Golden (eds.), *Framing Disease*. New Brunswick: Rutgers University Press, p. 277.

17 *Administration Reports, Ceylon (ARC)*, 1887, p. C52.

administrators and doctors over where exactly to mark the boundary between lawful and unlawful conditions of incarceration. The following are typical of the administrators' views: "the low animal natures of too many of the criminal class, and the admitted efficiency of reductions of food in cases of prison offences, render plain the value of diet as one form of penal correction"[18] and typical of the doctors' cautionary advice was "I have no hesitation in stating that diet should not be made an instrument of punishment for prisoners undergoing long periods of punishment."[19] Prison doctors drew on nutritional science to locate that boundary, while jailers called it into question. This boundary dispute was difficult to resolve, in large part because the principle that "no permanent harm be done" was difficult to assess. Prisoners fell ill for numerous reasons, other than diet, and the long-term effect of penal diets was unknown as the morbidity and mortality rates of former prisoners had never been recorded. As a consequence, prisoner diet became a site of ongoing struggle between prison administrators and doctors within the prisons and more widely between senior administrators and unofficial Europeans.

The calculation of hunger

The rancorous debate over prison food raged unresolved for the remainder of the century. The pre-reform diet in Ceylon was mocked as "luxuriant," although ironically it corresponded quite closely to what doctors and nutritionists in Britain believed at the time to be a minimally sound diet. But politics triumphed and a healthy diet was rejected over and over again as insufficiently harsh to act as a deterrent. In its place a new diet was introduced in 1867, which in its careful measurements of different foods gave the appearance of being scientifically sound, but was clearly not, even by the standards of the time. From 1867 onwards, extensive, but amateurish, experiments by doctors and jailers were conducted on the relationship between exertion during hard labour, diet and race in Ceylon prisons. This was typical of the colonies more generally, for as Herbst points out, "much of 'colonial science' was made up in the face of particular exigencies and often by the man on the spot rather than in the colonial capital, much less in Europe."[20] As we can see from Table 9.1, the prison diet of 1867 varied by race and stage of sentence. The authorities spent twice as much on food for Europeans, including Burghers, than they did for the other

---

18 Committee of the House of Lords on the State of Discipline in Gaols, and the Report and Evidence presented by the Royal Commission on Penal Servitude, in Digest and Summary of Information Respecting Prisons in the Colonies, supplied by the Governors of Her Majesty's Colonial Possessions in Answer to Mr. Secretary Cardwell's Circular Despatches of the 16th and 17th January 1865. Presented to both Houses of Parliament by Command of Her Majesty. Vol. LVII, Pt. 2, p. 73. London: George Edward Eyre and William Spottiswoode, 1867.

19 Dr. Kinsey, *ARC*, 1876, p. C118.

20 Herbst, J.I. 2000. *States and Power in Africa: Comparative Lessons in Authority and Control.* Princeton: Princeton University Press, p. 82.

prisoners.[21] An interesting development in terms of racial thinking can be seen in Table 9.1. By the end of the century, Burghers were grouped with "natives" in terms of diet. This change corresponded to a hardening of racial categories and increased concerns about miscegenation and tropical degeneration or biological devolution towards the end of the century.[22]

There were three levels of diet in prison: penal, given during the first stage of sentence, ordinary, given during the balance of the sentence, and short term punishment diet, for infractions of prison rules. Prisoners on ordinary diets got three meals per day, coffee and *appa* at 5.30 in the morning; a meal at 11.00 a.m., and the same meal again at 5.30 p.m. The ordinary diet of 1867 was grossly inadequate, in that a long-sentenced prisoner was expected to do rigorous labour on a diet of rice and a bit of dried fish. Even though the authorities were well aware that some vegetables and citrus were necessary, none were given. The ordinary diet for Europeans and Burghers, although generous in protein, was also lacking in vegetables. An even greater problem lay in the penal and punishment diets, which consisted of rice and salt, or bread for Europeans. The former was given during the first ten days of each month during the penal phase.[23] By the mid-1860s, Cornish and others had warned that rice alone was not enough to sustain health. His suggestion that rice be replaced by more nutritious grains such as wheat or ragi was ignored.[24] Some prisoners were able to withstand this diet for ten days each month, but many were placed on it additional times during the month as a punishment for non-completion of tasks or other violations of prison rules and so they ended up in hospital. That this diet was an experiment calculated to push bodies to the edge is very clear in the following injunction to prison officials.

> The local medical officer of each of the prisons is hereby directed to watch most carefully, and to report promptly and fully to the principal civil medical officer, what effects this diet produces on each prisoner's health; and the former is authorised to alter the diet of any prisoner if in his judgement the prisoner's state of health makes such an alteration necessary. Every such alteration, and the reasons for making it, and the effect produced by it, shall be specially reported to the principal civil medical officer.[25]

---

21 Return of the Gaols and Houses of Correction, *Ceylon Blue Books (CBB)*, 1862.

22 Roberts, M., Raheem, I. and Colin-Thome, P. 1989. *People Inbetween: Burghers and the Middle Class in the Transformations within Sri Lanka, 1790s–1960s, Volume 1*. Ratmalana: Sarvodaya Books, p. 148 point out that after the mid-nineteenth century Burghers were increasingly considered a coloured race.

23 Rules and Regulations of the Wellicadde Prison, 12 June 1867, reproduced in *Committee of the House of Lords*, p. 76.

24 Further research by Church in 1886 reinforced how dangerous a rice diet was. Arnold, "The 'discovery' of malnutrition," pp. 11–12.

25 *Committee of the House of Lords*, p. 78.

For a year and a half this harsh new diet was enforced, despite growing alarm among prison doctors that prisoners were growing weak and ill.

Of greater concern was the fact that against prison doctors' advice, jailers sometimes lowered the prisoners' rations even further, believing that life in prison was insufficiently harsh. In response to a report that the new penal stage diet greatly increased mortality, the secretary of state for the colonies appointed a local commission to examine it. The chairman of the commission sided with the jailers by claiming that disease in prisons was largely due to overcrowding rather than diet, and that the diet should be retained. The commission assured the government, as they had several years earlier, that "the attendance of medical men prevents the restriction of diet or hard labour discipline from being carried too far." But the mortality rate and the reports of the prison doctors told a different story.

In his annual report, the colonial surgeon protested against "reckless experiments" such as removing the morning meal for those on a penal diet to see if they could withstand it.[26] He noted that prisoners became so weakened by hard labour on so little food that rice gruel had to be given to them in the morning. He stated, "violent changes in dietary arrangements of prisons and jails, unsupported by medical recommendations have been generally attended with evil results in India, as in England."[27] He concluded,

> as a result of the operations of the penal diet in connection with penal labour ... we have had to deal with a class of diseases in the Kandy Jail that we had not observed there before, and which have been justly ascribed to defective nutrition[28]

The result, he claimed, was "a mortality not known before" in the jail.[29] The percentage of deaths in the jail increased from 1.67%, in the four years preceding the new diet, to 3.76% in the three years after it was introduced.[30]

26 *ARC*, 1870, p. 391. A more systematic version of this type of experiment was undertaken by Heinrich Kraut on concentration camp prisoners and forced labourers in Nazi Germany to attempt to ascertain the caloric minimum necessary to sustain life under different levels of physical activity. See Neswald, E. 2016. "Food fights: human experiments in late nineteenth-century nutrition physiology," in Dyck, E. and L. Stewart (eds.), *The Uses of Humans in Experiments: Perspectives From the Seventeenth to the Twentieth Centuries*. Leiden: Brill, Rodopi, p. 174.

27 *ARC*, 1870, p. 391.

28 *Ibid.*

29 *Ibid.*

30 *ARC*, 1868, p. 64. There were continual worries about the health of the prison population and in particular about the death rates. The latter varied widely from year to year as they did in Ceylon more generally. In some years cholera struck prisons hard, while in most years there were a substantial number of deaths from fevers (probably malaria) and dysentery. It was argued that improved sanitation and higher quality and better prepared food could reduce these death rates in prisons. In the mid-1870s the death rates of prisoners on the island was running at about 35 per 1,000 as opposed to 24 per 1,000 among those who were free. In England the death rate during that period was 12 per 1,000 for the free population and 16 per 1,000 for prisoners (*ARC*, 1878, p. 85).

The secretary of state for the colonies, taking advice from the prison doctors rather than the local commission, accepted that while overcrowding was a contributing factor, had it not been for the new penal diet, the "number of deaths would not have been so great."[31] He ordered that the penal diet be modified and reiterated that prison doctors should have the ultimate say on the effects of diet and hard labour. Henceforth, prison doctors assumed a much more activist role, and on a regular basis resisted regulations by ordering more liberal food allowances. In particular, after 1870, doctors in most prisons added a small amount of vegetables to the ordinary diet.[32]

Two years later, Whitehall backtracked on the issue of diet under pressure from the intransigent chairman of the prison commission, Du Cane, who claimed that a penal diet was a key component of punishment and that the specifics of a prison diet were a local issue. The constituents of diet necessarily vary, the dispatch read, "in different countries, as well as the climates and races of men, and equivalent measures of nourishment can only be approximately reached … through cautious experiment."[33] In effect, the government ignored current nutritional research in Europe and gave a green light for unregulated experiments on prisoners' diet to continue. And so the struggle between jailers and prison doctors continued.

In his 1874 report on health in Ceylon prisons, J.L. Vanderstraaten, assistant colonial surgeon, stated that prisoners' diet was lacking in protein, which at the time was referred to as nitrogen or albumen. The report further showed that the rations given to European prisoners in Ceylon had twice as much protein as the Ceylonese prisoner rations. Furthermore, the quantity of protein given to European prisoners in Ceylon, who did virtually no hard labour, was substantially higher than prisoners doing hard labour in English prisons. The extent to which European prisoners in Ceylon were privileged was presumably due to racial favouritism as well as anxieties about the difficulties faced by prisoners in an alien climate far from home. Vanderstraaten concluded that European prisoners were overfed while Ceylonese prisoners were seriously underfed in the ordinary as well as the penal diets. He strongly recommended that the prisoners' rations be increased from 2½ ounces to 6 or 7 ounces of fish or beef per day.[34] His proposals for the Ceylonese were dismissed as overgenerous, and predictably his claim that European prisoners were overfed was ignored.[35]

---

31 Despatch from The Earl Granville, K.G. to the Officer Administering the Government of Ceylon. January 29, 1869. Further Correspondence Respecting The Discipline and Management of Prisons in Her Majesty's Colonial Possessions. *Command Papers* 8th August 1870. Vol. LVII, pp. 63–64.; An occasional correspondent, "Ceylon," *The Times* (London) September 21, 1871, p. 10.

32 *CBB*, 1871, p. 596.

33 The Earl of Kimberley to the Governors of Colonies, April 15, 1871, in Discipline and Management of Prisons in Her Majesty's Colonial Possessions. *Command Papers* May 1871, Vol. LVIII, p. 3.

34 *ARC*, 1874, p. C24.

35 The estimated intake of "nitrogen," by which they meant protein, per day for inhabitants of Paris in 1896 was 19.3 grams, a figure in line with the ordinary diet of European prisoners in Ceylon (Billen, G., S. Barles, P. Chatzimpiros and J. Garnier. 2012. "Grain, meat and vegetables to feed Paris: where do they come from? Localising Paris food supply areas from the eighteenth to the twenty-first century." *Regional Environmental Change*, Vol. 12, No. 2, June, p. 326). Throughout

Table 9.1 A comparison of native, European and Burgher dietaries, 1867, 1897.

1867	"Native" Diet						
	Bread	*Rice*	*Fish*	*Plantain*	*Vegetable*	*Dhal*	*Sugar*
Ordin.	0	24	2.5	0	0	0	1
Penal.	0	26.5	0	0	0	0	0

1897	"Native" and Burgher Diet[36]						
	Bread	*Rice*	*Fish*	*Plantain*	*Vegetable*	*Dhal*	*Sugar*
Ordin.	6	16	4	2	2	2	0.5
Penal.	8	14	0	0	4	0	0
Punish	12	2	0	0	0	0	0

1867	European and Burgher Diet					
	Bread	*Rice*	*Meat*	*Plantain*	*Vegetable*	*Sugar*
Ordin.	28	0	16	4	0	1
Penal.	28	0	0	0	0	0

1897	European Diet					
	Bread	*Rice*	*Meat*	*Plantain*	*Vegetable*	*Sugar*
Ordin.	14	10	8	5	4	1
Penal.	24	8	0	0	4	0
Punish	24	0	0	0	0	0

In that same year, the Principal Civil Medical Officer reported that, although doctors had ordered that the official prison dietary be slightly increased, there were nevertheless many cases of scurvy in the jails throughout the country. He ordered that the diet should immediately be changed to include more vegetables.[37] The report of widespread scurvy caused a heated debate in the Legislative Council,

the rest of the British Empire, European prisoners were also much better fed than other prisoners (Pete, S. and Deven, A. 2005. "Flogging, fear and food: punishment and race in colonial Natal." *Journal of Southern African Studies*, Vol. 31, No. 1, March, pp. 3–21).

36 Again we see that in 1867 Burghers were classified with Europeans, but by 1897, with changing attitudes towards the mixing of races, they were classified with natives.

37 *ARC*, 1874, p. 6.

where non-European members accused the government of negligence regarding a preventable disease.[38] In fact, reports of scurvy and widespread sickness in the jails forced Whitehall to intervene once again in the experiments by jailers and to issue yet more detailed guidelines on diet. The experiments ordered by administrators in Ceylon jails were crude in comparison to those carried out by European nutritional scientists. The latter had been, since the 1860s, using input-output analyses, which measured food consumed and faeces and urine expelled. The cruel experiments in Ceylon prisons, on the other hand, simply measured the food consumed and then waited to see if prisoners collapsed or fell ill.[39]

The new scientific diet of 1876

As a consequence of the furore over widespread scurvy in the jails, the official guidelines for Ceylon prisons were more substantially revised in an attempt to force prison authorities to comply with the up-to-date knowledge of nutrition and to make it easier for prison doctors to fulfil their remit. The result was the Scientific Diet of 1876, which W.R. Kynsey, the principal civil medical officer, believed was greatly superior to the old diet, which he believed was grossly deficient in meat and vegetables and too high in carbohydrates.[40] After surveys had been made of the vegetables available in local markets, calculations were made for how much food was needed to perform work quotas at each stage of sentence and for each race.[41] The nutrients received were enumerated in albuminates, carbohydrates, fats, salts, nitrogen, and carbon, and the "dynamical value in foot/tons" of the daily allowance was estimated for each.[42] In this way it was argued, the amount of fuel used by the body was tailored to the exact amount of energy expended. For Ceylonese prisoners the amount of food consumed was calibrated to the amount of work done in foot-tons, but for European prisoners such measurements were irrelevant as they were usually not required to do hard labour.

38 *Ceylon Hansard (CH)*, 1875–76, p. 74. Scurvy had been known to be preventable through the consumption of citrus for 130 years.
39 Neswald, "Food fights," pp. 170–93. Anderson, C. 2004. *Legible Bodies: Race, Criminality and Colonialism in South Asia*. Oxford: Berg, p. 9, argues that the colonial prisons were not in fact used as laboratories more often than prisons in England, as has sometimes been assumed. The interchange of ideas between metropole and colony based on experimentation, she claims "was so great that the boundaries between them can be substantially blurred." The obsession with race was peculiarly colonial, however and furthermore, as far as nutrition is concerned, experimentation in Britain was much more sophisticated. I would agree with Anderson, in the sense that the experiments in Ceylon were crude, one is tempted to say parodies of scientific experimentation. But they were dangerous experiments, none the less.
40 *ARC*, 1876, p. C114.
41 On the "science" behind such calculations, see Rabinbach, *The Human Motor*.
42 *ARC*, 1876, pp. C117–119.

However, the tailoring of labour and food to the artificial norm of a strong 130-pound Ceylonese male continued to be a problem. Five years after the scientific diet was put in place, the inspector general admitted that while some prisoners would thrive at these "normal" levels, others became ill or died, and therefore diets should be tailored to that which would sustain the weakest of the healthy prisoners.[43] Dr. Kynsey concluded, "[t]he principles I have aimed at in proposing a new scale of dietaries is to give a sufficiency of nutritious food to maintain the prisoners in health without pampering them."[44] For all the language of rigorous, scientific measurement that accompanied this change of diet, the actual difference between the ordinary diet of 1867 and 1876 consisted of the addition of four ounces of vegetables and two ounces of dhal daily. Prisoners' bodies had been pushed so close to the edge of serious deprivation that this small adjustment made a noticeable difference. Kynsey wrote that, since the new diet was introduced, there had been "steady improvement in the physique and health of prisoners" which "has fully justified the increased expenditure."[45] He continued, however, with a warning to jailers: "I have no hesitation in stating that diet should not be made an instrument of punishment for prisoners undergoing long periods of punishment."[46] He argued that the old prison dietary requirement of enforcing the "penal diet" for the first ten days of every month throughout the penal stage was unlawful, as it exceeded the prisoners' sentences. He said that with recent advanced scientific knowledge of the body and nutrition the penal diet had been "improved in nutritive value and rendered as little injurious as possible" by adding some vegetables[47] and that it would be better coordinated with energy expenditure if it was given no more than twice a week, on Saturdays when prisoners only did a half day's labour and on Sundays when they rested.[48] Likewise, the punishment diet of rice and water was to be given for a maximum of three consecutive days.[49]

43 F.R. Ellis, *ARC*, 1880, p. C6.
44 *ARC*, 1876, p. C118.
45 *Ibid.*, p. C114. The annual cost per head rose from an average of 148 Rs. in the last two years of the old diet to 178 Rs. in the first four years of the new diet (*ARC*, 1884, p. C22).
46 *ARC*, 1876, p. C118. In fact, in the 1840s, the Home Secretary warned prison authorities against using diets "as an instrument of punishment." Ignatieff, M. 1978. *A Just Measure of Pain: The Penitentiary in the Industrial Revolution 1750–1850*. London: Penguin, pp. 176–77. This was interpreted to mean that diet should not permanently ruin a prisoner's health. Restriction of food for short periods of time was considered an acceptable form of punishment that did not violate the injunction.
47 In fact, Kynsey was being disingenuous in implying that before the mid-1870s it was not known that the absence of vegetables would affect health. The real reason for the addition of vegetables was that the Secretary of State for the Colonies had written to the governor in the previous year stating that, "prisoners at hard labour ought not for any period be confined to a diet of rice, salt and water, or bread, salt and water, as those articles do not in my judgement, contain the elements of nutrition necessary for the sustenance of prisoners at hard labour." The Earl of Carnarvon to the Right Hon'ble W.H. Gregory, *Ceylon Sessional Papers (SP)*, 1876–77, p. 20.
48 *ARC*, 1876, p. C114.
49 *Ibid.*

By the mid-1880s it appeared that the doctors, with the support of Whitehall, had won the battle over diet. As the Superintendent of the Convict Establishment wrote,

> The question of food is essentially one in which the opinion of the medical authorities must guide, if not rule, and unless a less generous dietary can be found capable of sustaining convicts in the performance of real hard labour, the present dietary should not, I consider, be interfered with.[50]

But in fact it was a temporary truce. Unfortunately, during the mid-to-late 1880s adverse economic conditions returned to Ceylon with a corresponding rise in crime. Under pressure from European residents, the Legislative Council once again appointed a prison committee to inquire into the diet and labour in prisons to see if they could be made more punitive and of even greater concern to them, if the cost of running the prison system could be reduced. But after comparing the dietary to English prisons and taking the advice of doctors, the Committee concluded that the cost of food could not be reduced without sending large numbers of prisoners to the hospital and that the lost labour would negatively impact public works.[51] A compromise was finally reached by agreeing that the food would be made less palatable by reducing the rations of chillies.

The search for the point of collapse

But in spite of the Committee's conclusion that food should not be reduced, in 1890 the newly appointed Inspector General of Prisons, F.R. Ellis, decided to once again attempt to limit food. To this end, the full diet was abolished on the grounds that it was "unnecessarily nutritive and expensive," and ordinary diet became the highest level of diet.[52] An effort was also made to make the food disagreeable by substituting bread, which the Ceylonese disliked, for rice. Despite the fact that doctors had stressed the importance of protein for decades, fish was removed from the ordinary diet during their first year in prison. Even though prisoners began to rapidly lose weight, the authorities were determined to seek the precise point where prisoners became too weak or sick to work.[53] The ill effects of this new penal regime were compounded by stricter discipline resulting in the placement of prisoners more frequently on the onerous punishment diet of rice and water. It is remarkable, given what was known about nutrition in the 1890s, that all Ceylonese prisoners, now including Burghers, were still given a month-long penal diet of sixteen ounces of bread and two ounces of rice *conjee* per day, thereby removing vegetables from the

---

50 *ARC*, 1885, p. C39.
51 Report of The Prison Committee, September 12, 1887, *SP*, 1887, p. 217.
52 Full diet was instituted in 1886. The only difference between full and ordinary diet was that the former included an extra 2 ½ ounces of fish and 2 ounces of plantain.
53 *ARC*, 1891, p. C6.

penal diet. By comparison, European prisoners were given twenty-four ounces of bread, six ounces of vegetable and eight ounces of rice per day.[54]

In the remaining two months of the penal phase, prisoners were given a crippling diet of four ounces of rice *conjee*, four ounces of vegetables, ten ounces of rice and eight ounces of bread per day.[55] The one-month penal diet while doing hard labour essentially ignored all medical evidence about its dangers and contravened what government in Britain had ruled in 1876. *The Ceylon Times* applauded the introduction of this greatly restricted diet, arguing that the convict establishment would finally become deterrent.[56] Within months of this new unofficial penal diet being introduced, O. Johnson, the Medical Officer in charge of Negombo Prison, wrote that prisoners were becoming debilitated due to a deficiency of proteins. He ordered the diet stopped.[57] By April of 1893 doctors at other prisons were reporting sickness due to the new penal diet, but rather than abandon it in the Convict Establishment, the Inspector General and Principal Civil Medical Officer agreed to conduct an experiment on all new prisoners admitted after April 1. One group had two ounces of dhal added to the penal stage diet, another three ounces of fish, and yet another four ounces of rice. Six months after this new experiment was begun, the illness and mortality rates continued to rise. These experiments revealed that the daily average sick rate per 1,000 for those with the supplement of rice was 140.76; for a supplement of dhal 126.57; and for a supplement of fish 95.55.[58] It is telling that the experiments were continued for as long as six months, given the obvious difference between the rice and fish supplements.

Prison officials tried to explain away this evidence by arguing that it must be a result of factors other than diet. They urged that further tests be undertaken by continuing to use fish and decreasing the labour quotas of prisoners.[59] But the Ceylon government had become impatient with Inspector General Ellis's harsh experiments and, following the advice of prison doctors, abandoned experimentation in 1895 for all inmates sentenced to more than one month. Henceforth, those prisoners went onto the ordinary diet of *conjee*, rice, dhal, fish, vegetable, bread and plantain immediately. Soon after this latest diet was abandoned, the health of prisoners improved and Ellis conceded that it had been "absolutely necessary to make some alteration; it was," he said, "undoubtedly better to increase the diet than to diminish the labour."[60]

As a concession to Ellis, the government allowed prisoners sentenced to a month or less to live on only rice, bread and water. But within a year this diet also came under attack by prison doctors. H. Huybertsz, the Medical Officer of

54 *CBB*, 1892, p. 308.

55 *Ibid.*

56 "Our prison system – the older system and the new," *The Ceylon Times*, October 15, 1891.

57 *ARC*, 1892, p. A35.

58 *ARC*, 1893, pp. C1–2.

59 *Ibid.*, p. C2.

60 *ARC*, 1894, p. C9.

the Convict Establishment, wrote in his annual report that the diet was "unphysiologic," producing diarrhoea, inflammatory dysentery and dyspepsia in prisoners.[61] His colleague in Kandy Prison reported that it was not possible for one-month prisoners to do pingo carrying on such as diet. He reported that prisoners were soon in hospital "with a quick small pulse, cold extremities, giddiness, high-coloured urine, all being signs of acute exhaustion."[62] The short term punishment diet of bread or *conjee* and water, which continued to be used, was also found to be problematic as prisoners were frequently sent to hospital.[63]

Prisoner resistance to the dietary regime

While it is clear that prisoners suffered greatly from these dietary experiments, they were also constantly looking for ways to subvert them. Ellis grumbled that the punishment diet was ineffective for the most "stubborn" prisoners, as it was only allowed to be given for three days in succession when a two-day break was required before it could be given again. The "stubborn" prisoner, he argued,

> was content to go on idling forever, if refractory [punishment] diet five days a week [including a break] was to be his only punishment. He knew well that if he got weak on this diet, he would either go to hospital, or at all events, would not get refractory diet for a while. When the doctor refused to pass a prisoner for refractory diet, the only mode of punishment was removed, and the prisoner remained master of the situation.

Consequently, over the objection of prison doctors and contrary to regulations, Ellis tried briefly to extend the refractory punishment diet to seven days with prisoners remaining in their cells, because they were too weak to work. But such an illegal practice was unsustainable given prisoner resistance and the opposition of doctors and in the end he concluded sourly that punishment diet was useless because, "the Ceylon prisoner is perfectly willing to live for any length of time on twelve ounces of bread if he is allowed, what to him is the inestimable privilege of complete idleness."[64] From this time on, it became clear that the punishment diet would only be employed on malleable prisoners and when it failed, as it often did, corporal punishment was resorted to in order to compel prisoners to work.[65] Ellis' remarks reveal that prisoners would resist work quotas by purposely making themselves ill, knowing that prison doctors

61 *ARC*, 1896, p. C11.
62 H.G. Tomasz, Medical Officer of the Kandy Jails, *ARC*, 1896, p. C14.
63 F.R. Ellis, Inspector General of Prisons, *ARC*, 1893, pp. C9, 7; *CBB*, 1892, pp. 308–09.
64 F.R. Ellis, *ARC*, 1891, pp. C10–11; *ARC*, 1892, p. C11.
65 R. E. Firminger, Superintendent, Convict Establishment, *ARC*, 1894, p. C14; *ARC*, 1895, p. C1; A. De Wilton, Inspector General of Prisons, *ARC*, 1903, p. C10.

were required to support them. So effective was this prisoner resistance that by 1900 all experiments in reducing food to starvation levels had been abandoned and it was widely accepted that the diet of prisoners could be superior to that of poor, free men.[66]

Not only did prisoners, as we have seen, resist the labour regime through foot-dragging and bribing or threatening guards, but an 1871 report claimed that penal diet was "a farce" because prisoners on penal fare working outside of prison regularly obtained extra food from friends.[67] While the sickness rates from lack of food show that the penal diet was in reality far from being a farce, it is clear that at times prisoners were able to evade it. For example, some prisoners, for a price, could get extra food that was kept hidden in the kitchens by warders and guards.[68] It appears that there was a regular underground economy within the prison walls based upon exchanges of money, food, information, protection and sexual favours. Those prisoners who were able to tap into this underground economy could sustain themselves. Those unable to were likely to become ill or die. Furthermore, as we have seen, there was a continual flow of prisoners to the prison doctors with real or feigned sickness and weakness, hoping that the doctors would increase their rations, reduce their work load, or better still, put them in hospital.[69]

All the debates over the scientific measurement of food and work in prison presupposed that jailers did, in fact, control the amount of labour done by convicts and the amount of food they ate. Despite the long drawn-out struggle and anxiety over prison diets in relation to labour, in truth, jailers had full control over neither. By the end of the century the influence of more humane attitudes coming from Britain toward prisoners coupled with prisoner resistance, managed to ensure that the prisoner existence, although still precarious, largely remained within lawful bounds, that is just above what Agamben has described as bare life.

66 A. Kalenberg, Medical Officer of the Kandy Jails, *ARC*, 1900, p. C11; LeMesurier, *ARC*, 1889, p. C6; Wickremeratne, L.A. 1973. "Grain consumption and famine conditions in late nineteenth century Ceylon." *The Ceylon Journal of Historical and Social Studies*, Vol. 3, pp. 28–53.

67 *ARC*, 1871, p. 356.

68 *Report of the Prisons*, p. 92.

69 *CBB*, 1887, p. 524.

10 Conclusion

Gyan Prakash described Britain's colonies as "underfunded and overextended laboratories of modernity. There," he stated, "science's authority as a sign of modernity was instituted with a minimum of expense and a maximum of ambition."[1] This study has explored attempts by the British to put in place a hybrid criminal justice system modelled upon a mix of systems from Ceylon, England and British India. One could argue that the court system, which often involved detectives and other members of the police, was the primary site where the peasantry encountered the state and, importantly, where the state encountered the peasantry. And although a much smaller percentage of the population experienced incarceration, the prison became the site of the most intense biopolitical-legal negotiation between individuals and the state. Drawing upon the writings of the "men on the spot," officials in the bureaucracies of the police, the courts and the prisons, I have examined the widespread resistance to the rule of law in Ceylon. Individuals at each level of these bureaucracies, from the governor down, as well as the general public, at times resisted the constraints imposed by the rule of law using different strategies and tactics according to their different perspectives, value systems and degrees of power within colonial society. Resistance was sometimes passive, sometimes active, sometimes backed by great power, sometimes coordinated, but perhaps more often the result of relatively powerless individuals acting alone. As Foucault claimed, "Power is everywhere; not because it embraces everything, but because it comes from everywhere."[2] As we have seen, the consequences and effectiveness of resistance varied greatly, at different scales and within the three parts of the justice system. Although power was, of course, unevenly distributed and highly differentiated, the total effect of all these points of resistance was "productive"

1 Prakash, G. 1999. *Another Reason: Science and the Imagination of Modern India.* Princeton: Princeton University Press, p. 13. Zinoman, on the other hand, found that the French in Indo-China showed little interest in such modernist experimentation in their prisons there. Zinoman, P. 2001. *The Colonial Bastille: A History of Imprisonment in Vietnam, 1862–1940.* Berkeley: University of California Press.
2 Foucault, M. 1990. *The History of Sexuality. Volume 1: An Introduction.* Translated by R. Hurley. New York: Vintage, p. 93.

of the rule of law in a Foucauldian sense, that is to say, as it developed on the ground. The cumulative impact was considerable, weighing down and distorting the system. While the government employed a range of surveillance strategies as well as legal and penal technologies to counter the various forms of resistance among the peasantry and within each bureaucracy, they were largely unsuccessful. All of this breaks down any sharp dichotomy between collaboration and resistance. For what this study has shown is that collaboration can be a powerful form of resistance through the undermining of colonial bureaucracies from within.

If rule of law was to prevail, it required a combination of effective state surveillance, accountability, transparency and a fair measure of hegemony, or what Foucault has termed the "conduct of conduct." This is most successful when the governed internalise the goals of the state and consequently to a large extent police themselves, producing a very opaque display of state authority. However, in nineteenth-century Ceylon the state lacked legitimacy among the people, and consequently there was little commitment to its institutions. Rule of law, so highly prized by the British, and invoked through liberal theories of jurisprudence, was at best a mixed blessing and at worst a tool of government oppression that prioritised the development of capitalist enterprise in Ceylon. It protected private property when it could be proven with legal documentation, but failed to protect the vast common lands which were so crucial to the welfare of the native population, especially during times of hunger.

Ironically, it often failed to protect the criminal justice system itself, and this is not surprising given the higher priority placed on capitalist infrastructure development. The questions that members of the revenue branch asked themselves foretold the neoliberal economic analysis of crime as described by Becker and commented upon by Foucault: what offences against society should be defined as crime worth the expense of punishment?[3] What expenditures are worth the social loss? These are utilitarian questions that the British contended with, although such economistic analysis clearly went against the rule of law and liberal ideals.[4]

As self-policing was ineffective under colonial conditions, the state was heavily dependent upon surveillance to enforce the rule of law. Unfortunately, from the British point of view, surveillance was also ineffective, for even in highly funded police states it is difficult to enforce laws using surveillance alone, for there are too many "dark zones" that the state cannot monitor. In a place like Ceylon, where there were thousands of small villages, few fully committed

3 Foucault shows great interest in, if not approval of, Becker's economistic approach to every aspect of human decision-making ranging from criminal behaviour to governmental practices. Foucault, M. 2008. *The Birth of Biopolitics: Lectures at the College de France, 1978–79*. Edited by M. Senellart. New York: Palgrave Macmillan, p. 262.
4 Becker, G. 1968. "Crime and punishment: an economic approach." *Journal of Political Economy*, Vol. 76, No. 2, pp. 169–70.

officials, and where the criminal justice system was run on a shoestring, "corruption" from the perspective of British bureaucratic norms was pervasive. As I have shown, British administrators were complicit in this corruption. In order to save money, they systematically exploited the Ceylonese, who occupied the lower ranks of the three bureaucracies. The police constables, *vidanes*, minor court officials and prison guards were paid the barest living wage while being expected to enforce the law. This created a virtual perfect storm of corruption in which bureaucrats exploited their fellow subjects in order to earn a living. Furthermore, even those who accepted the rule of law in principle often found it expedient to evade it in practice. Sometimes senior administrators and judges didn't have the power or the knowledge to counter the attempts of lower level bureaucrats to circumvent the law. They failed in their duty to enforce the law for lack of money, training, or simply lack of will due to their racist views of what was possible in a colonial society. They were further thwarted in their attempts to make the criminal justice system effective because they were faced with a population whose sense of justice did not sufficiently correspond to colonial law. Delegitimation was compounded by the widespread knowledge among the Ceylonese that too often the system criminalised the innocent and failed to punish the guilty.

It became clear from reading the official documents and newspaper editorials that the British often depoliticised resistance by seeing it as racial failure. Thus, I have argued that by naturalising the evasion tactics used by the Ceylonese including lower level officials, British members of the colonial government failed to fully comprehend the government's lack of legitimacy. I also argue that these naturalised evasion tactics can be considered a form of negotiation whereby the oppressed seek to lower the standard of labour and cooperation that the oppressor believes they can expect due to racial and cultural differences. The power of the subaltern to "reset" the norm, therefore, can be an effective way to mitigate the worst forms of oppression.

In part because the Ceylonese so rarely engaged in open defiance, the British took their acceptance largely for granted, albeit nervously, given the British experience of the Great Rebellion of 1857 in India and the smaller rebellion of 1848 in Ceylon. It seemed to most of the British that in day-to-day affairs the Ceylonese appeared to accept, and in the case of the courts, to actually embrace, the colonial system. In fact, as most of the Ceylonese, and the peasantry in particular, were living at a subsistence level, they were not inclined to political violence, which would have been dangerously disruptive to their ability to survive. Whatever sincere humanitarian concerns the British may have had for the protection and welfare of the Ceylonese, they were understood within the largely unquestioned *raison d'etat* that prioritised the success of British trade and capitalist enterprise in the colony over all other considerations.

Not surprisingly, the peasants and town dwellers rarely shared the goals of the state, and so they resisted the law when they could and used the courts extensively in ways that exasperated the British. They sought out the cracks in each bureaucratic structure—the physical and metaphorical spaces where the state had difficulty surveying them. As I have argued, some of these cracks were the result of insufficient funding, lack of local knowledge, or simply poor planning by the British, while others were created by the adherence of senior officials to the constraining force of the rule of law itself. In the latter case, as we have seen, officials expressed frustration that their hands were tied by laws that allowed those they believed to be guilty to escape justice.

I have examined four principal types of resistance. The first took the form of foot-dragging. This was quite literally the case when constables on the beat or prisoners doing hard labour walked or worked slowly. But it also took the form, in the case of the police, *vidanes* and court officials, of failing to do required tasks, falsely claiming to have misunderstood instructions or to be ignorant of the laws they were to enforce. As these claims were sometimes true, such ignorance cannot always be considered resistance. In the case of villagers and town dwellers it took the form of "knowing nothing," or giving very partial or misleading information when questioned by officials. Feigned ignorance was effective as it played into British racist stereotypes. As I have argued, a multitude of habitual individual actions some coordinated, some not, established low expectations and this could be advantageous to those resisting a bureaucracy to which they felt no real allegiance.

The second type of resistance took the form of invisibility and illegibility. This operated at a variety of geographical scales. As I have discussed, villages in the rural areas were relatively autonomous. Peasants were left to manage with little government help or interference. The police outstations were located far away from supervisory personnel and so bureaucratic norms could be largely ignored. There were also many out of the way spaces on police beats where constables could rest unobserved. Likewise, fiscal's peons operated free from supervision when sent out into the countryside to make arrests. And prisoners used the lightly monitored group wards to exchange information and forbidden items and occasionally to plot insurrection. Usually villagers could count on the remoteness of their villages as well as the silence of their fellow villagers to shield them from police inquiries, although as we saw in terms of prison escapes, it was rare for convicts to be protected indefinitely. And town dwellers and prisoners in extramural work gangs sometimes used the anonymity of crowds to hurl insults and even on rare occasions to violently attack police, court and jail officials. But the most pervasive use of invisibility was the bringing of false cases and the ubiquity of perjury in the courts. The courts throughout the nineteenth century were flooded with such cases because the remoteness of villages, the lack of detective work and insufficient commitment by regular police and *vidanes* meant that there were few ways for judges to check on the veracity of statements made in court. This was compounded by the power of translators who were by no means neutral conduits in the testimonial process. As translators and cultural interpreters, they were indispensable to judges, and therefore in a strong a position to distort facts and contextual information according to their personal biases or their stake in the outcome of cases. The broader problem was that most of the British, especially the newly arrived, found the Ceylonese culture to be relatively illegible. As I have shown, they were not only frustrated by their inability to understand the motivations of Ceylonese criminals and especially habitual criminals, but also the more general Ceylonese cultural values of freedom, work, truth and justice. They were unsure which penal technologies would be most onerous and thus deterrent and whether there was any possibility of rehabilitating criminals. Village communities, by remaining relatively "closed" to British understandings, were able to take advantage of this ignorance to give themselves more autonomy than they might otherwise have been granted.

The third type of resistance was what the British defined as corruption, but which in many cases was seen as ethical, or at least normal survival strategies by the Ceylonese. For example, some of the lower level bureaucrats accepted money, to which they felt entitled as a traditional form of payment. The bureaucrats' belief in their entitlement was reinforced by the lack of salaries for *vidanes* and the inadequate pay for other positions and so the low level of financial support for the criminal justice system was in large part to blame in this system of unofficial remuneration. Perhaps the major difference between the British and the Ceylonese understandings of what constituted corruption stemmed from the refusal of the Ceylonese to embrace British bureaucratic norms, which placed duty to the bureaucracy above family and community obligations.

The fourth type was violence. State-sanctioned violence, such as corporal punishment, denying food to prisoners or dispossessing people of their land and livelihoods, although deplorable, largely observed the rule of law. Resistance to the rule of law included much non-sanctioned individual violence. One was the illicit, but normalised, everyday violence committed by Europeans on the Ceylonese. Another was the everyday violence directed at the peasantry by members of the lower echelons of the police and prison bureaucracies as they beat confessions out of suspects in order to "solve" cases quickly and punish prisoners whose families failed to pay them bribes. Violence on the part of villagers against other villagers was viewed by the British as racial failure, including lack of self-control and innate viciousness. It was punished, but less harshly than it might have been because it was defined as non-political. Open resistance to the British right to rule was put down with extreme ruthlessness as it was considered potentially seditious and far more dangerous. The latter was in fact rare until the twentieth century when nationalist movements began in earnest.

I have traced in detail the repeated attempts to create elaborate systems of surveillance in order to counter the various types of resistance. As I discussed, superintendents timed the police on fixed routes in order to track their movements. Town police and those in outposts were subject to random surveillance. Those who surveyed were in turn themselves subject to surveillance. The police were moved periodically from route to route or from outstation to outstation to prevent them from participating in local crime. All these temporo-spatial techniques, as I have shown, met with very limited success.

Lack of trust between groups was a seemingly insurmountable problem because reports of abuse and corruption were unreliable and often motivated by animosity. *Vidanes* were known to resent the interference of the government in village affairs and were thought to be uncooperative with information and manpower. Consequently, attempts were made at having regular police supervise them; but as outsiders, the police found it nearly impossible to uncover village secrets. Villagers were questioned about police and *vidane* abuse, but officials were unable to know if villagers were covering for headmen either out of loyalty or fear or if they were trying to stir up trouble for the police, who were unwelcome outsiders.

Likewise, as we have seen, surveillance was very difficult to effect in the courts. Because of the pervasiveness of perjury, verbal evidence was always highly suspect. There were not enough detectives to check evidence, and the few there were were notorious for fabricating evidence. And so circumstantial evidence and rigorous cross-examination were relied upon in the struggle to discern the truth. But this process was undermined by the failure of many judges to understand local languages, leaving them dependent on translators who were known to accept bribes or to take matters of justice into their own hands by distorting testimony. And so at times, out of frustration due to the difficulty of uncovering the truth, judges simply passed judgement based upon their impressions of guilt or innocence. Such cases were frequently then overturned on appeal by the Supreme Court. It seemed nearly impossible to change the way the people used the courts. Even if the government had placed a higher priority on the criminal justice system and funded it accordingly, there was no getting around the fundamental lack of legitimacy of the whole system in the eyes of the majority of the population.

Surveillance was most effective within prisons as they at least approximated total disciplinary institutions. However even there, prisoners and guards found ways to resist regulations and the most extreme aspects of the penal regime. The primary goal of the prison, as I have argued, was to make incarceration as deterrent as lawfully possible. This is an example of biopower through the mathematisation of racialised bodies. In theory, no activity was to take place outside the confines of the meticulously devised space-time organisation. The amount of work prisoners were required to do and the food needed to maintain their health were, in theory at least, scientifically calculated based on what was deemed to be a "normal" Ceylonese male body. Those who were larger or weaker than the norm, and thus required more food, suffered even more than intended. I have described the decades-long experimentation process by which prison administrators attempted to refine their methods of keeping prisoners' bodies at the cliff-edge of permanent damage. However, the success or failure of this system of control depended to some degree upon the willingness of the guards to enforce it, and the guards had little incentive to enforce prison rules uniformly. The success of the administrators' efforts also depended on the extent to which prison doctors and prisoners themselves cooperated with the draconian measures. As we have seen, over and over again, the administrators' attempts to make the prisons effectively deterrent met with only limited success.

As was typical of nineteenth-century British bureaucratic practice, the Ceylon government's commissions and reports, which included the diaries of government agents and senior police officials, carefully documented in minute detail all of the forms of expertise that were brought to bear on the problems surrounding lawful rule. The archive the British officials left behind, biased and selective as it necessarily is, has allowed me to trace in detail the official development of the criminal justice system, including how it was consistently undermined by those it oppressed. As Legg writes in reference to the colonial police in New Delhi in the twentieth century, "it was often internal review that followed a period of problemisation that defined how disciplinary technologies were deployed, although each new development necessarily created as many spaces of resistance as those it made visible."[5] The colonial archive is a testament to what Richards has termed the "peculiarly Victorian confidence that knowledge could be controlled and controlling."[6] But, as Ann Stoler states in her commentary on the British colonial archive, "grids of intelligibility were fashioned from uncertain knowledge; disquiet and anxieties registered the uncommon sense of events and things; epistemic uncertainties repeatedly unsettled the imperial conceit that all was in order."[7] The frustrations and anxieties of the British are revealed, if not always with complete candour, in the archives, as are the activities of the Ceylonese. From these records I have been able document the myriad ways the Ceylonese resisted the bureaucracies and the rule of law. But, quite obviously, I can only speculate about the ideas and motivations of those who have little or no voice of their own in the recorded history. I hope these speculations, however, have been convincing enough to, at the very least, raise further questions about how the colonial justice system in Ceylon, and presumably elsewhere in the British Empire, was partially shaped through its undermining and creative use by those for whom the whole system was never fully legitimate.

5 Legg, S. 2007. *Spaces of Colonialism: Delhi's Urban Governmentalities*. Oxford: Blackwell, p. 85.
6 Richards, T. 1993. *The Imperial Archive: Knowledge and the Fantasy of Empire*. London: Verso, p. 7.
7 Stoler, A.L. 2009. *Along the Archival Grain: Epistemic Anxieties and Colonial Common Sense*. Princeton: Princeton University press, p. 1.

Bibliography

Official records

Addresses Delivered in the Legislative Council of Ceylon by Governors of the Colony together with Replies of the Council. 1880. Colombo: Government Printer.

Administration Reports, Ceylon (ARC). Colombo: Government Printer.

British Parliamentary Papers (BPP).

Ceylon Blue Books (CBB). Colombo: Government Printer.

Ceylon Hansard (CH), Debates of the Ceylon Legislative Council. Colombo: Government Printer.

Ceylon Sessional Papers (SP). Colombo: Government Printer.

Command Papers.

Committee of the House of Lords on the State of Discipline in Gaols, and the Report and Evidence presented by the Royal Commission on Penal Servitude, in Digest and Summary of Information Respecting Prisons in the Colonies, supplied by the Governors of Her Majesty's Colonial Possessions in Answer to Mr. Secretary Cardwell's Circular Despatches of the 16th and 17th January 1865. Presented to both Houses of Parliament by Command of Her Majesty. Vol. LVII, Pt. 2. London: George Edward Eyre and William Spottiswoode, 1867.

Correspondence Relating to the Police Establishment, Ceylon. 1866. Colombo: William Henry Herbert, Acting Government Printer, Ceylon.

Dickman, C. 1872. *The Ceylon Civil Service Manual.* Colombo: F. Fonseka, Government Printer.

Giles, A.H. 1889. *Report on the Administration of Police, Including the Actions of the Courts and the Punishment of Criminals in Ceylon.* Colombo: J.A. Skeen, Government Printer.

Instructions and Orders for the Regulation of the Police Force. 1857. Colombo: William Skeen, Government Printer.

Police Gazette, Colombo: Government Printer.

Public Record Office (PRO) (National Archives, Kew) CO 54.

Report of the Prisons Inquiry Commission, 1931. 1932. Colombo: Ceylon Government Press.

Select Committee on Ceylon Second Report 1850. London: HMSO.

Sessional Papers (SP). Colombo: Government Printer.

Sri Lanka National Archives (SLNA).

White, H. 1903. *The Ceylon Manual for the Use of Officials.* Colombo: GJA Skeen, Government Printer.

Newspapers

Ceylon Examiner.
Ceylon Independent.
Ceylon Times.
The Overland Times of Ceylon.
Ceylon Observer.
Overland Ceylon Observer.
Colombo Observer.
Colombo Overland Observer.
The Times (London).
Weekly Ceylon Observer.
Weekly Ceylon Observer and Summary of Intelligence.

Publications

Agamben, G. 1998. *Homo Sacer: Sovereign Power and Bare Life.* Stanford: Stanford University Press.

Ager, A.W. 2014. *Crime and Poverty in 19th Century England: The Economy of Makeshifts.* London: Bloomsbury.

Aguirre, C. 2007. "Prisons and prisoners in modernising Latin America," in F. Dikotter and I. Brown (eds.), *Cultures of Confinement: A History of the Prison in Africa, Asia and Latin America.* London: Hurst, pp. 14–54.

Ahmed, S. 2004. *The Cultural Politics of Emotion.* Edinburgh: Edinburgh University Press.

Alatas, S.H. 1977. *The Myth of the Lazy Native.* London: Frank Cass and Co.

Allen, C.J.W. 1997. *The Law of Evidence in Victorian England.* Cambridge: Cambridge University Press.

Ambirajan, S. 2008. *Classical Political Economy and British Policy in India.* Cambridge: Cambridge University Press.

Amin, S. 1995. *Event, Metaphor, Memory: Chauri Chaura 1922–1992.* Delhi: Oxford University Press.

Anderson, B. 1991. *Imagined Communities: Reflections on the Origins and Spread of Nationalism.* London: Verso.

Anderson, C. 2004. *Legible Bodies: Race, Criminality and Colonialism in South Asia.* Oxford: Berg.

Anderson, C. 2007. "Sepoys, servants and settlers: convict transportation in the Indian Ocean 1787–1945," in F. Dikotter and I. Brown (eds.), *Cultures of Confinement: A History of the Prison in Africa, Asia and Latin America.* London: Hurst, pp. 185–220.

Anderson, C. 2007. *The Indian Uprising of 1857–58. Prisons, Prisoners and Rebellion.* London: Anthem Press.

Anderson, C. 2008. "The politics of punishment in colonial Mauritius, 1766–1887." *Cultural and Social History,* Vol. 5, No. 4, pp. 411–22.

Anderson, C. 2012. *Subaltern Lives: Biographies of Colonialism in the Indian Ocean World, 1790–1920.* Cambridge: Cambridge University Press.

Anderson, C. 2014. "The power of words in nineteenth century prisons: colonial Mauritius, 1835–1887," in F. Paisley and K. Reid (eds.), *Critical Perspectives on Colonialism: Writing the Empire from Below.* New York: Routledge, pp. 199–218.

Anderson, C. and Arnold, D. 2007. "Envisioning the colonial prison," in F. Dikotter and I. Brown (eds.), *Cultures of Confinement: A History of the Prison in Africa, Asia and Latin America*. London: Hurst, pp. 304–31.

Anderson, C., Crockett, C.M., de Vito, C.G., Miyamoto, T., Moss, K., Roscoe, K. and Sakata, M. 2015. "Punishment space and place, c. 1750–1900," in K.M. Morin and D. Moran (eds.), *Historical Geographies of Prisons: Unlocking the Usable Carceral Past*. London: Routledge, pp. 206–33.

Arendt, H. 1968. "Truth and politics," in *In Between Past and Future*. New York: Viking, pp. 227–64.

Arford, T. 2016. "Prisons as sites of power/resistance," in D. Courpasson and S. Vallas (eds.), *The Sage Handbook of Resistance*. Los Angeles: Sage, pp. 385–417.

Arnold, D. 1976. "The police and colonial control in South India." *Social Scientist*, Vol. 4, No. 12, Jul., pp. 3–16.

Arnold, D. 1977. "The armed police and colonial rule in south India, 1914–1947." *Modern Asian Studies*, Vol. 2, No. 1, pp. 101–25.

Arnold, D. 1985. "Crime and crime control in Madras, 1858–1947," in A.A. Yang (ed.), *Crime and Criminality in British India*. Tucson: University of Arizona Press, pp. 62–88.

Arnold, D. 1985. "Bureaucratic recruitment and subordination in colonial India: the Madras Constabulary, 1859–1947," in R. Guha (ed.), *Subaltern Studies IV*. Delhi: Oxford University Press, pp. 1–53.

Arnold, D. 1986. *Police Power and Colonial Rule: Madras, 1859–1957*. Delhi: Oxford University Press.

Arnold, D. 1994. "The colonial prison: power, knowledge and penology in nineteenth century India," in D. Arnold and D. Hardiman (eds.), *Subaltern Studies VIII: Essays in Honour of Ranajit Guha*. Delhi: Oxford University Press, pp. 148–87.

Arnold, D. 1994. "The 'discovery' of malnutrition and diet in colonial India." *The Indian Economic and Social History Review*, Vol. 31, No. 1, pp. 1–26.

Arnold, D. 2007. "India: the contested prison," in F. Dikotter and I. Brown (eds.), *Cultures of Confinement: A History of the Prison in Africa, Asia and Latin America*. London: Hurst, pp. 147–84.

Bailey, V. 1981. "Introduction," in V. Bailey (ed.), *Policing and Punishment in Nineteenth Century Britain*. London: Croom Helm, pp. 11–24.

Bailkin, J. 2006. "The boot and the spleen: when was murder possible in British India?" *Comparative Studies in Society and History*, Vol. 48, No. 2, pp. 462–93.

Bandarage, A. 1983. *Colonialism in Sri Lanka: The Political Economy of the Kandyan Highlands, 1833–1886*. Berlin: Mouton Publishers.

Barkey, K. 2013. "Aspects of legal pluralism in the Ottoman empire," in L. Benton and R. J. Ross (eds.), *Legal Pluralism and Empires, 1500–1850*. New York: New York University Press, pp. 83–108.

Barrie, D.G. 2008. *Police in the Age of Improvement: Police Development and the Civic Tradition in Scotland, 1775–1865*. Cullompton: Willan.

Bashford, A. and Chaplin, J.E. 2016. *The New Worlds of Thomas Robert Malthus: Rereading the Principle of Population*. Princeton: Princeton University Press.

Basnayake, S. 1973. "The Anglo-Indian codes in Ceylon." *The International and Comparative Law Quarterly*, Vol. 22, No. 2, pp. 284–311.

Baxi, U. 2004. "Rule of law in India: theory and practice," in R. Peerboom (ed.), *Discourses of Rule of Law: Theories and Implementation of Rule of Law in Twelve Asian Countries, France and India*. New York: Routledge, pp. 324–45.

Bayly, C.A. 1996. *Empire and Information: Intelligence Gathering and Social Communication in India, 1780–1870*. Cambridge: Cambridge University Press.

Becker, G. 1968. "Crime and punishment: an economic approach." *Journal of Political Economy*, Vol. 76, No. 2, pp. 169–217.

Beckingham, D. 2017. *The Licensed City: Regulating Drink in Liverpool, 1830–1920*. Liverpool: Liverpool University Press.

Beirne, P. 1987. "Adolphe Quetelet and the origins of positivist criminology." *AUS*, Vol. 92, No. 5, pp. 1140–69.

Bellamy, R. 1992. *Liberalism and Modern Society: An Historical Argument*. Cambridge: Polity Press.

Bennett, T. 2011. "Habit, instinct, survivals: repetition, history, biopower," in S. Gunn and J. Vernon (eds.), *The Peculiarities of Liberal Modernity in Imperial Britain*. Berkeley: University of California Press, pp. 102–18.

Benton, L. 1999. "Colonial law and cultural difference: jurisdictional politics and the formation of the colonial state." *Comparative Studies in Society and History*, Vol. 41, No. 3, pp. 563–88.

Benton, L. 2002. *Law and Colonial Cultures: Legal Regimes in World History, 1400–1900*. Cambridge: Cambridge University Press.

Bernault, F. 2007. "The shadow of rule: colonial power and modern punishment in Africa," in F. Dikotter and I. Brown (eds.), *Cultures of Confinement: A History of the Prison in Africa, Asia and Latin America*. London: Hurst, pp. 55–94.

Berry, S. 1992. "Hegemony on a shoestring: indirect rule and access to agricultural land." *Africa*, Vol. 62, No. 3, pp. 327–55.

Bertolacci, A. 1817. *A View of the Agricultural, Commercial and Financial Interests of Ceylon*. London: Black, Parbury and Allen.

Bhandar, B. 2018. *Colonial Lives of Property*. Durham: Duke University Press.

Bhushan, V. 1970. *Prison Administration in India*. Delhi: S. Chand.

Billen, G., Barles, S., Chatzimpiros, P. and Garnier, J. 2012. "Grain, meat and vegetables to feed Paris: where do they come from? Localising Paris food supply areas from the eighteenth to the twenty-first century." *Regional Environmental Change*, Vol. 12, No. 2, June, pp. 325–35.

Blomley, N. 1994. *Law, Space, and the Geographies of Power*. New York: Guilford Press.

Blomley, N. 2007. "Making private property: enclosure, common right and the work of hedges." *Rural History*, Vol. 18, pp. 1–21.

Blomley, N. 2013. "Performing property, making the world." *Canadian Journal of Law and Jurisprudence*, Vol. 26, pp. 23–48.

Blomley, N., Delaney, D. and Ford, R., eds. 2001. *The Legal Geographies Reader*. Oxford: Blackwell.

Bohstedt, J. 2010. *The Politics of Provisioning, Food Riots, Moral Economy, and Market Transition in England, c. 1550–1850*. Farnham: Ashgate.

Braverman, I., Blomley, N. and Delaney, D., eds. 2014. *The Expanding Spaces of Law: A Timely Legal Geography*. Palo Alto: Stanford University Press.

Brogden, M. 1987. "The emergence of the police: the colonial dimension." *British Journal of Criminology*, Vol. 27, No. 1, pp. 4–14.

Brooks, C.W. 1998. *Lawyers, Litigation and English Society since 1450*. London: Hambleton Press.

Brooks, C.W. 2004. "The longitudinal study of civil litigation in England 1200–1996," in S.R. Anleu and W. Prest (eds.), *Litigation. Past and Present*. Sydney: University of New South Wales Press, pp. 24–43.

Brown, I. 2007. "South East Asia: reform and the colonial prison," in F. Dikotter and I. Brown (eds.), *Cultures of Confinement: A History of the Prison in Africa, Asia and Latin America*. London: Hurst, pp. 221–68.

Brown, M. 2001. "Race, science and the construction of native criminality in colonial India." *Theoretical Criminology*, Vol. 5, No. 3, pp. 345–68.

Brown, M. 2003. "Ethnology and colonial administration in nineteenth century British India: the question of native crime and criminality." *The British Journal for the History of Science*, Vol. 36, No. 02, pp. 201–19.

Brown, M. 2014. *Penal Power and Colonial Rule*. Abingdon: Routledge.

Brown, N. 2007. *Rule of Law in the Arab World: Courts in Egypt and the Gulf*. Cambridge: Cambridge University Press.

Brown, N. 2018. "Politics over doctrine: the evolution of Sharia-based state institutions in Egypt and Saudi Arabia." *James A. Baker III Institute for Public Policy of Rice University*, pp. 1–15.

Butler, J. 1997. *The Psychic Life of Power: Theories in Subjection*. Stanford: Stanford University Press.

Carpenter, K.J. 2006. "Nutritional studies in Victorian prisons." *The Journal of Nutrition*, Vol. 136, No. 1, pp. 1–8.

Carpenter, M. 1867. *Suggestions on Prison Discipline and Female Education in India*. London: Longmans.

Carroll-Burke, P. 2000. *Colonial Discipline: The Making of the Irish Convict System*. Dublin: Four Courts Press.

Casinader, I., Wijeyaratne, R.D. and Godden, L. 2018. "From sovereignty to modernity: revisiting the Colebrooke-Cameron reforms – transforming the Buddhist and colonial imaginary in nineteenth-century Ceylon." *Comparative Legal History*, Vol. 6, No. 1, pp. 34–64.

Chakrabarti, R. 2009. *Terror, Crime and Punishment: Order and Disorder in Early Colonial Bengal, 1800–1860*. Kolkata: Readers Service.

Chandra, U. 2015. "Rethinking subaltern resistance." *Journal of Contemporary Asia*, Vol. 45, No. 4, pp. 563–73.

Chandrachud, A. 2015. *An Independent, Colonial Judiciary: A History of the Bombay High Court during the British Raj, 1862–1947*. Oxford: Oxford University Press.

Chanock, M. 1985. *Law, Custom and the Social Order: The Colonial Experience in Malawi and Zambia*. Portsmouth: Heinemann.

Chanock, M. 1995. "Criminological science and the criminal law on the colonial periphery: perception, fantasy, and realities in South Africa, 1900–1930." *Law & Social Inquiry*, Vol. 20, No. 4, pp. 911–39.

Chatterjee, P. 1993. *The Nation and Its Fragments: Colonial and Postcolonial Histories*. Princeton: Princeton University Press.

Claeys, G. 2010. *Imperial Sceptics: British Critics of Empire, 1850–1910*. Cambridge: Cambridge University Press.

Clarence, L.B. 1896–1897. "Application of European law to natives of Ceylon." *Journal of the Society of Comparative Legislation*, Vol. 1, pp. 227–31.

Cohn, B. 1987. *An Anthropologist among the Historians*. Oxford: Oxford University Press.

Cole, S. 2002. *A History of Fingerprinting and Criminal Identification*. Cambridge, MA: Harvard University Press.

Collingham, E.M. 2001. *Imperial Bodies: The Physical Experience of the Raj, C. 1800–1947*. Cambridge: Polity Press.

Collins, C. 1951. *Public Administration in Ceylon*. London: Royal Institute of International Affairs.

Comaroff, J. 2006. "Colonialism, culture and law: a foreword." *Law and Social Inquiry*, Vol. 2, pp. 305–14.

Comaroff, J. and Comaroff, J. 1991. *Of Revelation and Revolution: Christianity, Colonialism and Consciousness*. Chicago: University of Chicago Press.

Cooper, F. and Stoler, A., eds. 1997. *Tensions of Empire: Colonial Cultures in a Bourgeois World*. Berkeley: University of California Press.

Cooper, R. 1981. "Jeremy Bentham, Elizabeth Fry, and English prison reform." *Journal of the History of Ideas*, Vol. 42, No. 4, pp. 675–90.

Coperehewa, S. 2011. "Colonialism and problems of language policy: formulation of a colonial language policy in Sri Lanka." *Sri Lanka Journal of Advanced Social Studies*, Vol. 1, No. 1, pp. 27–52.

Cornish, W.R. 1864. *Observations on the Nature of the Food of the Inhabitants of Southern India and on Dietaries in the Madras Presidency*. Madras: Gantz. pp. 1–72. Reprinted from the *Madras Quarterly Journal of Medical Science* no 15.

Courpasson, D. and Vallas, S. 2016. "Resistance studies: a critical introduction," in D. Courpasson and S. Vallas (eds.), *The Sage Handbook of Resistance*. Los Angeles: Sage, pp. 19–63.

Crimmins, J. 1996. "Contending interpretations of Bentham's utilitarianism." *Canadian Journal of Political Science*, Vol. 29, No. 4, pp. 751–77.

Critchley, T.A. 1967. *A History of Police in England and Wales, 900–1966*. London: Constable.

Crooks, P. and Parsons, T.H. 2016. "Empires, bureaucracy and the paradox of power," in P. Crooks and T.H. Parsons (eds.), *Empires and Bureaucracy in World History: From Late Antiquity to the Twentieth Century*. Cambridge: Cambridge University Press, pp. 1–20.

Darby, N. 2012. "A protestant purgatory: theological origins of the penitentiary act, 1779." *Journal for Eighteenth Century Studies*, Vol. 35, No. 4, pp. 617–18.

Darwin, C. 2003. (1871). *The Descent of Man*. London: Gibson Square Books.

Das, D.K. and Verma, A. 2003. *Police Mission: Challenges and Responses*. Oxford: Scarecrow Press.

Das, D.K. and Verma, A. 1998. "The armed police in the British colonial tradition." *Policing: An International Journal of Police Strategies and Management*, Vol. 21, No. 2, pp. 354–67.

Das, V. 2015. "Corruption and the possibility of life." *Contributions to Indian Sociology*, Vol. 49, No. 3, pp. 322–43.

Davie, N. 2005. *Tracing the Criminal: The Rise of Scientific Criminology in Britain, 1860–1918*. Oxford: Bardwell.

Dean, M. 2010. *Governmentality: Power and Rule in Modern Society*. 2nd edition. Los Angeles: Sage.

De Certeau, M. 1984. *The Practice of Everyday Life*. Berkeley: University of California Press.

Deflem, M. 1994. "Law enforcement in British colonial Africa: a comparative analysis of imperial policing in Nyasaland, the Gold Coast and Kenya." *Police Studies*, Vol. 17, No. 1, pp. 45–67.

Delaney, D. 2010. *The Spatial, the Legal, and the Pragmatics of World-Making: Nomospheric Investigations*. London: Glass House Books.

Den Otter, S. 2012. "Law, authority and colonial rule," in D.M. Peers and N. Gooptu (eds.), *India and the British Empire*. Oxford: Oxford University Press, pp. 168–90.

Dep, A.C. 1969. *A History of the Ceylon Police, Volume 2, 1866–1913*. Colombo: Police Amenities Fund.

De Silva, C.R. 1953. *Ceylon under the British Occupation, 1795–1833*, Volume 1. Colombo: Colombo Apothecaries.

De Silva, K.M. 1973. "The Development of the administrative system, 1833 to c. 1910," in K.M. De Silva (ed.), *University of Ceylon History of Ceylon, Volume 3*. Colombo: University of Ceylon Press Board, pp. 213–25.

De Silva, K.M. 1973. "The courts," in K.M. De Silva (ed.), *University of Ceylon. History of Ceylon. Volume 3*. Colombo: Colombo Apothecaries, pp. 317–26.

De Silva, K.M. 1979. "Resistance movements in nineteenth century Sri Lanka," in M. Roberts (ed.), *Collective Identities, Nationalisms and Protest in Modern Sri Lanka*. Colombo: Marga Institute, pp. 129–52.

De Silva, M.U. 2006. "Litigiousness in Sri Lankans: an examination of judicial change and its consequence during the late Dutch and early British administration in the Maritime Provinces of Sri Lanka." *Journal of the Royal Asiatic Society of Sri Lanka*, New Series, Vol. 52, pp. 127–42.

De Silva, P. 2016. *Leonard Woolf as a Judge in Ceylon*. Battaramulla: Neptune Publications.

Dewaraja, L.S. 1972. *The Kandyan Kingdom of Ceylon, 1707–1780*. Colombo: Lake House.

Dhillon, K.S. 1998. *Defenders of the Establishment: Ruler-Supportive Police Forces of South Asia*. Shimla: Indian Institute of Advanced Study.

Digby, W. 1879. *Forty Years of Official and Unofficial Life in an Oriental Crown Colony; Being the Life of Sir Richard Morgan, Kt., Queen's Advocate and Acting Chief Justice of Ceylon*. 2 vols. Madras: Higginbotham and Co.

Dikotter, F. 2004. "'A paradise for rascals': colonialism, punishment and the prison in Hong Kong (1841–1898)." *Crime, History and Societies*, Vol. 8, No. 1, pp. 49–63.

Dikotter, F. 2007. "Introduction," in F. Dikotter and I. Brown (eds.), *Cultures of Confinement: A History of the Prison in Africa, Asia and Latin America*. London: Hurst, pp. 1–13.

Dodsworth, F. 2012. "Men on a mission: masculinity, violence and self-presentation of policemen in England, c. 1870–1914," in D.G. Barrie and S. Broomhall (eds.), *A History of Police and Masculinities, 1700–2010*. London: Routledge, pp. 123–40.

D'Oyly, J. 1929. (1832). *A Sketch of the Constitution of the Kandyan Kingdom*. Colombo: Government Printer.

Driver, F. 1993. *Power and Pauperism: The Workhouse System, 1834–1884*. Cambridge: Cambridge University Press.

DuCane, E. 1885. *The Punishment and Prevention of Crime*. London: Macmillan.

Duncan, J.S. 2016. *In the Shadows of the Tropics: Climate, Race and Biopower in Nineteenth Century Ceylon*. London: Routledge.

Dwivedi, O.P. 1967. "Bureaucratic corruption in developing countries." *Asian Survey*, Vol. 7, No. 4, pp. 245–53.

Dwyer, P. and Nettelbeck, A. 2018. "'Savage wars of peace': violence, colonialism and empire in the modern world," in P. Dwyer and A. Nettelbeck (eds.), *Violence, Colonialism and Empire in the Modern World*. London: Palgrave Macmillan, pp. 1–22.

Elkins, J. 1996. "Legality with a vengeance: famines and humanitarian relief in 'complex emergencies'." *Millennium: Journal of International Studies*, Vol. 25, No. 3, pp. 547–75.

Emsley, C. 1996. *The English Police: A Political and Social History*. London: Routledge.

Emsley, C. 2014. "Policing the empire, policing the metropole: some thoughts on models and types." *Crime, Histoire & Sociétés/Crime, History & Societies*, Vol. 18, No. 2, pp. 5–25.

Engel, D.M. 1978. *Code and Custom in a Thai Provincial Court: The Interaction of Formal and Informal Systems of Justice*. Tucson: University of Arizona Press.

Engel, D.M. 2015. "Rights as wrongs: legality and sacrality in Thailand." *Asian Studies Review*, Vol. 39, No. 1, pp. 38–52.

Engel, D.M. 2011. "'The spirits were always watching': Buddhism, secular law, and social change, in Thailand," in W.F. Sullivan, R.A. Yelle and M. Taussig-Rubbo (eds.), *After Secular Law*. Palo Alto: Stanford University Press, pp. 242–60.

Evans, J. 2005. "Colonialism and the rule of law: the case of South Australia," in B.S. Godfrey and G. Dunstall (eds.), *Crime and Empire, 1840–1940: Criminal Justice in Local and Global Contexts*. Cullompton: Willan, pp. 57–75.

Evans, R. 1982. *The Fabrication of Virtue: English Prison Architecture, 1750–1840*. Cambridge: Cambridge University Press.

Eyler, J.M. 1992. "The sick poor and the state: Arthur Newsholme on poverty, disease, and responsibility," in C.E. Rosenberg and J. Golden (eds.), *Framing Disease*. New Brunswick: Rutgers University Press, pp. 275–96.

Fassin, D. 2009. "Another politics of life is possible." *Theory, Culture and Society*, Vol. 26, No. 5, pp. 44–60.

Fein, H. 1977. *Imperial Crime and Punishment: The Massacre at Jallianwala Bagh and British Judgement, 1919–1920*. Honolulu: University Press of Hawaii.

Felices-Luna, M. 2012. "Justice in the Democratic Republic of Congo: practicing corruption, practicing resistance." *Critical Criminology*, Vol. 20, pp. 197–209.

Ferguson, J. 1990. *The Anti-Politics Machine: 'Development,' Depoliticization and Bureaucratic Power in Lesotho*. Cambridge: Cambridge University Press.

Fernando, P.T.M. 1969–70. "The legal profession of Ceylon in the early twentieth century: official attitudes to Ceylonese aspirations." *The Ceylon Historical Journal*, Vol. 19, pp. 1–15.

Fernando, P.T.M. 1970. "The post riots campaign for justice." *The Journal of Asian Studies*, Vol. 29, No. 2, pp. 255–66.

Field, J. 1981. "Police, power and community in a provincial English town: Portsmouth 1815–1875," in V. Bailey (ed.), *Policing and Punishment in Nineteenth Century Britain*. London: Croom Helm, pp. 42–64.

Fischer-Tine, H. 2009. *Low and Licentious Europeans: Race, Class and White Subalternity in Colonial India*. New Delhi: Orient Longman.

Fitzpatrick, P. 2008. *Law as Resistance: Modernism, Imperialism, Legalism*. Aldershot: Ashgate.

Foucault, M. 1979. *Discipline and Punish: The Birth of the Prison*. New York: Vintage.

Foucault, M. 1990. *The History of Sexuality. Volume 1: An Introduction*. Translated by R. Hurley. New York: Vintage.

Foucault, M. 1991. "Governmentality," in G. Burchell, C. Gordon and P. Miller (eds.), *The Foucault Effect: Studies in Governmentality*. Chicago: University of Chicago Press, pp. 87–104.

Foucault, M. 2007. *Security, Territory, Population: Lectures at the College De France, 1977–78*. Edited by M. Senellart. New York: Palgrave Macmillan.

Foucault, M. 2008. *The Birth of Biopolitics: Lectures at the College De France, 1978–79*. Edited by M. Senellart. New York: Palgrave Macmillan.

Freitag, S.B. 1985. "Collective crime and authority in North India," in A.A. Yang (ed.), *Crime and Criminality in British India*. Tucson: University of Arizona Press, pp. 140–63.

Freitag, S.B. 1991. "Crime in the social order of colonial North India." *Modern Asian Studies*, Vol. 25, No. 2, pp. 227–61.

Fuller, C.J. and Benei, V. 2001. *The Everyday State and Society in Modern India*. London: Hurst.

Galanter, M. 1983. "Mega-law and mega-lawyering in the contemporary United States," in R. Dingwell and P. Lewis (eds.), *The Sociology of the Professions: Lawyers, Doctors and Others*. London: MacMillan, pp. 6–11.

Garland, D. 1985. *Punishment and Welfare: A History of Penal Strategies*. Aldershot: Gower.

Gatrell, V.A.C. 1980. "The decline of theft and violence in Victorian and Edwardian England," in V.A.C. Gatrell, B. Lenman and G. Parker (eds.), *Crime and the Law: The Social History of Crime in Western Europe since 1500*. London: Europa Publications, pp. 238–337.

Godfrey, B. 2014. *Crime in England, 1880–1945*. London: Routledge.

Goffman, E. 1961. *Asylums: Essays on the Social Situation of Mental Patients and Other Inmates*. Garden City: Anchor.

Gombrich, R. 1971. *Precept and Practice: Traditional Buddhism in the Rural Highlands of Ceylon*. Oxford: Clarendon Press.

Gooneratne, B. and Gooneratne, Y. 1999. *This Inscrutable Englishman: Sir John D'Oyly (1774–1824)*. London: Cassell.

Goonesekere, R.K.W. 1958. "Eclipse of the village court." *The Ceylon Journal of Historical and Social Studies*, Vol. 1, pp. 138–54.

Gordon, C. 1991. "Governmental rationality: an introduction," in G. Burchell, C. Gordon and P. Miller (eds.), *The Foucault Effect: Studies in Governmentality*. Chicago: University of Chicago Press, pp. 1–52.

Gorman, A. 2007. "Regulation, reform and resistance in the Middle Eastern prison," in F. Dikotter and I. Brown (eds.), *Cultures of Confinement: A History of the Prison in Africa, Asia and Latin America*. London: Hurst, pp. 95–146.

Graycar, A. and Jancsics, D. 2017. "Gift giving and corruption." *International Encyclopedia of Public Administration*, Vol. 4, No. 12, pp. 1013–23.

Griffin, E. 2018. "Diets, hunger and living standards during the British industrial revolution." *Past and Present*, No. 239, May, pp. 71–111.

Guha, R. 1989. *The Unquiet Woods: Ecological Change and Peasant Resistance in the Himalayas*. Delhi: Oxford University Press.

Guha, R. 1997. *Dominance without Hegemony: History and Power in Colonial India*. Cambridge, MA: Harvard University Press.

Guha, R. 1999. *Elementary Aspects of Peasant Insurgency in Colonial India*. Durham: Duke University Press.

Gupta, A. 2012. *Red Tape: Bureaucracy, Structural Violence and Poverty in India*. Durham: Duke University Press.

Guy, W.A. 1863. "On sufficient and insufficient dietaries, with special reference to the dietaries of prisoners." *Journal of the Statistical Society of London*, Vol. 26, No. 3, pp. 239–80.

Guy, W.A., Maitland, J. and Clarke, V.C. 1864. "Report of the committee on the dietaries of county and borough gaols." *Parliamentary Papers*, Vol. XLIX, pp. 563–719.

Hacking, I. 1975. *The Emergence of Probability: A Philosophical Study of Early Ideas about Probability, Induction and Statistical Inference.* Cambridge: Cambridge University Press.

Hacking, I. 1990. *The Taming of Chance.* Cambridge: Cambridge University Press.

Hargrove, J.L. 2006. "History of the calorie in nutrition." *The Journal of Nutrition,* Vol. 138, pp. 2957–61.

Harrison, M. 1992. "Tropical medicine in nineteenth century India." *British Journal of the History of Science,* Vol. 25, pp. 299–318.

Hay, D. 1975. "Property, authority and the criminal law," in D. Hay, P. Linebaugh, J. G. Rule, E.P. Thompson and C. Winslow (eds.), *Albion's Fatal Tree: Crime and Society in Eighteenth-Century England.* New York: Pantheon, pp. 17–64.

Hay, D. 1975. "Poaching and the game laws on Cannock Chase," in D. Hay, P. Linebaugh, J.G. Rule, E.P. Thompson and C. Winslow (eds.), *Albion's Fatal Tree: Crime and Society in Eighteenth-Century England.* New York: Pantheon, pp. 189–254.

Haynes, D.M. 1996. "Social status and imperial service: tropical medicine and the British medical profession in the nineteenth century," in D. Arnold (ed.), *Warm Climates and Western Medicine: The Emergence of Tropical Medicine, 1500–1900.* Amsterdam: Rodopi, pp. 208–26.

Heath, D. 2016. "Bureaucracy, power and violence in colonial India: the role of Indian subalterns," in P. Crooks and T.H. Parsons (eds.), *Empires and Bureaucracy in World History: From Late Antiquity to the Twentieth Century.* Cambridge: Cambridge University Press, pp. 364–90.

Herbst, J.I. 2000. *States and Power in Africa: Comparative Lessons in Authority and Control.* Princeton: Princeton University Press.

Herzfeld, M. 1990. "Pride and perjury: time and the oath in the mountain villages of Crete." *Man* (n.s.), Vol. 25, p. 305.

Hindess, B. 2001. "The liberal government of unfreedom." *Alternatives,* Vol. 26, pp. 93–111.

Hirsch, S. and Lazarus-Black, M. 1994. "Introduction: performance and paradox: exploring law's rule in hegemony and resistance," in M. Lazarus-Black and S. Hirsch, (eds.), *Contested States: Law, Hegemony and Resistance.* New York: Routledge, pp. 1–31.

Holmes, L. 2015. *Corruption: A Very Short Introduction.* Oxford: Oxford University Press.

Howard, J. 1784. *The State of Prisons in England and Wales.* London: William Eyres.

Howell, P. 2009. *Geographies of Regulation: Policing Prostitution in Nineteenth Century Britain and the Empire.* Cambridge: Cambridge University Press.

Hunter, I. and Dorsett, S. 2010. "Introduction," in S. Dorsett and I. Hunter (eds.), *Law and Politics in British Colonial Thought.* New York: Palgrave Macmillan, pp. 1–10.

Hussain, N. 2003. *The Jurisprudence of Emergency: Colonialism and the Rule of Law.* Ann Arbor: University of Michigan.

Hynd, S. 2008. "Killing the condemned: the practice and process of capital punishment in British Africa, 1900–1950s." *Journal of African History,* Vol. 49, pp. 403–18.

Hynd, S. 2012. "Murder and mercy: capital punishment in colonial Kenya, ca. 1909–1956." *The International Journal of African Historical Studies,* Vol. 45, No. 1, pp. 81–101.

Ibhawoh, B. 2009. "Historical globalization and colonial legal culture: African assessors, customary law, and criminal justice in British Africa." *Journal of Global History,* Vol. 4, pp. 429–51.

Ibhawoh, B. 2013. *Imperial Justice.* Oxford: Oxford University Press.

Ignatieff, M. 1978. *A Just Measure of Pain: The Penitentiary in the Industrial Revolution 1750–1850*. London: Penguin.

Inglis, H. 1835. *A Journey throughout Ireland during the Spring, Summer and Autumn of 1834, Vol. 1*. London: Whittaker and Co.

Jayawardena, K. 2000. *Nobodies to Somebodies: The Rise of the Colonial Bourgeoisie in Sri Lanka*. Colombo: Social Scientist's Association and Sanjiva Books.

Jeffrey, A. 2019. *The Edge of Law: Legal Geographies of a War Crimes Court*. Cambridge: Cambridge University Press.

Jeffrey, A. and Jakala, M. 2015. "Using courts to build states: the competing spaces of citizenship in transitional justice programmes." *Political Geography*, Vol. 47, pp. 43–52.

Jeffrey, C. 2002. "Caste, class, and clientelism: a political economy of everyday corruption in rural North India." *Economic Geography*, Vol. 78, No. 1, pp. 21–41.

Jennings, I. and Tambiah, H.W. 1952. *The Dominion of Ceylon: The Development of Its Laws and Constitution*. London: Stevens and Sons.

Joireman, S. 2006. "The evolution of common law: legal development in Kenya and India." *Journal of Commonwealth and Comparative Politics*, Vol. 44, No. 2, pp. 190–210.

Kamminga, H. and Cunningham, A. 1995. "Introduction," in H. Kamminga and A. Cunningham (eds.), *The Science and Culture of Nutrition, 1840–1940*. Amsterdam: Rodopi, pp. 1–14.

Kannangara, P.D. 1966. *The History of the Ceylon Civil Service, 1802–1833*. Dehiwala: Tisara Prakasakayo.

Kaplan, M. 1995. "Panopticon in Poona: an essay on Foucault and colonialism." *Cultural Anthropology*, Vol. 10, pp. 85–98.

Katz, C. 2004. *Growing up Global: Economic Restructuring and Children's Everyday Lives*. Minneapolis: University of Minnesota Press.

Kearns, G. 2006. "Bare life, political violence, and the territorial structure of Britain and Ireland," in D. Gregory and A. Pred (eds.), *Violent Geographies: Fear, Terror, and Political Violence*. New York: Routledge, pp. 7–35.

Kearns, G. and Nally, D. 2019. "An accumulated wrong: Roger Casement and the anticolonial moments within imperial governance." *Journal of Historical Geography*, Vol. 64, pp. 1–12.

Kemper, S. 1984. "The Buddhist monkhood, the law, and the state in colonial Sri Lanka." *Comparative Studies in Society and History*, Vol. 26, No. 3, pp. 401–27.

Killingray, D. 1986. "The maintenance of law and order in British colonial Africa." *African Affairs*, Vol. 85, No. 340, pp. 411–37.

Kirkby, D. and Coleborne, C., eds. 2001. *Law, History and Colonialism: The Reach of Empire*. Manchester: Manchester University Press.

Kolsky, E. 2005. "Codification and the rule of colonial difference: criminal procedure in British India." *Law and History Review*, Vol. 23, No. 3, Fall, pp. 631–83.

Kolsky, E. 2010. *Colonial Justice in British India: White Violence and the Rule of Law*. Cambridge: Cambridge University Press.

Latour, B. 1987. *Science in Action: How to Follow Scientists and Engineers through Society*. Cambridge: Harvard University Press.

Legg, S. 2007. *Spaces of Colonialism: Delhi's Urban Governmentalities*. Oxford: Blackwell.

Legg, S. 2015. *Prostitution and the Ends of Empire: Scale, Governmentalities and Interwar India*. Durham: Duke University Press.

Legg, S. 2016. "Empirical and analytical subaltern space? Ashrams, brothels and trafficking in colonial Delhi." *Cultural Studies*, Vol. 30, No. 5, pp. 793–815.

Legg, S. 2019. "Subjects of truth: resisting governmentality in Foucault's 1980s." *Environment and Planning D.: Society and Space*, Vol. 37, No. 1, pp. 27–45.

Legg, S. 2019. "Colonial and nationalist truth regimes: empire, Europe and the latter Foucault," in S. Legg and D. Heath (eds.), *South Asian Governmentalities: Michel Foucault and the Question of Post-Colonial Orderings*. Cambridge: Cambridge University Press, pp. 104–31.

Legg, S. and Jazeel, T. 2019. *Subaltern Geographies*. Athens: University of Georgia Press.

Lester, A. and Dussart, F. 2014. *Colonization and the Origins of Humanitarian Governance: Protecting Aborigines across the Nineteenth-Century British Empire*. Cambridge: Cambridge University Press.

Levine, P. 2003. *Prostitution, Race and Politics: Policing Venereal Disease in the British Empire*. New York: Routledge.

Li, T.M. 2007. *The Will to Improve: Governmentality, Development and the Practice of Politics*. Durham: Duke University Press.

Linebaugh, P. 2014. *Stop, Thief! The Commons, Enclosures, and Resistance*. Oakland: PM Press.

Livingstone, D.N. 1991. "The moral discourse of climate: historical considerations on race, place and virtue." *Journal of Historical Geography*, Vol. 17, pp. 413–34.

Livingstone, D.N. 1999. "Tropical climate and moral hygiene: the anatomy of a Victorian debate." *British Journal for the History of Science*, Vol. 32, pp. 93–110.

Lombroso, C. 2006. (1876). *Criminal Man*. Trans. M. Gibson and N.H. Rafter. Durham: Duke University Press.

Losurdo, D. 2011. *Liberalism: A Counter-History*. London: Verso.

Lucassen, L., Willems, W. and Cottar, A. 1998. *Gypsies and Other Itinerant Groups: A Socio-Historical Approach*. New York: Palgrave Macmillan.

Ludowyk, E.F.C. 1966. *The Modern History of Ceylon*. London: Weidenfeld and Nicolson.

Macaulay, T.B. 1867. *Speeches and Poems with the Report and Notes on the Indian Penal Code*. New York: Hurd and Houghton.

Makovick, N. and Henig, D. 2018. "Neither gift nor payment: the sociability of instrumentality," in A. Ledeneva (ed.), *The Global Encyclopedia of Informality*, Volume 1. London: University College London Press, pp. 125–212.

Mann, M. 1986. *The Sources of Social Power, Volume One*. Cambridge: Cambridge University Press.

Mann, M. 2004. "Dealing with oriental despotism: British jurisdiction in Bengal, 1772–93," in H. Fischer-Tine and M. Mann (eds.), *Colonialism as Civilizing Mission: Cultural Ideology in British India*. London: Anthem Press, pp. 29–48.

Mantena, K. 2009. "The crisis of liberal imperialism," in D. Bell (eds.), *Victorian Visions of the Global Order: Empire, and International Relations in Nineteenth Century Political Thought*. Cambridge: Cambridge University Press, pp. 113–35.

Mantena, K. 2010. *Alibis of Empire: Henry Maine and the Ends of Imperialism* Princeton: Princeton University Press.

Marasinghe, M.L. 1979. "Kandyan law and British colonial law: a conflict of tradition and modernity—an early stage of colonial development in Sri Lanka." *Verfassung und Recht in Übersee/Law and Politics in Africa, Asia and Latin America*, Vol. 12, No. 2, pp. 115–27.

Marquis, G. 1997. "The 'Irish model' and nineteenth-century Canadian policing." *Journal of Imperial and Commonwealth History*, Vol. 25, No. 2, pp. 193–218.

Mattei, U. and Nader, L. 2008. *Plunder: When the Rule of Law Is Illegal*. Oxford: Blackwell.

Mawani, R. 2014. "Law as temporality: colonial politics and Indian settlers." *University of California Irvine Law Review*, Vol. 4, No. 1, pp. 70–95.

Mayhew, H. 1862. *London Labour and the London Poor, Vol. 1*. London: Griffin, Bohn, and Company.

Mayhew, H. and Binny, J. 1862. *The Criminal Prisons of London and Scenes of Prison Life*. London: Griffin, Bohn, and Company.

Mayhew, R.J. 2014. *Malthus: Life and Legacies of an Untimely Prophet*. Cambridge: Harvard University Press.

Mayhew, R.J., ed. 2016. *New Perspectives on Malthus*. Cambridge: Cambridge University Press.

McConville, S. 1981. *A History of English Prison Administration. Volume 1. 1750–1877*. London: Routledge and Kegan Paul.

McConville, S. 2018. *English Local Prisons: 1860–1900. Next Only to Death*. London: Routledge.

McDonough, S.A. 2013. *Witnesses, Neighbors, and Community in Late Medieval Marseille*. New York: Palgrave Macmillan.

McGowen, R. 1998. "The well-ordered prison: England 1780–1865," in N. Morris and D. Rothman (eds.), *The Oxford History of the Prison: The Practice of Punishment in Western Society*. New York: Oxford University Press, pp. 71–99.

Mclane, J.R. 1985. "Bengali bandits, police and landlords after the permanent settlement," in A.A. Yang (ed.), *Crime and Criminality in British India*. Tucson: University of Arizona Press, pp. 26–47.

McLaren, J. 2010. "The uses of the rule of law in British colonial societies in the nineteenth century," in S. Dorsett and I. Hunter (eds.), *Law and Politics in British Colonial Thought*. New York: Palgrave Macmillan, pp. 71–90.

McLynn, F. 1991. *Crime and Punishment in Eighteenth-Century England*. Oxford: Oxford University Press.

Mehta, U. 1999. *Liberalism, and Empire: A Study in Nineteenth-Century British Liberal Thought*. Chicago: University of Chicago Press.

Mendis, G.C. 1948. *Ceylon Under the British*. Colombo: The Colombo Apothecaries.

Mendis, G.C., ed. 1956. *The Colebrooke-Cameron Papers: Documents in British Colonial Policy in Ceylon, 1796–1833*. Volume 1. Oxford: Oxford University Press.

Merry, S.E. 1990. *Getting Justice and Getting Even: Legal Consciousness among Working-Class Americans*. Chicago: University of Chicago Press.

Merry, S.E. 1992. "Anthropology of law and transnational processes." *Annual Review of Anthropology*, Vol. 21, pp. 357–79.

Merry, S.E. 2000. *Colonizing Hawaii: The Cultural Power of Law*. Princeton: Princeton University Press.

Merry, S.E. 2010. "Colonial law and its uncertainties." *Law and History Review*, Vol. 28, No. 4, November, pp. 1067–71.

Meyer, E. 1998. "Forests, chena cultivation, plantations and the colonial state in Ceylon 1840–1940," in R.H. Grove, V. Damodaran and S. Sangwan (eds.), *Nature and the Orient: The Environmental History of South and Southeast Asia*. Delhi: Oxford University Press, pp. 793–827.

Migdal, J.S. 1994. "The state in society: an approach to struggles for domination," in A. Kohl, V. Shue and J.S. Migdal (eds.), *State Power and Social Forces: Domination and Transformation in the Third World*. Cambridge: Cambridge University Press, pp. 7–32.

Miller, I. 2015. *Reforming Food in Post-famine Ireland: Medicine, Science and Improvement, 1845–1922*. Manchester: Manchester University Press.

Milles, D. 1995. "Working capacity and calorie consumption: the history of rational physical economy," in H. Kamminga and A. Cunningham (eds.), *The Science and Culture of Nutrition, 1840–1940*. Amsterdam: Rodopi, pp. 75–96.

Mills, L.A. 1964. *Ceylon under British Rule, 1795–1932*. London: Frank Cass and Co.

Mitchell, T. 1991. *Colonising Egypt*. Berkeley: University of California Press.

Moran, D. 2012. "Doing time in carceral space: timespace and carceral geography." *Geografiska Annaler B Human Geography*, Vol. 94, No. 4, pp. 305–16.

Moran, D. 2015. *Carceral Geography: Spaces and Practices of Incarceration*. Farnham: Ashgate.

Moran, D., Gill, N. and Conlon, D., eds. 2013. *Carceral Spaces: Mobility and Agency in Imprisonment and Migrant Detention*. Farnham: Ashgate.

Morin, K.M. 2018. *Carceral Spaces, Prisoners and Animals*. London: Routledge.

Morin, K.M. and Moran, D., eds. 2015. "Introduction," in *Historical Geographies of Prisons: Unlocking the Usable Carceral Past*. London: Routledge, pp. 15–32.

Muir, S. and Gupta, A. 2018. "Rethinking the anthropology of corruption: an introduction to supplement 18." *Current Anthropology*, Vol. 59, Supplement.18, pp. 4–15.

Nally, D. 2011. *Human Encumbrances: Political Violence and the Great Irish Famine*. South Bend: University of Notre Dame Press.

Nally, D. 2011. "The biopolitics of food provisioning." *Transactions. I.B.G.*, Vol. 36, pp. 37–53.

Nanaraja, T. 1972. *The Legal System of Ceylon in Its Historical Setting*. Leiden: E.J. Brill.

Neocleous, M. 2000. *The Fabrication of Social Order: A Critical Theory of Police Power*. London: Pluto Press.

Neswald, E. 2016. "Food fights: human experiments in late nineteenth-century nutrition physiology," in E. Dyck and L. Stewart (eds.), *The Uses of Humans in Experiments: Perspectives from the Seventeenth to the Twentieth Centuries*. Leiden: Brill, Rodopi, pp. 170–93.

Nijhar, P. 2009. *Law and Imperialism: Criminality and Constitution in Colonial India and Victorian England*. London: Pickering and Chatto.

Niti Niganduva or the Vocabulary of the Law as It Existed in the Days of the Kandyan Kingdom. 1880. Trans. T.B. Panabokke and C.J.R. Le Mesurier. Colombo: Ceylon Government Press.

Northwestern University School of Law. 1974. "Perjury: the forgotten offense." *The Journal of Criminal Law and Criminology*, Vol. 65, No. 3, pp. 361–72.

O'Brien, P. 1982. *The Promise of Punishment: Prisons in Nineteenth-Century France*. Princeton: Princeton University Press.

O'Brien, P. 1996. "Prison reform in France and other European countries in the nineteenth century," in N. Finzsch and R. Jutte (eds.), *Institutions of Confinement: Hospitals, Asylums and Prisons in Western Europe and North America, 1500–1950*. Cambridge: Cambridge University Press, pp. 285–300.

Ogborn, M. 1995. "Discipline, government and law: separate confinement in the prisons of England and Wales, 1830–1877." *Transactions of the Institute of British Geographers*, Vol. 20, pp. 295–311.

Oldham, J. 1994. "Truth-telling in the eighteenth-century English courtroom." *Law and History Review*, Vol. 12, No. 1, Spring, pp. 95–121.

Olund, E.N. 2002. "From savage space to governable space: the extension of United States judicial sovereignty over Indian Country in the nineteenth century." *Cultural Geographies*, Vol. 9, No. 2, pp. 129–57.

Osborne, T. 1994. "Bureaucracy as a vocation: governmentality and administration in nineteenth century Britain." *Journal of Historical Sociology*, Vol. 7, No. 3, pp. 289–313.

Pardo, I. 2004. "Introduction: corruption, morality and the law," in *Between Morality and the Law: Corruption, Anthropology and Comparative Society*. New York: Routledge, pp. 1–18.

Parkin, F. 1971. *Class Inequality and Political Order*. New York: Praeger.

Parsons, T.H. 2016. "The unintended consequences of bureaucratic 'modernization' in post-World War II British Africa," in P. Crooks and T.H. Parsons (eds.), *Empires and Bureaucracy in World History: From Late Antiquity to the Twentieth Century*. Cambridge: Cambridge University Press, pp. 412–33.

Peebles, P. 1981. "Governor Arthur Gordon and the administration of Sri Lanka, 1883–1890," in R.T. Crane and N.G. Barrier (eds.), *British Imperial Policy in India and Sri Lanka, 1858–1912*. Columbia: South Asia Books, pp. 84–106.

Peebles, P. 1995. *Social Change in Nineteenth Century Ceylon*. New Delhi: Navrang.

Pennington, K. 2013. "Innocent until proven guilty: the origins of a legal maxim." *The Jurist*, Vol. 63, pp. 106–24.

Perera, N. 1998. *Society and Space: Colonialism, Nationalism and Postcolonial Identity in Sri Lanka*. Oxford: Westview.

Pete, S. and Deven, A. 2005. "Flogging, fear and food: punishment and race in colonial Natal." *Journal of Southern African Studies*, Vol. 31, No. 1, March, pp. 3–21.

Phillips, R. 2017. *Sex, Politics and Empire: A Postcolonial Geography*. Oxford: Oxford University Press.

Pierce, S. 2016. "The invention of corruption: political malpractice and selective prosecution in colonial northern Nigeria." *Journal of West African History*, Vol. 2, No. 2, pp. 1–28.

Pieris, A. 2009. *Hidden Hands and Divided Landscapes: A Penal History of Singapore's Plural Society*. Honolulu: University of Hawai'i Press.

Pieris, R. 1956. *Sinhalese Social Organization*. Colombo: Ceylon University Press Board.

Pippet, G.K. 1938. *A History of the Ceylon Police, Volume 1, 1795–1870*. Colombo: The Times of Ceylon.

Pitts, J. 2005. *A Turn to Empire: The Rise of Imperial Liberalism in Britain and France*. Princeton: Princeton University Press.

Planters' Association. *Proceedings of the Planters' Association*. 1883–84. Kandy, Ceylon: Planters' Association.

Porter, B. 1968. *Critics of Empire: British Radicals and the Imperial Challenge*. London: Bloomsbury.

Prakash, G. 1999. *Another Reason: Science and the Imagination of Modern India*. Princeton: Princeton University Press.

Pratt, M.L. 1992. *Imperial Eyes: Travel Writing and Transculturation*. New York: Routledge.

Priestley, P. 1985. *Victorian Prison Lives: English Prison Biography 1830–1914*. New York: Methuen.

Rabinbach, A. 1992. *The Human Motor: Energy, Fatigue and the Origins of Modernity*. Berkeley: University of California Press.

Radhakrishna, M. 2008. "Laws of metamorphosis: from nomad to offender," in K. Kannabiran and R. Singh (eds.), *Challenging the Rule(s) of Law: Colonialism, Criminology and Human Rights in India*. New Delhi: Sage, pp. 3–27.

Radzinowicz, L. and Hood, R. 1990. *A History of English Criminal Law and Its Administration from 1750, Vol. 5. The Emergence of Penal Policy in Victorian and Edwardian England*. Oxford: Clarendon Press.

Raman, K.K. 1994. "Utilitarianism and the criminal law in colonial India: a study of the practical limits of Utilitarian jurisprudence." *Modern Asian Studies*, Vol. 28, No. 4, pp. 739–91.

Rao, A. and Dube, S. 2013. "Questions of crime," in S. Dube and A. Rao (eds.), *Crime through Time*. Oxford: Oxford University Press, pp. xii–lxiii.

Redfield, P. 2005. "Foucault in the tropics: displacing the panopticon," in J.X. Inda (ed.), *Anthropologies of Modernity: Foucault, Governmentality and Life Politics*. Oxford: Blackwell, pp. 50–79.

Rejali, D. 1994. *Torture and Modernity: Self, Society and State in Modern Iran*. Boulder: Westview Press.

Richards, T. 1993. *The Imperial Archive: Knowledge and the Fantasy of Empire*. London: Verso.

Roberts, D.F. 2002. *The Social Conscience of the Early Victorians*. Stanford: Stanford University Press.

Roberts, M. 1965. "The master servant laws of 1841 and the 1860s and immigrant labour in Ceylon." *Ceylon Journal of Historical and Social Studies*, Vol. 8, pp. 24–37.

Roberts, M. 1982. *Caste Conflict and Elite Formation: The Rise of the Karava Elite in Sri Lanka, 1500–1931*. Cambridge: Cambridge University Press.

Roberts, M., Raheem, I. and Colin-Thome, P. 1989. *People Inbetween: Burghers and the Middle Class in the Transformations within Sri Lanka, 1790s–1960s*, Volume 1. Ratmalana: Sarvodaya Books.

Robinson, F.B. 1985. "Bandits and rebellion in nineteenth century western India," in A. A. Yang (ed.), *Crime and Criminality in British India*. Tucson: University of Arizona Press, pp. 48–61.

Rogers, J.D. 1987. *Crime, Justice and Society in Colonial Sri Lanka*. London: Curzon Press.

Rogers, J.D. 1991. "Cultural and social resistance: gambling in colonial Sri Lanka," in D. Haynes and G. Prakash (eds.), *Contesting Power: Resistance and Everyday Social Relations in South Asia*. Berkeley: University of California Press, pp. 175–212.

Rose, N. and Miller, P. 2010. "Political power beyond the state: the problematics of government." *British Journal of Sociology*, Vol. 61, No. 1, pp. 271–303.

Ross, J.I. 2009. "Resisting the carceral state: prisoner resistance from the bottom up." *Social Justice*, Vol. 36, No. 3, pp. 28–45.

Ross, J.I. 2013. "Deconstructing correctional officer deviance: towards typologies of actions and controls." *Criminal Justice Review*, Vol. 38, No. 1, pp. 110–26.

Saha, J. 2011. "Histories of everyday violence in South Asia." *History Compass*, Vol. 9, No. 11, pp. 844–53.

Saha, J. 2013. "Colonization, criminalization and complicity: policing gambling in Burma c. 1880–1920." *South East Asia Research*, Vol. 21, No. 4, Special Issue: Colonial Histories in South East Asia—Papers in Honour of Ian Brown, pp. 655–72.

Saha, J. 2013. *Law, Disorder and the Colonial State: Corruption in Burma c. 1900*. London: Palgrave Macmillan.

Saho, B. 2018. *Contours of Change: Muslim Courts, Women, and Islamic Society in Colonial Bathurst, the Gambia, 1905–1965*. East Lansing: Michigan State University Press.

Samaraweera, V. 1974. "The Ceylon charter of Justice of 1833: a Benthamite blueprint for judicial reform." *The Journal of Imperial and Commonwealth History*, Vol. 2, No. 3, pp. 263–77.

Samaraweera, V. 1978. "The 'village community' and reform in colonial Sri Lanka." *The Ceylon Journal of Historical and Social Studies*, New Series, Vol. VIII, No. 1, January–June, pp. 68–75.

Samaraweera, V. 1979. "Litigation and legal reform in colonial Sri Lanka." *South Asia: Journal of South Asian Studies*, Vol. 2, No. 1–2, pp. 78–90.

Samaraweera, V. 1981. "British justice and the 'oriental peasantry'; the working of the colonial legal system in nineteenth century Sri Lanka," in R.I. Crane and G. Barrier (eds.), *British Imperial Policy in India and Sri Lanka, 1858–1912*, Columbia: South Asia Books, pp. 107–41.

Samaraweera, V. 1985. "The legal system, language and elitism: the colonial experience in Sri Lanka," in L.M. Marasinghe and W.E. Conklin (eds.), *Law, Language and Development*. Colombo: Lake House Investments, pp. 93–110.

Sarat, A. 1990. "Law is all over: power, resistance and the legal consciousness of the welfare poor." *Yale Journal of Law and the Humanities*, Vol. 2, pp. 343–79.

Sauvain, P. 1987. *British Economic and Social History 1700–1870*. Cheltenham: Stanley Thornes.

Schneider, W.E. 2015. *Engines of Truth: Producing Veracity in the Victorian Courtroom*. New Haven: Yale University Press.

Schuller, K. 2016. "Biopower before and below the individual." *GLQ: A Journal of Lesbian and Gay Studies*, Vol. 22, No. 4, pp. 629–36.

Schuller, K. 2018. *The Biopolitics of Feeling: Race, Sex and Science in the Nineteenth Century*. Durham: Duke University Press.

Scott, D. 1995. "Colonial governmentality." *Social Text*, Vol. 43, pp. 191–220.

Scott, J.C. 1985. *Weapons of the Weak: Everyday Forms of Peasant Resistance*. New Haven: Yale University Press.

Scott, J.C. 1986. "Introduction," in J.C. Scott, J. Benedict and T. Kerkvliet (eds.), *Everyday Forms of Peasant Resistance in South-East Asia*, Library of Peasant Studies no. 9. London: Frank Cass and Co, pp. 5–35.

Scott, J.C. 1990. *Domination and the Arts of Resistance*. New Haven: Yale University Press.

Scott, J.C. 1992. "Domination, acting and fantasy," in C. Nordstrom and A.A. Martin (eds.), *The Paths to Domination, Resistance and Terror*. Berkeley: University of California Press, pp. 55–84.

Scott, J.C. 1998. *Seeing like a State: How Certain Schemes to Improve the Human Condition Have Failed*. New Haven: Yale University Press.

Scott, J.C. 2013. *Decoding Subaltern Politics: Ideology, Disguise, and Resistance in Agrarian Politics*. London: Routledge.

Sen, A. 1981. *Poverty and Famine: An Essay on Entitlement and Deprivation*. Oxford: Clarendon Press.

Sen, M. 2007. *Prisons in Colonial Bengal 1838–1919*. Kolkata: Thema.

Sen, S. 2000. *Disciplining Punishment: Colonialism and Convict Society in the Andaman Islands*. New Delhi: Oxford University Press.

Sharafi, M. 2014. *Law and Identity in Colonial South Asia: Parsi Legal Culture, 1772–1947*. New York: Cambridge University Press.

Sharp, J., Routledge, P., Philo, C. and Paddison, R., eds. 2000. "Introduction," in *Entanglements of Power: Geographies of Domination and Resistance*. New York: Routledge, pp. 1–40.

Shaw, K. 2004. "Creating/negotiating interstices: indigenous sovereignties," in J. Edkins, V. Pin-Fat and M.J. Shapiro (eds.), *Sovereign Lives: Power in Global Politics*. London: Routledge, pp. 165–88.

Sherman, T.C. 2009. "Tensions of colonial punishment: perspectives on recent developments in the study of coercive networks in Asia, Africa and the Caribbean." *History Compass*, Vol. 7, No. 3, pp. 659–77.

Sherman, T.C. 2010. *State Violence and Punishment in India*. London: Routledge.

Shinar, A. 2013. "Dissenting from within: why and how public officials resist the law." *Florida State University Law Review*, Vol. 40, No. 3, pp. 602–57.

Shutt, S.K. 2007. "'The natives are getting out of hand': legislating manners, insolence and contemptuous behaviour in Southern Rhodesia, c. 1910–1963." *Journal of Southern African Studies*, Vol. 33, No. 3, pp. 653–72.

Sinclair, G. 2008. "The 'Irish' policeman and the empire: influencing the policing of the British Empire/Commonwealth." *Irish Historical Studies*, Vol. 36, No. 142, pp. 173–87.

Singha, R. 1998. *A Despotism of Law: Crime and Justice in Early Colonial India*. New Delhi: Oxford University Press.

Sirr, H.C. 1850. *Ceylon and the Cingalese*. Volume 2. London: William Shoberl.

Sivasundaram, S. 2007. "Tales of the land: British geography and Kandyan resistance in Sri Lanka, C. 1803–1850." *Modern Asian Studies*, Vol. 41, No. 5, pp. 925–65.

Sivasundaram, S. 2013. *Islanded: Britain, Sri Lanka and the Bounds of an Indian Ocean Colony*. Chicago: University of Chicago Press.

Skinner, T. 1974. (1891). *Fifty Years in Ceylon: An Autobiography*. Dehiwala: Tisara Prakasakayo.

Smith, D.J. 2007. *A Culture of Corruption: Everyday Deception and Popular Discontent in Nigeria*. Princeton: Princeton University Press.

Sneath, D. 2006. "Transacting and enacting: corruption, obligation and the use of monies in Mongolia." *Ethnos: Journal of Anthropology*, Vol. 71, No. 1, pp. 89–112.

Spivak, G.C. 1994. "Can the subaltern speak?" in P. Williams and L. Crisman (eds.), *Colonial Discourse and Postcolonial Theory*. New York: Columbia University Press, pp. 66–111.

Stokes, E. 1959. *The English Utilitarians and India*. Delhi: Oxford University Press.

Stoler, A.L. 1995. *Race and the Education of Desire: Foucault's History of Sexuality and the Colonial Order of Things*. Durham: Duke University Press.

Stoler, A.L. 2002. *Carnal Knowledge and Imperial Power: Race and the Intimate in Imperial Rule*. Berkeley: University of California Press.

Stoler, A.L. 2009. *Along the Archival Grain: Epistemic Anxieties and Colonial Common Sense*. Princeton: Princeton University Press.

Tamanaha, B.Z. 2004. *On the Rule of Law: History, Politics, Theory*. Cambridge: Cambridge University Press.

Tambiah, H.W. 1954. *The Laws and Customs of the Tamils of Ceylon*. Colombo: Tamil Cultural Society of Ceylon.

Tambiah, H.W. 1968. *Sinhala Laws and Customs*. Colombo: Lake House.

Taylor, D. 1997. *The New Police in Nineteenth Century England: Crime, Conflict and Control*. Manchester: Manchester University Press.

Thomas, M. 2012. *Violence and Colonial Order: Police, Workers and Protest in the European Colonial Empires, 1918–1940*. Cambridge: Cambridge University Press.

Thompson, E.P. 1975. *Whigs and Hunters: The Origin of the Black Act*. London: Allen Lane.

Tijsterman, S.P. and Overeem, P. 2008. "Escaping the iron cage: Weber and Hegel on bureaucracy and freedom." *Administrative Theory & Praxis*, Vol. 30, No. 1, pp. 71–91.

Tobias, J.J. 1967. *Crime and Industrial Society in the Nineteenth Century*. London: Batsford.

Tomlinson, M. 1978. "Not an instrument of punishment: prison diet in the mid-nineteenth century." *Consumer Studies*, Vol. 2, No. 1, pp. 15–26.

Travers, R. 2007. *Ideology and Empire in Eighteenth Century India*. Cambridge: Cambridge University Press.

Udagama, D. 2012. "The Sri Lankan legal complex and the liberal project: only thus far and no more," in T.C. Halliday, L. Karpik and M.M. Feeley (eds.), *Fates of Political Liberalism in the British Post-Colony the Politics of the Legal Complex*. Cambridge: Cambridge University Press, pp. 219–44.

Valverde, M. 1996. "'Despotism' and ethical liberal governance." *Economy and Society*, Vol. 25, pp. 357–72.

Valverde, M. 2003. *Law's Dream of a Common Knowledge*. Princeton: Princeton University Press.

Valverde, M. 2008. "Police, sovereignty and law. Foucauldian reflections," in M. Dubber and M. Valverde (eds.), *Police and the Liberal State*. Stanford: Stanford University Press, pp. 15–32.

Vanden Driesen, L.H. 1957. "Land sales policy and some aspects of the problem of tenure, 1836–86 Part 2." *University of Ceylon Review*, Vol. 15, pp. 36–52.

Vernon, J. 2007. *Hunger: A Modern History*. Cambridge: Harvard University Press.

Véron, R., Corbridge, S., Williams, G. and Srivastava, M. 2003. "The everyday state and political society in Eastern India: structuring access to the employment assurance scheme." *The Journal of Development Studies*, Vol. 39, No. 5, pp. 1–28.

Vinthagen, S. and Johansson, A. 2013. "Everyday resistance: exploration of a concept and its theories." *Resistance Studies Magazine*, Vol. 1, No. 1, pp. 1–46.

Vold, G.B., Bernard, P.J. and Snipes, J.B. 2002. *Theoretical Criminology*. New York: Oxford University Press.

Von Benda-Beckmann, F., Von Benda-Beckmann, K. and Griffiths, A. 2009. "Space and legal pluralism: an introduction," in F. Von Benda-Beckmann, K. Von Benda-Beckmann and A. Griffiths (eds.), *Spatializing Law: An Anthropological Geography of Law in Society*. Farnham: Ashgate, pp. 1–30.

Warnapala, W.A.W. 1974. *Civil Service Administration in Ceylon*. Colombo: Department of Cultural Affairs.

Weinberger, B. 1981. "The Police and the public in mid-nineteenth century Warwickshire," in V. Bailey (ed.), *Policing and Punishment in Nineteenth Century Britain*. London: Croom Helm, pp. 65–93.

Weindling, P. 2017. "Introduction: a new historiography of the Nazi medical experiments and coerced research," in P. Weindling(ed.), *From Clinic to Concentration Camp: Reassessing Nazi Medical and Racial Research, 1933–45*. London: Routledge, pp. 3–33.

Wener, R.E. 2012. *The Environmental Psychology of Prisons and Jails: Creating Humane Spaces in Secure Settings*. Cambridge: Cambridge University Press.

Wickremeratne, L.A. 1973. "Grain consumption and famine conditions in late nineteenth century Ceylon." *The Ceylon Journal of Historical and Social Studies*, Vol. 3, pp. 28–53.

Wiener, M.J. 1990. *Reconstructing the Criminal: Culture, Law and Policy in England, 1830–1914*. Cambridge: Cambridge University Press.

Wiener, M.J. 2009. *An Empire on Trial: Race, Murder and Justice under British Rule, 1870–1935*. Cambridge: Cambridge University Press.

Wilson, J. 2017. "Re-appropriation, resistance, and British autocracy in Sri Lanka, 1820–1850." *The Historical Journal*, Vol. 60, No. 1, pp. 47–69.

Wood, E. 1999. *The Origin of Capitalism*. New York: Monthly Review Press.

Woolf, L. 1961. *Growing: An Autobiography of the Years 1904–1911*. New York: Harcourt Brace Jovanovich.

Woolf, L. 1963. *Diaries in Ceylon (1908–1911): Records of a Colonial Administrator*. London: The Hogarth Press.

Yang, A.A. 1985. "Dangerous castes and tribes: the criminal tribes act and the Magahiya Doms of Northeast India," in A.A. Yang (ed.), *Crime and Criminality in British India*. Tucson: University of Arizona Press, pp. 108–27.

Yang, A.A. 1987. "Disciplining 'natives': prisons and prisoners in early nineteenth century India." *South Asia*, Vol. 10, No. 2, pp. 29–45.

Zinoman, P. 2001. *The Colonial Bastille: A History of Imprisonment in Vietnam, 1862–1940*. Berkeley: University of California Press.

Index

Note: References in *italics* are to figures, those in **bold** to tables; 'n' refers to chapter notes.:

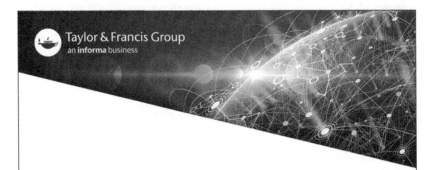

For Product Safety Concerns and Information please contact our EU
representative GPSR@taylorandfrancis.com Taylor & Francis Verlag GmbH,
Kaufingerstraße 24, 80331 München, Germany

Printed and bound by CPI Group (UK) Ltd, Croydon, CR0 4YY
01/05/2025
01858588-0002